Control of Cellular Growth in Adult Organisms

Control of Cellular Growth in Adult Organisms

A SIGRID JUSÉLIUS FOUNDATION SYMPOSIUM

Edited by

HARALD TEIR

and

TAPIO RYTÖMAA

II Department of Pathology, University of Helsinki, Finland

1967

ACADEMIC PRESS

London and New York

ACADEMIC PRESS INC. (LONDON) LTD
Berkeley Square House
Berkeley Square
London, W.1.

U.S. Edition published by
ACADEMIC PRESS INC.
111 Fifth Avenue
New York, New York 10003

Library of Congress Catalog Card Number: 67-21444

PRINTED IN GREAT BRITAIN BY
THE UNIVERSITY PRESS
ABERDEEN

List of Contributors in the Symposium

P. Alexander, Chester Beatty Research Institute, Institute of Cancer Research, Royal Cancer Hospital, London, England.

A. Alho, II Department of Pathology, University of Helsinki, Helsinki, Finland.

E. G. Bade, Instituto de Patología General y Experimental, Universidad Nacional de Cuyo, Mendoza, Argentina.

T. E. Bolden, Department of Oral Pathology and Oral Medicine, School of Dentistry, Meharry Medical College, Nashville, Tennessee, U.S.A.

M. W. Brenner, Radiotherapy Clinic, University Central Hospital, Helsinki, Finland.

W. S. Bullough, Department of Zoology, Birkbeck College, University of London, London, England.

A. Claude, Laboratoire de Cytologie et de Cancérologie Expérimentale, Université Libre de Bruxelles, Brussels, Belgium.

C. G. Craddock, Department of Medicine, University of California at Los Angeles, and Hematology Research Laboratory, Veterans Administration Center, Los Angeles, California, U.S.A.

J. F. Duplan, Laboratoire Pasteur de l'Institut du Radium, Paris, France.

J. M. Echave Llanos, Instituto de Patología General y Experimental, Universidad Nacional de Cuyo, Mendoza, Argentina.

K. Elgjo, Institute for General and Experimental Pathology, University of Oslo, Oslo, Norway.

K. E. Fichtelius, Department of Histology, University of Uppsala, Uppsala, Sweden.

K-J. Forsell, II Department of Pathology, University of Helsinki, Helsinki, Finland.

A. D. Glinos, Department of Cellular Physiology, Walter Reed Army Institute of Research, Walter Reed Army Medical Center, Washington, D.C. and Department of Zoology, University of Maryland, College Park, Maryland, U.S.A.

R. J. Goss, Division of Biological and Medical Sciences, Brown University, Providence, Rhode Island, U.S.A.

E. Grundmann, Institut für experimentelle Pathologie der Farbenfabriken Bayer AG, Wuppertal-Elberfeld, German Federal Republic.

V. I. Guelstein, Institute of Experimental and Clinical Oncology, Academy of Medical Sciences, Moscow, USSR.

A. Haddow, F.R.S. Chester Beatty Research Institute, Institute of Cancer Research: Royal Cancer Hospital, London, England.

L. R. Holsti, Radiotherapy Clinic, University Central Hospital, Helsinki, Finland.

P. Holsti, Radiotherapy Clinic, University Central Hospital, Helsinki, Finland.

O. H. Iversen, Institute for General and Experimental Pathology, University of Oslo, Oslo, Norway.

E. Jalavisto, Department of Physiology, University of Helsinki, Helsinki, Finland.

M. KEKOMÄKI, II Department of Pathology, University of Helsinki, Helsinki, Finland.

K. KIVINIEMI, II Department of Pathology, University of Helsinki, Helsinki, Finland.

A. LAHTIHARJU, II Department of Pathology, University of Helsinki, Helsinki, Finland.

L. G. LAJTHA, Paterson Laboratories, Christie Hospital & Holt Radium Institute, Manchester, England.

P. LARSEN, Botanical Laboratory, University of Bergen, Bergen, Norway.

E. B. LAURENCE, Department of Zoology, Birkbeck College, University of London, London, England.

J. S. LAWRENCE, Department of Medicine, University of California at Los Angeles, and Hematology Research Laboratory, Veterans Administration Center, Los Angeles, California, U.S.A.

J. LINNA, Department of Histology, University of Uppsala, Uppsala, Sweden.

B. LITTBRAND, Department of Tumour Biology, Karolinska Institute Medical School, Stockholm, Sweden.

Y. MATSUYUKI, Department of Medicine, University of California at Los Angeles, and Hematology Research Laboratory, Veterans Administration Center, Los Angeles, California, U.S.A.

H. MODIG, Department of Tumour Biology, Karolinska Institute Medical School, Stockholm, Sweden.

Y. PERTTALA, Radiotherapy Clinic, University Central Hospital, Helsinki, Finland.

H. RABES, German Cancer Research Center, Heidelberg; Department of Experimental Pathology and Department of Pathology, University of Munich, Munich, German Federal Republic.

T. RÄSÄNEN, II Department of Pathology, University of Helsinki, Helsinki, Finland.

K. O. RENKONEN, Department of Serology and Bacteriology, University of Helsinki, Helsinki, Finland.

L. RÉVÉSZ, Department of Tumour Biology, Karolinska Institute Medical School, Stockholm, Sweden.

T. RYTÖMAA, II Department of Pathology, University of Helsinki, Helsinki, Finland.

D. W. SMITHERS, Radiotherapy Department, The Royal Marsden Hospital and Institute of Cancer Research, Royal Cancer Hospital, London, England.

H. TEIR, II Department of Pathology, University of Helsinki, Helsinki, Finland.

G. D. TUMANISHVILI, Institute of Experimental Morphology, Academy of Sciences of the Georgian SSR, Tbilissi, USSR.

J. M. VASILIEV, Institute of Experimental and Clinical Oncology, Academy of Medical Sciences, Moscow, USSR.

M. VIROLAINEN, II Department of Pathology, University of Helsinki, Helsinki, Finland.

P. WEISS, Graduate School of Biomedical Sciences, University of Texas, Houston, Texas, U.S.A.

B-J. P. Wikgren, Åbo Academy, Åbo (Turku), Finland.

A. Winkelstein, Department of Medicine, University of California at Los Angeles, and Hematology Research Laboratory, Veterans Administration Center, Los Angeles, California, U.S.A.

H. Wrba, German Cancer Research Center, Heidelberg; Department of Experimental Pathology and Department of Pathology, University of Munich, Munich, German Federal Republic.

List of Additional Participants in the Symposium

A. DE LA CHAPELLE, Minerva Foundation, Institute for Medical Research, Helsinki, Finland.

K. DAMMERT, Department of Pathology, University of Oulu, Oulu, Finland.

A. R. G. GRÄSBECK, Minerva Foundation, Institute for Medical Research, Helsinki, Finland.

J. HONDIUS BOLDINGH, N.V. Organon, Oss, Netherlands.

O. JÄRVI, Department of Pathology, University of Turku, Turku, Finland.

A. KMENT, Institute of Physiology, Tierärztliche Hochschule, Vienna, Austria.

L. G. KOSS, Sloan-Kettering Institute for Cancer Research, New York, N.Y., U.S.A.

S. MUSTAKALLIO, Radiotherapy Clinic, University Central Hospital, Helsinki, Finland.

N. OKER-BLOM, Department of Virology, University of Helsinki, Helsinki, Finland.

K. PENTTINEN, Department of Virology, University of Helsinki, Helsinki, Finland.

R. PETERSON, Department of Pediatrics, University of Minnesota, Minneapolis, Minneapolis, U.S.A.

B. POOLE, The Rockefeller Institute, New York, N.Y., U.S.A.

M. F. RAJEWSKY, Max Planck-Institute of Biophysics, Frankfurt a/M., German Federal Republic.

A. F. RIECK, Department of Physiology, Maguette University School of Medicine, Milwaukee, Wisconsin, U.S.A.

J. J. SAUKKONEN, State Serum Institute, Helsinki, Finland.

E. SAXÉN, III Department of Pathology, University of Helsinki, Helsinki, Finland.

L. SAXÉN, II Department of Pathology, University of Helsinki, Helsinki, Finland.

J. STEIN, Heidelberg, German Federal Republic.

S. TOIVONEN, Department of Zoology, University of Helsinki, Helsinki, Finland.

A. VOUTILAINEN, Department of Radiology, University of Oulu, Oulu, Finland.

C. WASASTJERNA, Department of Medicine, Vaasa Central Hospital, Vaasa, Finland.

C. O. WEGELIUS, IV Department of Internal Medicine, University of Helsinki, Helsinki, Finland.

R. WOODS, Miles College, Birmingham, Alabama, U.S.A.

FOREWORD

THE theme of this symposium—"Control of Cellular Growth in Adult Organisms"—presents a fundamental event in life: reproduction of a cell. This phenomenon has been a central concern of biology since the beginning of cell research.

Events in cell reproduction may be considered either as kinetic or dynamic problems: the main difference between the approaches lies in the consideration of forces which regulate cell behaviour under normal and pathological conditions. Technical advances of the last two decades have led to many important discoveries in the field of cellular kinetics. This has provided a necessary structure to which the dynamic data can be referred. Thus the primary purpose of this symposium is to provide a basic insight into the problems of cellular growth control and possibly to stimulate further research in this complex field of biology.

The trend within the field of Natural Sciences has been towards specialization. However, the division of the fundamental sciences has advanced in such a way that workers in the different branches experience some difficulty in easily comprehending the language employed by colleagues, with the result that a reaction has set in which lends to emphasis to the need to create contact across the frontiers of the various fields of knowledge. The theme of this meeting presents a problem which has brought together scientists engaged in disciplines within biology and medicine from all over the world, thereby enabling them to cross the frontiers, to exchange views and to discuss experiences.

The symposium was held in Helsinki, Finland, October 4th–6th, 1965 and financed by The Sigrid Jusélius Foundation for Medical Research; this volume is the first concrete manifestation of its international activity. The Foundation extends its respectful thanks to the numerous prominent research workers who attended the symposium and by co-operation guaranteed its high scientific standard; in addition the Foundation would like to express its thanks to all those who actively contributed to the planning and successful outcome of the meeting.

Ambassador Helge von Knorring, Director of The Sigrid Jusélius Foundation, made all the practical arrangements of the symposium. The scientific programme was planned in collaboration with the members of the steering committee: Professor W. S. Bullough, Professor C. G. Craddock, Professor A. Haddow, Professor H. Teir (*chairman*), Professor P. Weiss and Professor H. Wrba. Dr. A. Lahtiharju and Dr. T. Rytömaa acted as secretaries of the symposium.

BERTEL VON BONSDORFF, M.D.

Member of the Advisory Medical Committee of Sigrid Jusélius Foundation

CONTENTS

Section I. General Aspects (A)

Section II. General Aspects (B)

Section III. Haematopoiesis

SECTION IV. Lymphoid Cells

Section V. Liver Regeneration

Section VI. Miscellaneous Aspects

Section VII. Aspects in Malignancy

SECTION VIII. General Discussion

Section 1

General Aspects (A)

The Strategy of Growth *

RICHARD J. GOSS

Division of Biological and Medical Sciences, Brown University, Providence, Rhode Island, U.S.A.

I. Introduction

As an animal grows, its parts continually adjust their histology to the expanding dimensions of the body. If an organ is to enlarge without sacrificing its physiological efficiency, it must increase either the size (hypertrophy) or number (hyperplasia) of the structural units upon which its function depends (Goss, 1966). Larger anatomical structures such as tissues and organs usually grow in size, while smaller units (cells, molecules) grow in number. In this sense, then, hypertrophy of a tissue or an organ is achieved only by hyperplasia of the smaller sub-units of which it is composed.

Common usage has tended to restrict the term hyperplasia to the cellular level of organization, thus equating it with mitosis. Indeed, there is a tendency to regard cell division as equivalent to growth, when in fact it represents only one of many levels of organization at which hyperplasia can occur. Thus, the truly remarkable way in which division occurs at the cellular level has given rise to a "mitotic mystique", which attaches more significance to the event than it may deserve. In the present context, therefore, the terms hyperplasia and hypertrophy will be used in a general sense to refer respectively to increases in number or size of units at any level of organization, from the molecule to the organ.

During the course of development, the generative potential in different kinds of organs declines to varying degrees. Some organs lose all capacity for proliferation except of subcellular entities, while others retain remarkable powers of regeneration even at histologically complex levels. Thus, it is possible to classify the various tissues and organs of the body according to the highest level of organization at which parts are capable of being added or replaced. This level can best be determined by studying the phenomenon of compensatory growth, which has recently been reviewed by Liosner (1963) and Goss (1964), and upon which much of the present analysis is based.

* Supported by a research grant (HD-00192) from the National Institute of Child Health and Human Development, U.S. Public Health Service.

II. Turnover and Levels of Organization

A. *Organ renewal*

There are a few cases in which substantial portions of organs are renewed from time to time. These include: the deciduous teeth by various vertebrates, which are capable of being replaced one or more times by successive development of enamel organs; the antlers of deer which are regrown annually from frontal pedicles; and portions of the uterine endometrium in primates which are regularly shed and regenerated during each menstrual cycle. The recurrent formation of ovarian follicles and Haversian systems may also be included in this category. Inasmuch as these instances of renewal at the organ level all take place following a kind of spontaneous autotomy or regression, they may be considered to be "intentional", as opposed to the more "accidental" cases of regeneration commonly encountered in the amputated appendages of lower vertebrates.

In somewhat the same category are the various epidermal appendages of vertebrates which may be periodically molted and replaced. The repeated development of such structures as hairs and feathers is a familiar example of spontaneous regeneration at the histological level of organization. Equally impressive is the shedding and replacement of the colorful bills of puffins, the keratinized "keel" on the dorsal beaks of male pelicans, and the horn sheaths of pronghorn antelopes. These last three cases are secondary sexual characters which are dispensed with once the mating season is over. Accordingly, their growth cycles are hormonally co-ordinated.

The foregoing examples of turnover at the histological level of organization all require rather elaborate mechanisms for preserving the potential for development into the adult stages of life. This is achieved by not committing *all* cells to the fully differentiated condition, but by reserving a subpopulation of partially undifferentiated cells which constitute a germinative zone to provide for future regeneration of the organ in question.

B. *Tissue growth*

In 1894, Bizzozero proposed a classification of tissues and organs based on mitotic potentialities. Along similar lines, Messier and Leblond (1960) have more recently designated Bizzozero's three kinds of tissues as static, renewing, and expanding, according to whether they lack the capacity for mitosis altogether, undergo cell division to replenish lost cells, or resort to mitosis only to augment a population of permanent cells.

Embryologically, all tissues may be regarded as originally of the expanding type. Only when they undergo certain kinds of differentiation which are incompatible with the retention of mitotic potentialities are some of them converted to static or renewing tissues (Fig. 1).

1. Static and renewing tissues

Static tissues are formed when all cells of a given kind become so fully differentiated that they can no longer divide. If only a fraction of the total population of cells becomes this specialized, while the rest remain incompletely differentiated and mitotically competent, then a renewing tissue may be produced. Thus, static and renewing tissues differ in that the latter retain a subpopulation of immature stem cells from which those of the differentiated compartment are recruited. In static tissues (nerve cells, striated muscle fibers), cells have the potentiality for living as long as the organism survives.

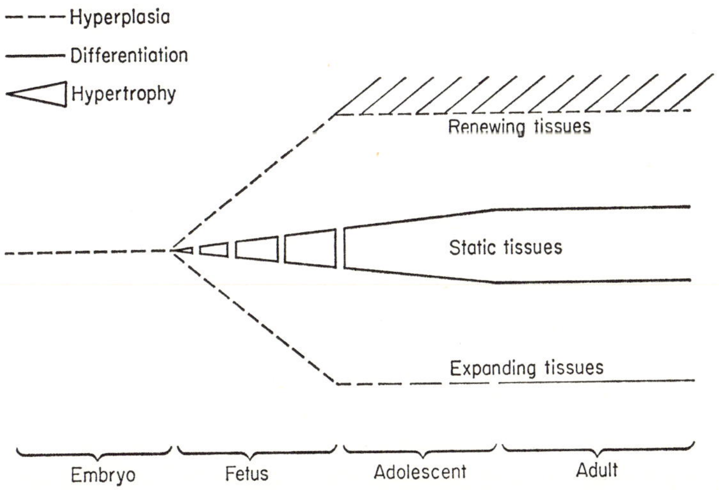

FIG. 1. Alternate pathways of growth and differentiation in tissues and organs. Renewing tissues continue to proliferate throughout life, giving off descendants that differentiate into nonmitotic cells. Static tissues permanently lose the capacity to divide as they differentiate early in life; thereafter they grow by hypertrophy. Expanding tissues do not stop dividing until body growth ceases, despite their fully differentiated condition.

A germinative zone is thus superfluous. The differentiated cells of renewing tissues, however, have relatively short life spans. In some cases, cells are not completely differentiated until they are dead. Expendable cells are dispensed with by being sloughed from body surfaces (epidermis, villous epithelium), extruded via ducts (holocrine glands, gametes), or destroyed internally (blood cells). To the extent that such cells are lost, replacements are differentiated from the stem cell population.

2. Expanding organs

The inability of fully differentiated cells in static and renewing tissues to divide distinguishes them from the cells of expanding organs, which never lose the capacity for mitosis. In organs such as the liver, kidney, and many of the endocrine and exocrine glands, cell division normally enables the organ to add to its mass in keeping pace with somatic growth. It is also a means of replacing accidentally lost cells should the need arise. Inasmuch as maximal differentiation is not incompatible with cellular proliferation in such tissues, these two processes are not spatially segregated. Expanding organs, therefore, do not have growth zones as do renewing tissues.

3. Generalities and exceptions

In general, the three categories of tissues may be characterized as follows: Expanding tissues are capable of cell division when fully differentiated, but do not have growth zones. Renewing tissues can proliferate only in growth zones where differentiation has not yet gone to completion. Static tissues, owing to the total differentiation of all cells, have neither growth zones nor mitotic potential. In most instances, the differentiated cells of expanding and static tissues are long-lived and are therefore not subject to turnover. Those of renewing tissues have short life spans and are constantly being replaced.

Although the above generalizations hold true in most cases, certain organs do not conform to such a simple pattern. The adrenal cortex, for example, does not meet all of the requirements of an expanding organ as do other endocrine glands. Although differentiation and mitosis are not mutually exclusive, proliferation is confined principally to a growth zone between the *zona glomerulosa* and the *zona fasciculata*. Furthermore, the occasional loss of cells from the *zona reticularis* testifies to the definite, albeit slow, turnover of cells (Brenner 1963). Thus, attributes of both expanding and renewing tissues are combined in the adrenal cortex.

Smooth muscle cells resemble static and expanding tissues because of their longevity and lack of a growth zone. Although they contain specialized structural proteins which perform a physical function, they are nevertheless capable of mitosis (Pollister 1932), at least in amphibians. It has been frequently claimed that the mammalion smooth muscle fibers can divide mitotically in such organs as the intestine, bladder and pregnant uterus, but indisputable demonstrations of such cells actually in mitosis are as rare as the event itself appears to be.

Still another exception is the lens of the eye, which shares characteristics of static and renewing tissues. Like static tissues, its fully differentiated cells (lens fibers) survive as long as the organism does. Yet the equatorial growth zone to which mitotic activity is restricted is more typical of renewing tissues. Since growth occurs not by the replacement, but by the addition of cells, the lens also resembles an expanding organ. However, certain biochemical considerations to be discussed later suggest that the lens should properly be regarded as a renewing tissue in which the differentiated cells are neither shed nor destroyed, but are sequestered.

These and other difficult-to-classify structures render the static, renewing, and expanding categories inadequate to account for all of the modes of tissue growth as indicated by the overlapping areas of the diagram in Fig. 2. This is because the growth of *tissues* cannot always be explained in terms of *cellular* phenomena, namely, mitosis and differentiation. A more consistent and satisfactory scheme must therefore relate to individual cell types, instead of tissues, because the latter often consist of many kinds of cells in varying stages of maturation and physiological activity.

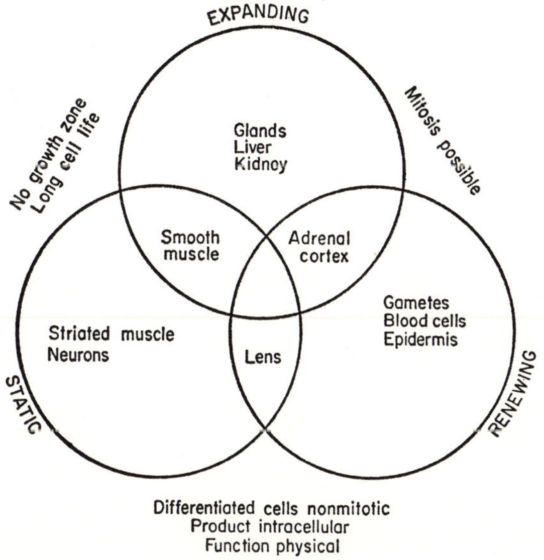

FIG. 2. Diagram illustrating classification of tissues according to modes of growth and types of differentiation. Exceptions to the general rule are listed in overlapping areas.

C. *Cellular hyperplasia*

The inverse relationship between mitotic activity and differentiation has long been defended and disputed (Berrill and Huskins 1936, Dawson 1937, 1940, Trinkaus 1956, Swann 1958, Grobstein 1959, Holtzer 1963). One reason for the confusion lies in the failure to recognize that there are two general categories of cellular differentiation, one compatible with cell division and the other not.

1. Differentiation v. mitosis

As indicated above, differentiation of cells in static and renewing tissues precludes mitosis, while the cells of expanding organs can do both. Such cells as neurons, retinal rods and cones, striated muscle fibers, lens fibers, keratinized epidermis, ameloblasts, red blood corpuscles, and sperm, all are incapable of division and all are associated with either static or renewing tissues. The fact that some of them have shorter life spans than others, and must therefore be renewed by proliferation from germinative zones, is of secondary importance. All differentiated cells are derived ultimately from progenitors capable of division, whether they be neuroblasts and myoblasts found only in immature organisms, or lens epithelium, stratum germinativum, hemocytoblasts, and spermatogonia in adults. These cells have differentiated to the extent that their fates have become considerably restricted, but not necessarily unipotential. Eventually, however, they must pursue the course of differentiation beyond the point of mitotic competence. Whether or not this transition occurs when the destiny of a cell is limited to only one irreversible pathway of differentiation is a facet of the problem which will be interesting to explore. In doing so, however, it may be well to keep in mind the possibility of reversible differentiation.

2. *Dedifferentiation*

Though not common, there exist several cases in which cells already differentiated to the extent of becoming nonmitotic, may revert to a more immature condition and recover their ability to divide. The Schwann cell is a case in point. When it invests an axon with myelin it is a highly differentiated cell and is not known to divide. Yet if the associated nerve fibers undergo Wallerian degeneration, Schwann cells dedifferentiate into mitotically competent phagocytes.

In amphibians there are two well-documented examples of comparable changes. In the regenerating limb, muscle fibers undergo fragmentation into mitotically active mononucleate cells lacking all overt indications of their former differentiated state (Hay 1959). In the newt eye, melanocytes of the iris and retina lose their pigment granules as they regain their reproductive capacities in the course of regenerating lens and retina (Eguchi 1963). Except under very unusual conditions, however, the differentiation of cells in static and renewing tissues is irreversible. The recovery of mitotic competence by such cells is therefore a rather extraordinary phenomenon.

The cells of expanding organs can divide despite the full extent of their differentiation. Yet even these cells cannot divide at all times, for they must temporarily suspend certain other processes (e.g. secretion and synthesis of secretory products) when engaged in mitosis. Cell division therefore cannot take place in the *functioning* cells of expanding organs, nor in the *differentiated* cells of static and renewing tissues.

3. *Structural proteins and mitotic competence*

How is it possible to reconcile these two categories of cellular differentiation with the fact that one permits mitosis and the other does not? To answer this, it is necessary to re-examine exactly what happens when a cell differentiates. Basically, the cell engages in the synthesis of a specific pattern of enzymes which catalyze the synthesis of products (usually, but not always, proteins) either to be retained or secreted. Those cells in which differentiation involves the retention of their specialized products as structural proteins usually forfeit the capacity to divide. Those which secrete their products seldom lose their mitotic potentialities.

This correlation finds support in the following examples. Various differentiated cells of mitotically static and renewing tissues contain in their cytoplasm such specialized molecules as actin, myosin, rhodopsin, hemoglobin, keratin, or lens crystallins. These and other molecules may be organized into such cytological structures as myofilaments, neurofibrillae, outer rod segments, or tonofibrils. Although intracellular structures of these kinds would be expected to create certain difficulties for a cell during division, it is not known if their presence in the cell is what actually blocks mitosis. Presumably it is not, because cytokinesis is only the final act in a sequence of events with beginnings at the biochemical level. Since replication of DNA apparently does not occur in such cells, the repression of mitosis may be inherent in the original synthesis of the cell's differentiated structure.

Characterized by functions of a *physical* nature, the cells of static and renewing tissues perform such services as impulse conduction, reaction to light, contraction, oxygen transport, or various mechanical functions. In these cells, differentiation brings about the formation of intracellular machinery, not to manufacture a product but to carry out a function.

The cells of expanding organs, however, perform *chemical* functions in that their principal physiological activity is to secrete end products which then exert effects outside the cell. These effects may be of a chemical nature, as in the case of hormones, antibodies, and enzymes, or of a mechanical nature, as may be illustrated by collagen, bone, cartilage, and dentine. Differentiation in these cells involves the formation of intracellular machinery to make exportable products.

Thus, we may distinguish two categories of fully differentiated cells. In one, mitosis does not occur, the specific cellular product is an internal, structural protein, and the cell performs a physical function. In the other, mitosis can occur, the specific product is extracellular, and function is chemical in nature.

D. *Turnover at subcellular levels of organization*

The various organelles in the cytoplasm and nucleus, and the molecules of which they are composed, exhibit a wide range of turnover rates. At the molecular level, probably the greatest longevity is represented by DNA, which persists indefinitely, and by collagen, which Thompson and Ballou (1956) described as being "uniquely inert". Most body components, however, turn over at variable rates measured in days or weeks (see review by Neuberger and Richards 1964). Perhaps the shortest life span is represented by messenger RNA, with molecules sometimes surviving only a matter of minutes. It is interesting to note that molecules of mRNA last longer (many hours) in cells which continue to manufacture specific proteins for some time after the nucleus has become inactive. Thus, Scott and Bell (1964) have demonstrated that in the lens, as in the down feather and reticulocyte, molecules of mRNA persist for up to 24 hours. In contrast to other organs with short-lived mRNA, these cell types all lose their nuclei by inactivation or extrusion during the latter stages of differentiation.

1. Mitosis, differentiation and function

In the life of a cell, the conditions of cytoplasmic entities frequently reflect various physiological upheavals, not the least of which is cell division. Some of the organelles manage to pass intact from mother to daughter cells, while other less stable structures disappear during mitosis and must be reconstructed after each division is completed. In the latter category, for example, are the Golgi apparatus and nuclear membrane (Robbins and Gonatas 1964), while the endoplasmic reticulum (Porter and Machado 1960) and mitochondria (Tahmisian *et al.* 1956) persist in the peripheral cytoplasm from one generation to the next, presumably because their services may be required during mitosis.

The fate of intracellular structures during mitosis is particularly important in the case of those organelles which carry out the specific functions of differentiated cells. By studying how they behave, we may learn the extent to which cells must temporarily suspend other physiological activities in order to give their undivided attention to mitosis.

The pros and cons of function v. mitosis have been reviewed by Berrill and Huskins (1936), who emphasized the importance of morphological dedifferentiation in many kinds of dividing cells, and by Dawson (1937, 1940), who in contrast stressed the persistence of many specialized structures in the course of cell division. Judging from the numerous cases on record, it is clear that mitosis is often preceded by the loss of such attributes of differentiation as the motile organs of protozoans, cilia and basal bodies of epithelium, and glycogen and fat in liver cells. Yet there are even more instances in which equally specialized structures have been observed to remain intact in dividing cells. Thus, proximal tubule cells retain their brush borders during mitosis, endocrine and exocrine cells may divide without losing their secretory granules, and chloroplasts and pigment granules may survive mitosis.

Apparently it does not matter whether cellular organelles dedifferentiate or persist during mitosis. In either case at least some new ones must ultimately be produced if the daughter cells are to be provided with their full complement of structures. What is important is the functional state of the dividing cell. Despite the persistence of differentiated organelles in many kinds of dividing cells, there is no conclusive proof that specialized functions are carried on throughout mitosis. The disappearance of the Golgi apparatus probably indicates that the elaboration of new secretory material is interrupted during the mitotic episode, and that such secretory granules as may be present at the time of division are only residual. Whether or not cilia continue to beat in mitotic epithelium is not known, but a negative answer may reasonably be predicted.

2. *Turnover in non-dividing cells*

Aside from mitotically related changes, certain intracellular entities are known to undergo turnover in the normal course of events, but it is important to distinguish morphological reconstruction from chemical renewal. Mitochondria, for example, have been claimed to have a half life of 10·3 days in rat livers. This was based upon measurements by Fletcher and Sanadi (1961) on turnover rates of mitochondrial protein, lipid, and cytochrome c as determined by ^{35}S-methionine and ^{14}C-acetate tracer studies. Since all three components have the same half lives, it was concluded that whole mitochondria, rather than their constituent molecules, were probably destroyed and reconstituted *de novo* in rat liver cells. Along similar lines, Meerson *et al.* (1964) have reported increased rates of mitochondrial destruction and production in hypertrophying rabbit heart muscle. The centrifugal movement of mitochondria along nerve fibers, as observed by Weiss *et al.* (1962), also suggests their degeneration in the distal regions of axons and dendrites.

Other organelles, though less thoroughly investigated, are also believed to undergo turnover. The variability with which endoplasmic reticulum and ribosomes appear and disappear in the cytoplasm suggests that they are subject to destruction and reconstruction depending upon physiological conditions. It is probably safe to assume that the Golgi apparatus is likewise renewed at a rate dictated by functional demands, but its morphological persistence in non-dividing cells suggests that the turnover is probably at the molecular or ultra-structural levels.

In addition to the above cytoplasmic entities common to nearly all cells, differentiated cells contain a variety of highly specialized structures which at least in some cases are known to exist in a state of dynamic equilibrium. It is self-evident that the products of secretory cells are transient in nature, but what of the intracellular products of cells in which differentiation is incompatible with mitosis? In renewing tissues, it is probably reasonable to conclude that such proteins as keratin and hemoglobin are not resynthesized intracellularly, especially since ribosomes are lost in the process of differentiation. Similarly, the crystallin proteins of lens fibers, which may be augmented but not replaced, are so stable that the innermost ones are in a sense "fossils" of proteins synthesized in the embryonic lens. In these cells, too, the apparatus for protein synthesis is lost during the course of differentiation (Eguchi 1964).

In mitotically static tissues, renewal must take place at subcellular levels of organization. In the brain, Thompson and Ballou (1956) detected two principal classes of protein with half lives of 16 and 150 days, respectively. This may account in part for the distal flow in axoplasm in neurons which Weiss *et al.* (1962) found to occur at the rate of about 1 mm per day. Although the soluble components of muscle fibers are replaced at independently variable rates, Dreyfus *et al.* (1960) found that the myosin of rat muscle has a life span of approximately 30 days.

Retinal rod cells also replace the substance of their differentiated elements, for recent studies by Droz (1963) and Young (1965) have shown that the outer rod segments are apparently in a continual state of growth. When labeled amino acids were administered to rats and studied autoradiographically in the retina, the label was found to be located at first in the inner rod segments, but it subsequently moved distally through the outer rod segments, eventually to disappear at the outer ends after a week. What this means in terms of the stability of such morphological structures as the multi-layered membranes of the outer rod segments is not yet known. Such provocative studies as these, however, make one wonder if other specialized cytological structures (e.g. cilia, flagella, myelin sheaths) might not also grow in comparable manner, and if so, where their generative zones may be situated.

III. The Limitations of Growth

Growth is most commonly thought of in connection with the maturation of the organism, but most organs in the body are capable of growing beyond the size limits generally considered normal for the adult. When they do, it may be in compensation for pathological deficiencies, for surgical ablations, or for aging processes. It is legitimate to ask, then, if growth under different compensatory circumstances can be attributed to a single basic cause. Indeed, we must also consider if enlargement in the maturing organism might not be a special case of compensatory growth, and if so, in compensation for what? Finally to what extent does the mechanism of growth in a given organ, tissue, or cell differ according to the circumstances under which size increases? In exploring the last question, we may find answers to the others.

A. *Organs of limited growth potential*

The mechanism by which structures turn over in adults is an extension of how they grow during normal development. In both cases it is necessary to produce new parts at various levels of organization, whether it be to supplement what is already present (growth) or to replenish what has been lost (turnover). Yet in the growing organism, hyperplasia can occur at a higher level of organization than in the adult, although the extent to which this is true varies from one organ or tissue to another. This accounts for the general decline in regenerative potential which most organs and tissues undergo with increasing size and maturity. It may even explain in part why the overall growth of the body slows down or stops altogether when a certain size is reached.

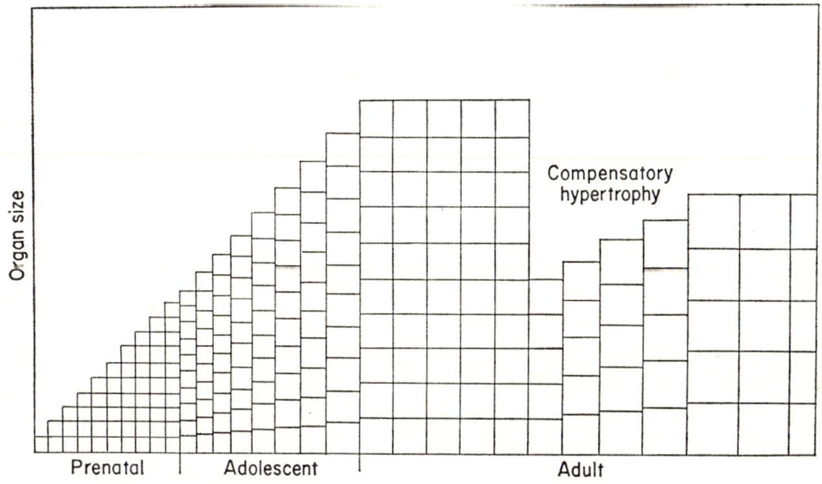

FIG. 3. Graphic representation of how an organ of limited growth potential enlarges. Each square indicates *n* functional units, which can proliferate only during early developmental periods. Subsequent growth occurs solely by hypertrophy, whether it be normal maturation to adult dimensions or compensatory growth after partial ablation (cf. Fig. 4).

The most limited organs in this respect are those in which a pre-determined number of functional units is formed. In striated muscle, the nervous system, and the kidneys, the capacity to produce new muscle fibers, neurons, or nephrons, respectively, is lost during early developmental stages. Subsequent growth must take place solely by enlargement of these basic structural units (Fig. 3) and it is this which accounts for the serious consequences of hypertrophy when carried beyond physiologically advantageous size limits. Hence, muscle fibers, neurons, and nephrons can all enhance their physiological efficiencies by becoming larger only up to a point, for the declining surface: volume ratio in structures that become too big tends to reduce functional competence.

1. Muscle

In cardiac and skeletal muscle, growth occurs during the normal maturation of the individual, and in addition, may be pushed beyond normal limits by increased functional demands. Under both conditions, however, the tissues respond with the only growth mechanism they are capable of undertaking. Thus, hypertension (Carney and Brown 1964) or aortic stenosis (Meerson *et al.* 1964) promotes hypertrophy of cardiac muscle fibers, and administration of testosterone to mice or rats (Venable 1964) stimulates hypertrophy of the *levator ani* muscle. Whatever the cause, hypertrophy can only occur by means of hyperplasia at some lower level of organization.

Studies have shown that in hypertrophic human hearts a twofold average increase in organ weight and cross-sectional fiber area is *not* accompanied by enlargement of individual myofilaments (Richter and Kellner 1963). Apparently these elements cannot alter their dimensions without affecting their functional competence. Myofibrils, which are composed of many myofilaments, were found to be 60% more numerous in hypertrophic fibers than in normal ones, indicating that the optimal dimensions of these organelles may also be limited. Richter and Kellner (1963) suggest that hypertrophy of cardiac muscle fibers may be achieved by the formation of extra myofibrils, as well as by the addition of new myofilaments to the periphery of preexisting ones. Carney and Brown (1964) concluded that new myofilaments must be added peripherally to the myofibrils in order to "augment, rather than disturb, the existing (hexagonal) array". Venable (1964) has proposed that the sarcoplasmic reticulum of *levator ani* muscles may repartition myofibrils as a means of increasing their numbers during fiber hypertrophy. Although the mechanism is not entirely understood, there is agreement that in both cardiac and skeletal muscles the fibers enlarge by increasing the number of fibrils and filaments within them. Thus, the dimensions of the fundamental units of function are held constant during the course of normal ontogeny as well as under conditions of functionally stimulated hypertrophy.

2. Nervous system

Even the brain is capable of growing beyond its normal dimensions under appropriate conditions. Bennett *et al.* (1964) showed that in rats exposed to challenging experiences for 80 days, the cortical regions of their brains were 3·2 to 5·8% larger than in animals maintained in restricted environments. Accompanied by increases in acetylcholinesterase activity, this effect was most marked in the cerebral cortex and especially in the visual regions.

Whether or not the neurons become larger has not yet been determined, but this might be expected in view of how the brain grows during normal ontogeny.

3. Kidneys

In the kidney, the only way growth can occur is by increasing the lengths and diameters of nephrons. This involves both hypertrophy and hyperplasia of the tubular cells not only during normal growth of the individual but also as a result of increased demands for function as caused by high protein diets, excess sodium, potassium depletion, partial nephrectomy, or reductions in the number of nephrons with advancing age. No matter what the cause, the affected nephrons can only respond by hypertrophy. Moderate hypertrophy can alleviate the physiological distress, but excessive enlargement of nephrons is functionally disadvantageous and leads to renal insufficiency.

4. Control of body size

In view of the very limited growth potentials of heart, brain, and kidney, it is possible that the overall size to which the body can grow may be determined by how big these organs can become without sacrificing functional efficiency. Any tendency for the body to grow beyond a mass which can be comfortably handled by these vital organs may logically be expected to diminish their relative physiological performance and thus limit any further somatic enlargement. Thus, body size may be predetermined during early developmental periods when the ultimate number of muscle fibers, neurons, or nephrons is being fixed.

In the case of the kidney, the number of glomeruli increases with the species size, as indicated by Rytand's (1938) studies of miscellaneous animals from mouse to elephant. Kunkel (1930) showed that the number of nephrons in rabbits of different sizes and ages was constant, but that in adult dogs of different breeds the number of glomeruli was proportional to body size. Even when one kidney is congenitally absent (Boycott 1910), or removed from newborn rats (Moore 1929), the remaining organ develops no more than its normal complement of glomeruli. Therefore, what determines how many of these functional units are to be produced appears to be species specific. However, it is possible that such genetic factors may be subject to experimental alteration, perhaps by appropriate manipulation of the inductive or endocrine conditions under which development occurs. Indeed, it might be predicted that if the number of functional units in organs of limited growth potential were artificially increased, then the organism as a whole might be permitted to grow to a larger size commensurate with the greater functional capacities of some of its vital organs.

B. *Organs capable of partial growth*

In the foregoing cases, the inability to augment the normal adult complement of functional units at the cell or tissue levels of organization explains why compensatory growth is achieved only by hypertrophy of these structures. In other organs, however, the capacity to produce new structural units of function at the histological level persists to varying degrees into postnatal or adult life. Consequently, it is possible for such organs to grow beyond their normal adult dimensions by increasing both the numbers and sizes of their basic structural units.

1. Lungs

In the lungs, for example, the number of alveoli increases with age since their dimensions do not change appreciably during maturation. When one lung is removed, however, the way in which compensatory hypertrophy is achieved depends upon the age of the individual. Studies on cats (Bremer 1937) and dogs (Longacre and Johansmann 1940) have revealed that after unilateral pneumonectomy in immature animals the remaining lung enlarges by forming additional alveoli, but in adults compensatory growth occurs by expansion of pre-existing alveoli. Clearly the decline in the capacity of an animal to develop new alveoli after attaining adulthood seriously restricts its ability to compensate efficiently for increased respiratory demands.

2. Exocrine glands

The body's exocrine glands are generally capable of regeneration in adult animals, although the extent to which this happens varies considerably from one gland to another. The acinus is the fundamental histological structure without which normal secretory activity cannot be carried on. During normal ontogeny, growth is achieved by the differentiation of additional acini, for their dimensions are remarkably constant throughout the maturation of the individual. When compelled to undergo compensatory growth, these organs may either produce extra acini, or enlarge those already present. Sometimes they do both.

The salivary glands do not readily enlarge after partial ablation (Alho 1961, Babaeva 1964, Wells and Peronace 1964), but their cells proliferate (Alho 1961). Hypertrophy of the submandibular gland has been reported, however, after denervation of the contralateral gland (Lacassagne and Caussé 1941), ablation of the opposite gland and ligation of the parotid ducts (Wells and Peronace 1964), repeated amputation of the ipsilateral incisor (Well *et al.* 1959, Handelman and Wells 1963), administration of the sympathomimetic drug, isoproterenol (Selye *et al.* 1961), or feeding diets containing proteolytic enzymes (Ershoff and Bajwa 1963, Wells and Voelkel 1963). In all of these cases, enlargement occurs by hypertrophy of the acini and their secretory cells. Conversely, acinar atrophy results from denervation (Lacassagne and Caussé 1941, Wells and Peronace 1964), treatment with adrenergic blocking agents (Handelman and Wells 1963), duct ligation (Standish and Shafer 1957) or liquid diets (Hall and Schneyer 1964). Although the salivary glands readily adjust their size to these various conditions by increasing or decreasing the dimensions of their acini, their capacity to regenerate new acini cannot be ignored in view of Milstein's (1950) description of how new acini bud off severed ducts in partially excised glands. Indeed, Babaeva and Bugrilova (1961) have shown that after removal of 1½ parotid glands from rats the residual parotid enlarges to normal dimensions but the sizes of its acini and cells are preserved. Therefore, new acini must have been formed in the course of regenerative hyperplasia.

It would appear that under conditions of direct injury to a gland, or during the course of normal ontogeny, salivary gland growth is achieved by *de novo* formation of acini. In a variety of other circumstances, characterized by alterations in the demands for functional activity, the response is one of acinar hypertrophy. How this discrepancy is to be explained we do not yet know.

C. *Organs of unlimited growth capacity*

Organs vary in the stage of development at which they lose the capacity to make new functional units. Those which retain their hyperplastic abilities throughout life never have to resort to hypertrophy in their compensatory responses (Fig. 4). They are therefore capable of unlimited growth potentials. All renewing tissues, plus many of the glands in the body, are included in this category.

1. Pancreas

Evidently the pancreas never loses the capacity to form new acini, for although this organ readily enlarges under a variety of experimental and pathological conditions, the diameters of its acini are not known to increase appreciably. The method by which regeneration occurs was first studied after partial pancreatectomy, duct ligation, or transplantation, and found to involve the sprouting of new acini from the duct epithelium (see review by Grauer 1926). It has been shown more recently that if dl-ethionine is injected into rats, the pancreatic acinar cells are selectively and totally destroyed within two to three weeks. Nevertheless, their complete anatomical and physiological restoration takes place after cessation of treatment, although the exact origin of the regenerated acini is not yet clear (Goldberg and Chaikoff 1951, Fitzgerald and Alvizouri 1952, Wachestin and Meisel 1953, Edwards and Brehme 1961). A different effect was investigated by Chernick *et al.* (1948) who discovered that pancreatic hypertrophy occurred when chicks were fed raw soybeans. Attributable to the presence of a trypsin inhibitor in soybean, the growth of the pancreas was accompanied by excessive secretion of all pancreatic enzymes (trypsin, amylase, lipase)

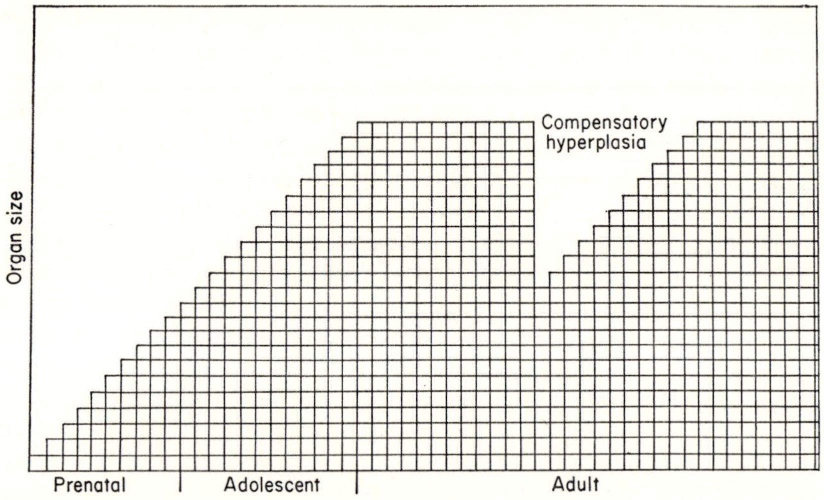

FIG. 4. Growth profile of an organ capable of unlimited enlargement. Each square represents *n* functional units, which can be indefinitely reproduced as long as the body grows. Should the number of such units be reduced, complete regeneration is possible by further increasing the number, not the size, of the functional units (cf. Fig. 3).

(Lyman 1957). Since Booth *et al.* (1960) noted no enlargement of acini in rats fed soybean, despite a 52% increase in relative organ weight, it must be concluded that growth takes place in the pancreas solely by the development of new histological units of function.

2. *Mammary glands*

Mammary glands also retain the ability for acinar regeneration into adult life, a phenomenon which may be observed whenever an animal prepares for lactation. Thus, under the appropriate hormonal influences new acini are budded off the terminal ducts, and during lactation they become enlarged. If one mammary gland is removed during pregnancy or lactation, the other gland grows extra large. Whether or not this involves acinar hypertrophy is not yet settled, for Linzell (1963) reported that in unilaterally mastectomized goats the acini were of normal dimensions despite physiological and anatomical hypertrophy of the remaining gland. Kuosaite (1965), however, noted that enlargement of the mammary acini did occur in similarly treated guinea pigs.

3. Liver

The unlimited regenerative potential of the liver is perhaps best illustrated by the demonstration of Simpson and Finckh (1963) that regeneration can occur at least five times in succession following repeated partial hepatectomies. They estimated that the final restored organ must have been derived from only 4% of the original liver cells. Other conditions which promote overgrowth of the liver include high protein diets (MacKay *et al.* 1928, Walter and Addis 1939), pregnancy (Doljanski 1960), and nephrosis (Lippman *et al.* 1952, Marsh and Drabkin 1958). The last condition may be explained by the excessive loss of plasma proteins due to albuminuria and the resultant stimulation of protein synthesis in the liver. The extent of hepatic enlargement under such conditions is proportional to the rate of protein excretion.

There are probably as many causes of liver regeneration as there are functions performed. Whatever the nature of the stimulus, however, the mechanism by which the liver grows is by the formation of parenchymal cords, which are the basic histological units of liver function. There also occurs some elongation of the cords, for Sidorova (1959) has shown that the hepatic lobules increase almost twofold in cross-sectional area during regeneration. Since this cannot account entirely for the full extent to which the liver mass is capable of enlarging, especially after repeated hepatectomies (cf. Simpson and Finckh 1963), the lobules must grow in length as well as width, a process which would necessitate the differentiation of additional hepatic cords. It is the indefinite retention of this capacity to augment the number of its basic functional units that makes it possible for the liver to grow as much as functional demands may dictate.

4. Endocrine glands

At least some of the endocrine glands also appear to be capable of unlimited growth, for their function depends only upon the integrity of individual cells, and these are all capable of proliferation. The adrenal cortex, which is to some extent a renewing organ, can undergo either compensatory hypertrophy or morphogenetic regeneration under appropriate conditions. The endocrine cells of the ovary are normally reproduced with each new crop of follicles, and as long as estrous cycles continue they are never depleted. The various endocrine cells of the testis, islets of Langerhans, adenohypophysis, and parathyroid also retain the capacity for indefinite multiplication, even though this potential may seldom be expressed.

The thyroid gland is somewhat unique inasmuch as its function depends upon the integrity of its histological, as well as its cytological structure. Yet the thyroid gland can undergo compensatory growth after partial ablation, or can develop into a goitre as a result of iodine deficiency or treatment with thiourea. In either case, its enlargement is attributable to overstimulation by excess thyrotrophic hormone.

Whatever the circumstances under which the thyroid is called upon to enlarge, the mechanism of growth appears to be the same as in the normal ontogeny of the individual. If the thyroid glands of young and old rats are compared, it is evident that growth involves both hypertrophy and hyperplasia of follicles. Similar processes are involved in compensatory growth following removal of one lobe. Gibadulin (1962) has described cellular proliferation, as well as increase in the height of the follicular epithelium, in residual portions of rat thyroids, both reactions being indicative of hypersecretory activity. However, the most striking change is the production of new microfollicles, which are formed by fragmentation or budding from the walls of the larger follicles, and by development from aggregates of interfollicular cells. In these ways, the thyroid gland is always able to proliferate its functional units and can therefore grow indefinitely.

IV. Size, Age and Growth

When forced to work harder, a functional unit first tends to grow in size. This phenomenon may be exemplified by enlargement of thyroid follicles, expansion of pulmonary alveoli, elongation of renal nephrons, or by cellular hypertrophy. Since excessive enlargement ultimately reduces physiological efficiency, it is to the organism's advantage to multiply its functional units whenever possible. This it can sometimes do at histological levels (thyroid and ovarian follicles, hepatic cords), usually does at cellular levels (except in nerves and striated muscle), and must always do at macromolecular levels of organization (because structures such as myofilaments, tropocollagen, and ribosomes have fixed dimensions).

For each kind of tissue or organ, there is a level of organization below which hyperplasia occurs, and above which hypertrophy is the rule. Up to this critical level, functional units are of constant dimensions, indicating the importance of size in their physiological activities. Indeed, the factors which counteract deviations from optimal size limits are so universal as to have operated during phylogeny as well as ontogeny. In both evolution and development, the ways in which organs cope with the problem of how to become bigger without becoming less useful can be compared over a wide range of body sizes.

A. *Evolutionary size adaptations*

Phylogenetically, it is possible to explore these adaptations from mouse to whale (see review by Hosokawa and Sekino 1958). Mitotically competent cells (e.g. hepatic parenchyma, adrenal cortex, parietal cells of gastric glands) are approximately the same size regardless of the species, but neurons and skeletal muscle fibers are both larger and more numerous the bigger the animal. Cardiac muscle fibers, however, are approximately the same size in all adult mammals although the diameters of Purkinje fibers vary with body size (Davies and Francis 1952). Myofibrils and myofilaments, of course, show little or no change from animal to animal.

Histological units of function are sometimes of constant dimensions in all mammals (e.g. pancreatic acini), but commonly grow bigger in larger species. Pulmonary alveoli, for example, range from diameters of about 25 μ in the bat to 277 μ in the right whale. The dimensions of nephrons also increase with species size, but the greatest differences are found among the smaller mammals (Sperber 1944).

As small species evolve into large ones, their functional units tend to enlarge as much as is compatible with the physiological needs of the animal. Along with this trend, there usually occurs an increase in the number of functional units. The latter tendency constitutes a selective advantage only when further enlargement of functional units fails to accommodate for the mounting physiological demands of greater body size. When phylogenetic enlargement occurs exclusively by increments in the numbers of functional units, then it can only be concluded that the uniform dimensions of such structures must be very critical indeed.

Why the phylogenetic enlargement of functional units can be tolerated in some instances (e.g. renal nephrons, pulmonary alveoli, Purkinje fibers) and not in others (e.g. most cells of the body, cardiac muscle fibers, pancreatic acini) must remain a matter for conjecture. It might possibly be related to the fact that the rates of some physiological activities are correlated with body surface area, others with body mass. Increases in functional demands with somatic size would rise less steeply in the former case than in the latter, which may explain why some functional units can afford to sacrifice physiological efficiency for the sake of becoming larger, while others must meet their functional obligations by numerical increments. These same principles may apply to ontogeny as well as phylogeny.

B. *Size changes during development*

In ontogeny, different sized individuals are not of the same ages; in phylogeny, they are of different species. In the former case, as in the phylogenetic examples already discussed, functional units are found to increase in number and/or size during the course of development. When the ability to make new units is lost early in life, only by hypertrophy can further growth be achieved. The determinate number of functional units in the nervous, muscular, and renal systems thus accounts for their very limited capacities for compensatory growth. When the addition of new units can take place up to maturity, somewhat greater powers of regeneration are evident, as in the lungs. Prolongation of hyperplastic competence into adulthood permits unlimited growth in such organs as renewing tissues, the liver, and certain endocrine and exocrine glands in which the number of functional units is potentially indeterminate.

In view of these considerations, it is reasonable that the dimensions of functional units in young v. old animals should be more uniform in organs capable of unlimited growth than in those with restricted potentials. Thus, liver lobules, exocrine acini, thyroid and ovarian follicles, and the cells of renewing tissues all tend to maintain approximately constant dimensions from infancy through adult life. Neurons, nephrons, and the fibers of skeletal and cardiac muscles, on the other hand, undergo hypertrophy as the body grows.

Maturation leads inevitably to aging, which involves the occasional atrophy of individual functional units. Thus, there is a 30% decline in the number of nerve cells in human spinal ganglia with advancing age (Gardner 1940). In senile rats, skeletal and cardiac muscle fibers degenerate (Simms and Berg 1957), and the number of nephrons in the kidney is reduced one-third (Arataki 1926). If this "normal" loss of tissue is to be compensated, it is necessary for most organs to continue growing even after the adult body size has been attained. This process further overloads the already hypertrophic functional units of such organs as the heart and kidney, and may contribute to physiological inadequacies in older individuals. In the case of muscle fibers, Wright (1963) has taken the view that chronic wear and tear may necessitate the constant repair of sarcoplasm and myofibrils with increasing age, but that the progressive depletion of irreplaceable muscle nuclei may ultimately restrict the capacity for fiber regeneration, leading eventually to atrophy and degeneration.

From the foregoing account, it is clear that during development as well as evolution, similar changes occur in the dimensions of functional units when those of small or young animals are compared with those of large or adult animals, respectively. That is, when the sizes of such units increase with phylogenetic enlargement in a series of species, they undergo hypertrophy during the growth of the individual organism. This correlation is illustrated by the previously cited cases of nerve cells, skeletal muscle fibers, Purkinje fibers, and nephrons. Conversely, structures which are approximately the same size in all adult mammals tend to have nearly uniform dimensions during ontogeny (e.g. exocrine acini, thyroid and ovarian follicles, cells of renewing tissues). There are exceptions, however. Cardiac muscle fibers, which are the same diameter in species of different sizes, exhibit considerable enlargement during maturation of the individual. Pulmonary alveoli, on the other hand, enlarge only slightly in the course of normal ontogeny, yet their adult dimensions vary in direct proportion with phylogenetic changes in body size. What basic laws are obeyed in the above cases we do not know.

C. *Comparative kinetics of growth*

In view of how the dimensions of functional units are affected by variations in body size, it is also of interest to explore how the *rates* of growth and turnover vary in ontogeny and phylogeny. In spite of the fragmentary state of our knowledge in this field, it may be possible to reach some general conclusions from a few selected systems.

Turnover in the blood occurs at the biochemical and cytological levels, and in each case there are correlations with species size. McFarlane (1957) has shown that plasma proteins have half lives of 1 to 2 days in the mouse, and 8 to 25 days in humans. Similarly, the reticulocyte percentage, which may be taken as an inverse indication of red cell life span, is higher in the rat than in the dog, and higher in the dog than in man (Boycott 1934). Therefore, in smaller animals plasma proteins and erythrocytes are used up and replaced at faster rates than in larger animals. Such relationships, however, cannot be attributed entirely to the influence of body size, for in human beings the normal reticulocyte percentage is fairly constant from infant to adult.

Most physiological processes, however, seem to go on at approximately the same rate regardless of the size or age of the animal. Such diverse phenomena as hair growth, immunological responses, nerve regeneration, enzyme activities, and wound healing are not known to occur at rates proportional to body size although age changes inevitably occur. Indeed, the life spans of villous epithelial cells or of epidermal cells are fairly constant among mammals, perhaps because the sizes of the villi and the thickness of the epidermis do not vary much with animal size. Although the duration of mitosis is likewise relatively uniform among warm blooded vertebrates, the incidence of cell division in various organs may depend upon their mass. Unfortunately, we do not have enough comparative data to learn whether or not the mitotic index in such organs as the liver or kidney vary from smaller to larger species. In the course of development of the individual, however, it is well known that cellular proliferation declines with advancing age, especially in expanding organs. Moreover, studies of compensatory hyperplasia in the livers and kidneys of rats have shown that maximal mitotic activity declines in magnitude as the animal grows larger (Goss 1965). Whether these effects are due to age or to size, however, will have to be studied in other kinds of animals that do not grow throughout life as does the rat

V. General Conclusions

The fundamental unit of function may be defined as the smallest structure that is still capable of performing the particular task for which a cell or organ is specialized. This unit is commonly represented by the cell itself, as in neurons, blood cells, and in many of the endocrine glands. The muscle fiber, or perhaps even the myofibril, is the indivisible functional unit of heart and striated muscle. At histological levels of organization, specialized physiological activity depends upon the anatomical integrity of such structures as nephrons, hepatic parenchymal cords, thyroid follicles, exocrine acini, pulmonary alveoli, seminiferous tubules, and ovarian follicles.

There is reason to believe that these units function on an all-or-nothing basis, with the rate of output being determined more by the length than the intensity of their activity periods. A fundamental unit of function thus has two speeds, stop and go. The overall activity of most organs is governed by how many of their constituent units are working at a given moment, and not by activity gradations throughout the population of these units. This being the case, whenever an organ is called upon to work harder it will press into service more of its functional units. The ratio of resting to working units, or the ratio of resting to working intervals for an average unit, represents the physiological reserve characteristic of all organs.

An organ can heighten its functional activity only to the extent that all of its units are working all of the time. It is this immediate physiological compensation which insures survival in emergencies until the original margin for safety can be restored by the more dilatory responses of growth. When the number of functional units can be increased without limit, then the organ in question (e.g. liver, endocrine glands, renewing tissues) can adjust indefinitely to the demands for increased physiological activity. It is when there are restrictions to the proliferation of functional units that chronic insufficiencies develop. For example, neurons are capable of regenerating fibers, and of establishing extra synaptic connections via the growth of new collaterals, but their capacities along these lines are limited by the fixed number of nerve cells in the body. Similarly, the augmentation of myofibrils is restricted by the limited size to which striated muscle fibers can grow and still function efficiently. In the kidneys and lungs of adult animals, the number of nephrons and alveoli cannot be increased. Hence, compensation in these organs can only be carried as far as the hypertrophy of their functional units will permit. Whenever the functional demands exceed these limits of compensation, serious pathological consequences ensue.

It is a curious fact that some of the most vitally essential organs of the body (brain, heart, kidneys, lungs) are most lacking in regenerative potential. Indeed, such less necessary organs as endocrine and exocrine glands exhibit greater capacities for growth. How could natural selection have accounted for the evolution of this seemingly disadvantageous situation? It may have been more important for an organism to be healthy and well-adjusted during its reproductive period than to live longer. Natural selection is only advantageous when it contributes to reproductive efficiency whereby the species is perpetuated. Except in humans, longevity beyond the reproductive age can only detract from the common benefit to the population (but sees Brues and Sacher 1965). In this sense, then, it is not how long, but how efficiently an animal lives, that is important. The singular distribution of regenerative potentials in favor of organs that control the body's physiological equilibrium, as opposed to those which are necessary for its very survival, is not so unreasonable when it is realized that the body was never meant to last forever.

REFERENCES

Alho, A. (1961). *Acta path. microbiol. scand. Suppl.* **149**, 1–84.

Arataki, M. (1926). *Am. J. Anat.* 36, 399–436.

Babaeva, A. G. (1964). *In* "Regeneration and Cell Division in Animals" (L. D. Liosner and V. N. Dobrokhotov eds.), *Trans. Moscow Soc. Nat.* **11**, 78–94.

Babaeva, A. G. and Bugrilova, R. S. (1961). *Bull. exp. Biol. Med. U.S.S.R.* **51**, 92–94.

Bennett, E. L., Diamond, M. C., Krech, D. and Rosenzweig, M. R. (1964). *Science* **146**, 610–619

Berrill, N. J. and Huskins, C. L. (1936). *Am. Nat.* **70**, 257–261.

Bizzozero, G. (1894). *Brit. med. J.* **1**, 728–732.

Booth, A. N., Robbins, P. J., Ribelin, W. E., and DeEds, F. (1960). *Proc. Soc. exp. Biol. Med.* **104**, 681–683.

Boycott, A. E. (1910). *J. Anat. Physiol., Lond.*, **45**, 20–22.

Boycott, A. E. (1934). *Trans. R. Soc. trop. Med. Hyg.* **27**, 529–532.

Bremer, J. L. (1937). *J. thorac. Surg.* **6**, 336–343.

Brenner, R. M. (1963). *Am. J. Anat.* **112**, 81–96.

Brues, A. M. and Sacher, G. A. (editors) (1965). "Aging and Levels of Biological Organization." The University of Chicago Press, Chicago and London. 353 pp.

Carney, J. A. and Brown, Jr., A. L. (1964). *Am. J. Path.* **44**, 521–529.

Chernick, S. S., Lipkovsky, S., and Chaikoff, I. L. (1948). *Am. J. Physiol.* **155**, 33–41.

Davies, F. and Francis, E. T. B. (1952). *J. Anat.* 86, 302–309.

Dawson, A. B. (1937). *Am. Nat.* **71**, 605–609.

Dawson, A. B. (1940). *Growth*, Suppl. pp. 91–106.

Doljanski, F. (1960). *Int. Rev. Cytol.* **10**, 217–241.

Dreyfus, J. C., Kruh, J. and Schapira, G. (1960). *Biochem. J.* 75, 574–578.

Droz, B. (1963). *Anat. Rec.* **145**, 157–167.

Edwards, E. L. and Brehme, H. C. (1961). *Growth* 25, 79–87.

Eguchi, G. (1963). *Embryologia* 8, 45–62.

Eguchi, G. (1964). *Embryologia* 8, 247–287.

Ershoff, B. H. and Bajwa, G. S. (1963). *Proc. Soc. exp. Biol. Med.* **113**, 879–881.

Fitzgerald, P. J. and Alvizouri, M. (1952). *Nature* **170**, 929–930.

Fletcher, M. J. and Sanadi, D. R. (1961). *Biochim. biophys. Acta* **51**, 356–360.

Gardner, E. (1940). *Anat. Rec.* 77, 529–536.

Gibadulin, R. A. (1962). *Bull. exp. Biol. Med. U.S.S.R.* **54**, 84–87.
Goldberg, R. C. and Chaikoff, J. L. (1951). *Archs. Path.* **52**, 230–238.
Goss, R. J. (1964). "Adaptive Growth." Academic Press, London and New York. 360 pp.
Goss, R. J. (1965). *Q. Rev. Biol.* **40**, 123–146.
Goss, R. J. (1966). *Science* **153**, 1615–1620.
Grauer, T. P. (1926). *Am. J. Anat.* **38**, 233–253.
Grobstein, C. (1959). *In* "The Cell" (J. Brachet and A. E. Mirsky eds.), vol. 1, pp. 437–496. Academic Press, New York.
Hall, H. D. and Schneyer, C. A. (1964). *Proc. Soc. exp. Biol. Med.* **117**, 789–793.
Handelman, C. S. and Wells, H. (1963). *Am. J. Anat.* **112**, 65–79.
Hay, E. D. (1959). *Devel. Biol.* **1**, 555–585.
Holtzer, H. (1963). *In* "General Physiology of Cell Specializations" (D. Mazia and A. Tyler eds.), pp. 80–90. McGraw-Hill, New York.
Hosokawa, H. and Sekino, T. (1958). *Scient. Rep. Whales Res. Inst., Tokyo* **13**, 269–301.
Kunkel, P. A., Jr. (1930). *Johns Hopkins Hosp. Bull.* **47**, 285–291.
Kuosaite, B. A. (1965). *Bull. exp. Biol. Med. U.S.S.R.* **59**, 97–100.
Lacassagne, A. and Caussé, R. (1941). *C. r. Séanc. Soc. Biol.* **135**, 1017–1020.
Linzell, J. L. (1963). *Q. J. exp. Physiol.* **48**, 34–60.
Liosner, L. D. (ed.) (1963). "Compensatory Hypertrophy of Mammalian Organs." Medgiz, Moscow. 318 pp. (In Russian.)
Lippman, R. W., Marti, H. U. and Jacobs, E. E. (1952). *Archs. Path.* **54**, 169–184.
Longacre, J. J. and Johansmann, R. (1940). *J. thorac. Surg.* **10**, 131–149.
Lyman, R. L. (1957). *J. Nutr.* **62**, 285–294.
MacKay, E. M., MacKay, L. L. and Addis, T. (1928). *Am. J. Physiol.* **86**, 459–465.
Marsh, J. B. and Drabkin, D. L. (1958). *J. biol. Chem.* **230**, 1063–1071.
McFarlane, A. S. (1957). *Progr. Biophys. biophys. Chem.* **7**, 115–163.
Meerson, F. Z., Zaletayeva, T. A., Lagutchev, S. S. and Pshennikova, M. G. (1964). *Expl. Cell Res.* **36**, 568–578.
Messier, B. and Leblond, C. P. (1960). *Am. J. Anat.* **106**, 247–285.
Milstein, B. B. (1950). *Br. J. exp. Path.* **31**, 664–669.
Moore, R. A. (1929). *J. exp. Med.* **50**, 709–712.
Neuberger, A. and Richards, F. F. (1964). *In* "Mammalian Protein Metabolism" (H. N. Munro and J. B. Allison eds.), vol. 1, pp. 243–296. Academic Press, New York.
Pollister, A. W. (1932). *Anat. Rec.* **53**, 11–17.
Porter, K. R. and Machado, R. D. (1960). *J. Cell Biol.* **7**, 167–180.
Richter, G. W. and Kellner, A. (1963). *J. Cell Biol.* **18**, 195–206.
Robbins, E. and Gonatas, N. K. (1964). *J. Cell Biol.* **21**, 429–463.
Rytand, A. D. (1938). *Am. J. Anat.* **62**, 507–520.
Scott, R. D. and Bell, E. (1964). *Science* **145**, 711–714.
Selye, H., Veilleux, R. and Cantin, M. (1961). *Science* **133**, 44–45.
Sidorova, V. F. (1959). *Bull. exp. Biol. Med. U.S.S.R.* **48**, 99–104.
Simms, H. S. and Berg, B. N. (1957). *J. Geront.* **12**, 244–252.
Simpson, G. E. C. and Finckh, E. S. (1963). *J. Path. Bact.* **86**, 361–370.
Sperber, I. (1944). *Zool. Bid. Upps.* **22**, 249–432.
Standish, S. M. and Shafer, W. G. (1957). *J. dent. Res.* **36**, 866–879.
Swann, M. M. (1958). *Cancer Res.* **18**, 1118–1160.
Tahmisian, T. N., Powers, E. L. and Devine, R. L. (1956). *J. biophys. biochem. Cytol.* (suppl.) **2**, 325–330.
Thompson, R. C. and Ballou, J. E. (1956). *J. biol. Chem.* **223**, 795–809.
Trinkaus, J. P. (1956). *Am. Nat.* **90**, 273–289.

Venable, J. (1964). *Anat. Rec.* **148**, 347.
Wachstein, M. and Meisel, E. (1953). *Lab. Invest.* **2**, 253–260.
Walter, F. and Addis, T. (1939). *J. exp. Med.* **69**, 467–483.
Weiss, P., Taylor, A. C. and Pillai, P. A. (1962). *Science* 136, 330.
Wells, H. and Peronace, A. A. (1964). *Am. J. Physiol.* **207**, 313–318.
Wells, H. and Voelkel, E. F. (1963). *Am. J. Physiol.* **205**, 1117–1121.
Wells, H., Zackin, S. J., Goldhaber, P. and Munson, P. L. (1959). *Am. J. Physiol.* **196**, 827–830.
Wright, G. P. (1963). *In* "Research in Muscular Dystrophy", pp. 63–72. J. B. Lippincott Co. Philadelphia.
Young, R. W. (1965). *Anat. Rec.* **151**, 484.

Epigenetic Mitotic Control

W. S. BULLOUGH and E. B. LAURENCE

Department of Zoology, Birkbeck College, University of London, England

I. Introduction

It has long been evident that in any typical adult mammalian tissue the dynamic equilibrium between cell gain, cell function, and cell loss must be controlled by some precise homeostatic mechanism. Recently Bullough (1965) and Bullough and Laurence (1966) have reviewed what is now known concerning this mechanism and have concluded that one basic component may be a tissue-specific chalone. This substance, which is produced within the cells on which it acts, evidently determines the programme of protein synthesis in which the cells indulge, one programme leading to mitosis and the other to tissue function. This theory of tissue autoregulation is in close agreement with the views especially of Osgood (1959), Glinos (1960), and Iversen (1961).

However, if such chalone action is indeed a basic feature of tissue control, it is obvious that different tissues may be expected to show variations on this theme. Thus, for instance, in nervous tissue there is no replacement of those cells that are steadily lost throughout life, while in certain target tissues, such as the vaginal epithelium, cell production, function, and loss are all controlled by specific mitogenic hormones. The reactions of adult tissue cells to chalones and hormones form part of the rapidly growing subject of epigenetics (see Harris 1964, Hechter and Halkerston 1965). It is now commonly believed that any change in protein synthesis within a cell must be dictated at the gene level, and that any change in gene activity must itself be related to some change in the chromosomal environment. Although these mechanisms have been analysed primarily in micro-organisms (see Monod and Jacob 1961), it is becoming increasingly apparent that similar mechanisms must also exist in metazoan cells. In the first place, tissue differentiation may be explicable in terms of gene activation and repression by specific inducing substances, and in the second, chalone-controlled gene activation and repression may determine whether a differentiated cell indulges in mitotic activity or in tissue function (see Bullough 1965).

Also relevant at this point are the recent intensive studies on histones (see Bonner and Ts'o 1964). These basic proteins are principal constituents of the chromosomes and it is believed that when the DNA is bound to them it is incapable of directing the synthesis of mRNA. There seems little doubt that in this way the histones must play an important role in the mechanism of gene control, but it also seems probable that the binding and unbinding of DNA regions of specific informational content may be mediated by other molecules, which may perhaps compete with the histones. Since considerable emphasis has been placed on the evident ionic nature of the DNA-histone association, it is interesting to find that the epidermal chalone is itself basic and might therefore be able to act competitively.

II. Tissue Autoregulation

A. *The epidermal chalone*

Substances which inhibit mitotic activity selectively in their tissues of origin have been extracted from liver, kidney (Saetren 1956, 1963), epidermis (Bullough and Laurence, 1964b, Iversen, Aandahl and Elgjo, 1965), and granulocytes (see Bullough and Rytömaa 1965). Most of the available information concerns the epidermal chalone, and indeed for a study of mitotic and functional homeostasis the epidermis has distinct advantages. In particular it is stratified in such a way that when a cell lies in the basal layer it is most likely to synthesise the enzymes needed for mitosis and when it migrates distally it begins instead to synthesize typical epidermal enzymes (see Mercer 1962). It has been suggested more than once that mitotic activity in the basal cells may be controlled by negative feedback information emanating from the maturing distal cells (Iversen 1961, Mercer 1962, Bullough and Laurence 1964a). Seeking for a diffusible chemical with the necessary characteristics, Bullough and Laurence (1964b) prepared extracts of macerated epidermis. They found that a mitotic inhibitor, evidently specific for epidermis, exists in a water extract, and after some preliminary purification (Bullough, Hewett and Laurence 1964) a major effort has been made by Homan and Hondius-Boldingh (1965) to determine its nature. Briefly, it now appears that the epidermal chalone is a basic glyco-protein with a molecular weight of perhaps 40,000. More details of its physical and chemical nature will be published in due course.

B. *The involvement of the adrenal hormones*

Although it is probable that the epidermal chalone is able to operate by itself, it is also evident that its power is considerably augmented by the presence of adrenalin and of a glucocorticoid hormone (see Bullough 1965). One result of the co-operation of adrenalin in the suppression of mitotic activity is the epidermal diurnal mitotic cycle (Bullough and Laurence 1961) and it has been suggested that the epidermal chalone and adrenalin may form an active but unstable complex. It is significant that similar diurnal mitotic cycles are common in other tissues (see Bullough 1965) and it is possible that all of them are dependent on the co-operation between adrenalin and the tissue chalone.

Regarding the glucocorticoid hormones, most of the evidence, though extensive, is indirect (see Bullough 1965). It has long been known that these hormones can inhibit mitotic activity, at least in some tissues, but the doses used have been more pharmacological than physiological (see Bullough 1952). The time has now come when this subject should be re-examined in detail.

The full meaning of the adrenal hormone involvement in the chalone mechanism is still obscure, but Bullough (1965) has suggested that in wild animals there may be some survival value in the depression of cell production and the maintenance of cell function in any stressful situation, and especially during starvation.

C. *The chalone concentration*

Both Iversen (1961) and Bullough and Laurence (1964b) have assumed that the highest rate of chalone production is in the maturing cells, which in epidermis may extend from the stratum spinosum to the stratum granulosum. Bullough (1965) has also suggested that in any tissue the rate of chalone production is genetically determined and is typical of the tissue. The reduced chalone content of a cell after damage may be due to a reduced rate of synthesis, to an increased rate of loss, or to both. Certainly it is clear that changes in cell wall permeability do occur after wounding and during regeneration. Thus after epidermal wounding the adjacent cells show what is evidently a rapid water uptake leading to an increase of about $\times 2{\cdot}5$ in cell volume (Bullough and Laurence, unpublished), and during liver regeneration the cells are able to absorb intact serum protein molecules which cannot enter normal liver cells (Ghose and Tso 1964).

In this connection the effects of glucocorticoid hormones on cell wall permeability may also be relevant. Takeda *et al.* (1964) have recently described how these hormones inhibit the loss of enzymes from rat liver cells *in vitro*, and if the chalone loss is similarly inhibited then the evident glucocorticoid hormone co-operation with the chalone mechanism might be at least partly explicable in these terms.

Any reduction in chalone concentration due to an increased rate of loss would also be expected to result in a reduced rate of chalone production since the previously mature cells would tend to revert towards mitotic activity. Thus chalone loss and chalone gain may both be self-promoting.

Chalone production may also be reduced by interference, for instance by the escaping contents of damaged lysosomes, in any part of the DNA→ mRNA→ enzyme→ chalone pathway. Again it is interesting that the gluco-corticoid hormones are regarded as preventing the escape of lysosome enzymes by strengthening the lysosome wall (de Duve *et al.* 1962).

D. *The gene mechanism*

The situation in adult epidermis may be summarized as follows. All the cells are fully differentiated, which implies that most of the potentiality originally possessed by the genome has been masked or lost. The most that can remain either actually or potentially active are, first, that group of genes which directs the synthesis of essential metabolic enzymes, and second, those two groups of genes which direct the alternative programmes of protein synthesis for mitosis or for keratin production. A discussion of this situation has already been published (Bullough 1965) and only a brief summary of these alternative programmes need be given here.

1. Synthesis for mitosis

The phases of the mitotic cycle have been redefined (Bullough 1963, 1965) and evidence has been reviewed to suggest that enzyme synthesis may be specially active in early prosphase (which is the whole period of preparation

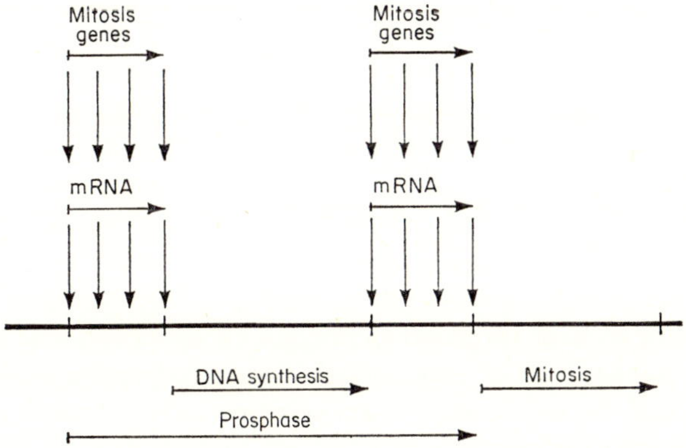

FIG. 1. Diagram illustrating the genetic control of the preparations for mitosis. Gene dependent syntheses are most active during early and late prosphase and are reduced during DNA duplication and during mitosis.

for mitosis), relatively inactive during DNA synthesis, active again in late prosphase (including antephase or G_2), and inactive again in mitosis. This situation is summarized in Fig. 1, and it may explain why the mitotic process is readily inhibited only in early and late prosphase. This theory receives support from work on kidney cortex cells *in vitro* by Kishimoto and Lieberman (1964), who have also listed earlier literature. They conclude that "to acquire competency to form DNA, a kidney cell must undergo a sequence of quantitative and perhaps qualitative changes in RNA and protein metabolism", and that the mRNA synthesis needed for passage through antephase into mitosis is probably mainly completed late in the phase of DNA synthesis. In the latter case the consequent protein synthesis evidently occurs an hour or two later in antephase. It may be noted that results obtained from studies of mRNA synthesis in micro-organisms (see Mittermayer *et al.* 1964) may not be directly relevant to conditions in the adult mammal since the mode of DNA synthesis may differ in detail.

2. Synthesis for tissue function

It has been shown that the non-mitotic cells in any tissue such as epidermis do not all have the same potentialities, and this is illustrated in Fig. 2. Briefly, the cells (*I*) may be immature and preparing for function, or mature

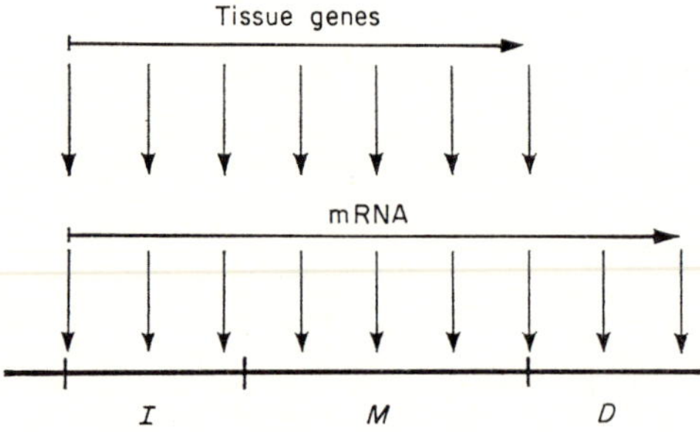

FIG. 2. Diagram illustrating the control of syntheses for tissue function. During the immature (*I*) and mature (*M*) phases these syntheses are gene dependent, but during the dying phase (*D*) they are directed by long-lived messenger RNA.

(*M*) but still capable of reversion to mitosis, or dying (*D*) and incapable of reversion to mitosis. In the epidermis the dying cells belong to the stratum granulosum and the dead cells remain attached as the stratum corneum. It would appear that during the immature and mature stages a cell must continually produce the mRNA appropriate to tissue function. If for any reason it fails to do this it rapidly reverts to the non-functional state in which other mRNA is produced to direct the cell towards mitosis. In dying cells tissue function is evidently maintained without the production of mRNA (see p. 37).

It is the present thesis that a cell progresses towards tissue function in the presence of sufficient chalone and that if the necessary genes are still potentially active, as in immature and mature cells, it reverts towards mitosis if the chalone concentration falls below a critical level (see Fig. 3). The latter situation is most obvious after tissue damage (Bullough and Laurence 1960a). Although there is at the moment no direct evidence linking chalone concentration with tissue function, there is a certain amount of indirect evidence (see Bullough 1965). Also in this connection it is interesting that

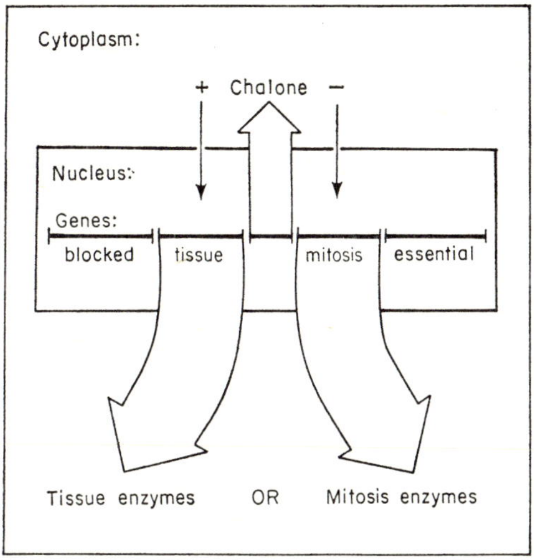

FIG. 3. Diagram illustrating the possible mechanism of chalone control within a cell. With low chalone concentration the mitosis genes are activated and with high chalone concentration the tissue genes are activated.

Sekeris and Lang (1964) and Lang and Sekeris (1964) have recently shown that in liver cells cortisol promotes the production of mRNA leading to liver protein synthesis, which is what would be expected if the glucocorticoid hormones strengthen the chalone action.

E. *The flow of cells through a tissue*

It is evident that, as the chalone concentration in a tissue decreases, the less intense is the cells' preparation for and maintenance of tissue function and the more intense and rapid is their preparation for mitosis; when the chalone concentration increases this situation is reversed.

When epidermal mitotic activity increases, for instance in relation to wounds, not only do many cells revert from preparation for keratin synthesis to preparation for mitosis (see Fig. 4) but so also do all the mitotically active cells pass through the mitotic cycle more quickly than normal. This shorter cycle is especially due to a much shorter interval between the end of the previous mitosis and the beginning of the preparations for the next mitosis. To a much lesser extent it is also due to a shorter prosphase and mitosis (Bullough and Laurence 1966). In wounded ear epidermis the whole mitotic cycle may perhaps be as short as 12 hours instead of the usual 30 days, while mitosis itself occupies only about 1·5 hours instead of the usual 2·6 hours (in sleep) or 3·5 hours (when awake).

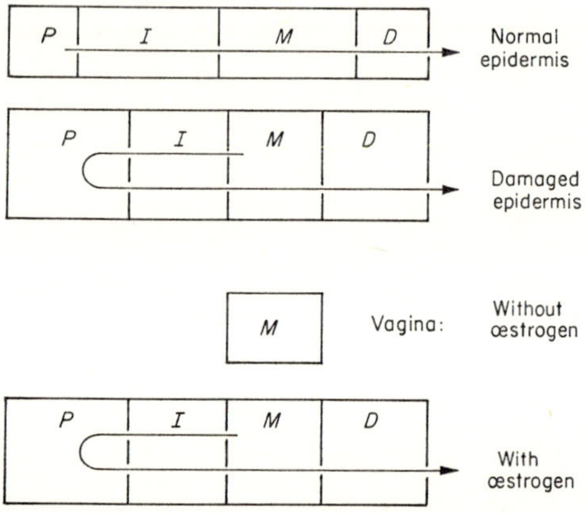

FIG. 4. Diagrams illustrating the manner in which tissue mass may be increased. After tissue damage the mitotically active progenitor cells (P) are augmented by cells which had ceased mitosis and were preparing for tissue function ($I+M$); the reaction of a target tissue to a mitogenic hormone is similar but more dramatic. In both cases the new cells pass through the dying phase (D) to death with unusual rapidity.

The effects of a reduced chalone concentration are also strikingly evident in those cells which remain or manage to become functional. Such cells begin to pass through the various phases of their life ($I+M+D$) in a much shorter time than normal. In epidermis the fastest time recorded from mitosis through maturity to death (in psoriasis) is 4 days instead of the usual 27 days (Scott and Ekel 1963). It is for this reason that a greatly increased mitotic rate does not result in a greatly increased tissue mass. The increased tissue mass becomes stabilized as soon as the newly formed functional cells begin to die.

With increased chalone concentration or efficiency these trends are reversed. Thus in epidermis it has been shown that when, through stress, the mitotic rate is reduced to one quarter of normal, the life expectancy of the maturing cells increases from about the usual 1 month to about 4 months (Bullough and Ebling 1952). In a mitotically inert tissue the life expectancy of the functional cells is extremely long. In mouse liver it has been estimated to be about 3000 days (MacDonald 1961), which is beyond the life expectancy of the animal itself.

Thus increased chalone concentration or efficiency not only seems to result in mitosis suppression and functional promotion but also in some way to reduce the flow of the functional cells towards death. With full mitotic suppression the flow of cells ceases and conditions in the tissue become static, with most if not all cells apparently in the mature stage.

F. *Conclusions*

Mitotic and functional homeostasis is evidently determined by the interaction between the severely limited gene potentiality of the differentiated cell and the micro-environment in which it lies. In the micro-environment the most important factors so far discovered are the chalones, the adrenal stress hormones, and in some cases certain mitogenic hormones. Although any change in the behaviour of a tissue following some alteration in the micro-environment may appear superficially to be simple (as, for example, when there is a shift of emphasis towards mitotic activity) the actual details are complex and involve a number of checks and balances. The most important of these acts to prevent any great change in tissue mass resulting from a change in the mitotic rate.

It is also important to notice that in different tissues the functional cells may be in different phases of cell life. Thus, in tissues with a very low mitotic rate (e.g. liver and kidney), and therefore with a very slow flow of cells towards death, it is the mature cells that are functional, while in tissues with a moderate or high mitotic rate it is the dying cells (e.g. erythrocytes and granulocytes) or even the dead cells (e.g. epidermis and hair) that are functional.

III. Tissue Heteroregulation

A. *Mitogenic hormones*

A number of tissues, especially of the reproductive system, only show high mitotic activity and grow to full size when stimulated by specific mitogenic hormones, the best known of which are the androgens and oestrogens. It appears that while autoregulation by chalones is typical of tissues which are continually functional, heteroregulation by mitogenic hormones is typical of tissues which are only occasionally functional. It is also evident that when, during evolution, a tissue becomes hormone-dependent this is the result of some modification in the tissue reaction; the hormone itself seems to be modified only rarely. This is particularly obvious in the special sensitivity to androgens shown by such epidermal structures as the thumb pad of the male frog and the comb and wattles of the cock (see Bullough 1946); these are mere augmentations of the much lesser sensitivity of the general epidermis (Montagna and Hamilton 1949; Bullough and van Oordt 1950).

It is interesting to note that just as the high mitotic activity of damaged tissues is offset by the reduced life expectancy of the functional cells, so also the high mitotic activity induced by mitogenic hormones is offset in the same way (see Bullough 1965). It is this which limits the growth of reacting tissues and ensures that, when stimulated, an organ merely progresses from one stable point of balance to another.

It seems to be a universal characteristic of such hormone-dependent tissues that the mature cells are non-functional. It is only when a flow of cells towards death is initiated by a raised mitotic rate that cells enter the dying phase and become functional (see Fig. 4).

B. *The gene reaction*

In recent years it has been increasingly suspected that hormones may act as effector-like substances at the gene level. The most striking evidence in favour of this comes from insect material containing giant chromosomes (see Clever 1964, Burdette and Anderson 1965). In these chromosomes the activity of certain genes, as recognized by their "puffing", can be seen to be determined by such hormones as ecdysone. Similar visible proof of hormone action on genes is not available for vertebrates, but there is much indirect evidence to suggest that the situation is similar (see Hechter and Halkerston 1965).

In considering the question of the mitogenic hormones, Bullough (1965) has suggested that their relationship with their target tissues may, at least in part, depend on their ability to neutralize the tissue chalones. This agrees with Kroeger's (1963) conclusion that insect hormones "cannot act directly on the genetic loci, and that there must be an intermediate system which relates the hormonal stimulus to the loci", and with Jensen's (1962) suggestion that in mammals an "estrogen may inhibit some growth-restraining factor, for many aspects of its action are more in keeping with an inhibitory rather than a stimulating role". It is therefore particularly interesting that Talwar *et al.* (1964) have extracted from uterus a factor which is possibly chalone-like and which inhibits RNA polymerase activity in the absence, but not in the presence, of oestradiol. They show, both *in vivo* and *in vitro*, that this factor binds selectively to any available oestradiol.

If it is true that a mitogenic hormone neutralizes the chalone of its target tissue, then in such a tissue the mitogenic effects of hormone action and of wound action should be additive. Although a full hormone reaction and a full wound reaction are both so extreme that no additional response is possible (see Bullough 1946), some evidence of an additive reaction in a subthreshold situation has been obtained using mammary gland *in vitro* (Prop 1964).

C. *Conclusions*

Although ordinary adult tissues may prove to be controlled by a homeostatic mechanism of which a chalone forms an integral part, some tissues are also subject to a second and higher level of control of which a mitogenic hormone forms an integral part. From this latter situation there emerges the impression of a hierarchy of controls, which may commonly be linked to some external factor. Thus in annually breeding female mammals the increasing day length in spring, acting through the sense organs and the brain, triggers the production of gonadotropic hormones by the pituitary, and thus of oestrogenic hormones by the ovary, to modify the genetic control mechanism of a target tissue in an accessory sexual organ.

IV. General Conclusions

It appears that many, if not most, of the maturing and functional cells within a tissue are in an unstable state in the sense that if the chalone mechanism weakens they tend to revert to mitotic activity. This indicates that within such cells those parts of the genome controlling mitosis and tissue function both remain actually or potentially active. When, however, a cell passes into the dying state reversion to mitosis becomes impossible and this may imply that the whole genome has finally ceased to function and that the remaining protein syntheses may all be directed from the ribosome level.

Support for this suggestion comes from studies of lens cells, down feather cells, and reticulocytes (Bishop *et al.* 1961, Humphreys *et al.* 1964, Scott and Bell 1964). In the words of Scott and Bell "in all three cell types the nucleus is effectively 'turned off' as . . . the cell progresses towards its terminal state", and the final period of intense protein synthesis is possible only because of the repeated use of pre-formed and relatively long-lived mRNA. It is possible that at this time some control of the synthesis rate might be exercised through end-product repression (see Donachie 1965).

That this may also be the situation existing in such non-mitotic tissues as striped muscles and nerves has been indicated by Yaffe and Feldman (1964). In striped muscle it appears that the entire genome is finally blocked, that consequently no cell replacement is possible, and that the necessary molecular replacement is directed by long-lived mRNA. Thus both neurones and striped muscle cells may be regarded as being in the dying stage, which in these cases is greatly prolonged. In man it is now known that large numbers of neurones, and probably also of striped muscle cells, are steadily lost throughout life (Curtis 1963).

In apparent contradiction to these conclusions, Harris (1965) and Harris and Watkins (1965) have recently shown that in artificial heterokaryons both

DNA and RNA synthesis can sometimes be restarted. Thus the nuclei, for instance, of bird erythrocytes "can be induced to resume RNA or DNA synthesis, or both, when they are incorporated into the cytoplasm of a cell which normally synthesizes RNA and DNA", although it is not clear from their descriptions whether this activity ever resulted in actual mitosis. In an apparently similar manner RNA and DNA synthesis can recommence in peripheral leucocytes in the presence of phytohaemagglutinin (see Rigas and Johnson 1964). Such observations imply that the blockage of the genome in the dying stage is at least sometimes reversible, although this conclusion may prove to apply only to those cells which are just entering the dying stage and in which the gene blockage is not yet complete.

The work on heterokaryons has also indicated that since "in a common cytoplasm DNA synthesis may, at any one time, be taking place in some nuclei and not in others", preparation for mitosis may not be entirely controlled by the micro-environment of the nucleus. However, it is clear that different nuclei may be controlled by different chalones, that their sensitivity to chalone concentrations may be different, and that their time-lag in reverting to preparation for mitosis may vary (for similar differing time-lags in wounded tissues, see Bullough and Laurence 1960b). In any case, in these short term experiments with heterokaryons it is clear that the nuclei had little time to synchronize their reactions before death occurred after about the third day. Such doubts are strengthened by the well-known fact that in any natural syncytium the nuclei always divide in unison.

Perhaps the most important point arising from this survey is the indication that the mechanisms controlling mitotic activity and tissue function in mammalian tissues, even including nerves and striped muscle, may all conform to a common pattern. This pattern includes, firstly, the closure of the greater part of the genome in response to the micro-environment during differentiation (see also Fujita 1965); secondly, the selection between alternative gene programmes in response to chalonal, and sometimes hormonal, action; and thirdly, during the ageing of the functional cell, the final closure of the genome with function being directed for shorter (e.g. epidermis) or longer (e.g. striped muscle) periods by relatively stable mRNA. Another important conclusion is that in those tissues which retain the power of mitosis excessive growth is prevented by the inverse relationship between mitotic activity and the life expectancy of the mature cells.

REFERENCES

Bishop, J., Favelukes, G., Schweet, R. and Russel, E. (1961). *Nature* 191, 1365–1368.

Bonner, J. and Ts 'o, P. O. P. (editors) (1964) "*The nucleohistones.*" Holden-Day, San Francisco.

Bullough, W. S. (1946). *Phil. Trans. R. Soc.* B **231**, 453–516.
Bullough, W. S. (1952). *J. Endocr.* 8, 265–274.
Bullough, W. S. (1965). *Cancer Res.* 25, 1683–1727.
Bullough, W. S. (1963). *Nature* **199**, 859–862.
Bullough, W. S. and Ebling, F. J. (1952). *J. Anat.* 86, 29–34.
Bullough, W. S., Hewett, C. L. and Laurence, E. B. (1964). *Expl. Cell Res.* **36**, 192–200.
Bullough, W. S. and Laurence, E. B. (1960a). *Proc. R. Soc.* B **151**, 517–536.
Bullough, W. S. and Laurence, E. B. (1960b). *Expl. Cell Res.* **21**, 394–405.
Bullough, W. S. and Laurence, E. B. (1961). *Proc. R. Soc.* B **154**, 540–556.
Bullough, W. S. and Laurence, E. B. (1964a). *Symp. zool. Soc. Lond.* **12**, 1–23.
Bullough, W. S. and Laurence, E. B. (1964b). *Expl. Cell Res.* **33**, 176–194.
Bullough, W. S. and Laurence, E. B. (1966). *In* "Advances in biology of skin. Carcinogenesis" (Montagna, W. ed.). Pergamon Press. New York.
Bullough, W. S. and Oordt, G. J. van. (1950). *Acta Endocr. Copenh.* **4**, 291–305.
Bullough, W. S. and Rytömaa, T. (1965). *Nature* **205**, 573–578.
Burdette, W. J. and Anderson, R. (1965). *Genetics* **51**, 625–633.
Clever, U. (1964). *In* "The nucleohistones" (Bonner, J. and Ts'o, P. O. P. eds.). Holden-Day, San Francisco.
Curtis, H. J. (1963). *Science* **141**, 686–694.
Donachie, W. D. (1965). *Nature* **205**, 1084–1086.
Duve, C. de, Wattiaux, R. and Wibo, M. (1962). *Biochem. Pharmac.* **9**, 97–116.
Fujita, S. (1965). *Nature*, **206**, 742–744.
Ghose, T. and Tso, S. C. (1964). *Nature* **204**, 1210–1211.
Glinos, A. D. (1960). *Ann. N.Y. Acad. Sci.* **90**, 592–602.
Harris, H. (1965). *Nature* **206**, 583–588.
Harris, H. and Watkins, J. F. (1965). *Nature* **205**, 640–646.
Harris, M. (1964). "Cell culture and somatic variation." Holt, Rinehart and Winston. San Francisco.
Hechter, O. and Halkerston, I. D. K. (1965). *Ann. Rev. Physiol.* **27**, 133–162.
Homan, J. D. H. and Hondius-Boldingh, W. (1965). Personal communication (from N.V. Organon, Oss, The Netherlands).
Humphreys, T., Penman, S. and Bell, E. (1964). *Biochem. biophys. Res. Commun.* **17**, 618–623.
Iversen, O. H. (1961). *Acta path. microbiol. scand.* suppl. **165**, 1–74.
Iversen, O. H., Aandahl, E. and Elgjo, K. (1965). *Acta path. microbiol. scand.*, **64**, 506–510.
Jensen, E. V. (1962). *Perspect. Biol. Med.* **6**, 47–59.
Kishimoto, S. and Lieberman, I. (1964). *Expl. Cell Res.* **36**, 92–101.
Kroeger, H. (1963). *Nature* **200**, 1234–1235.
Lang, N. and Sekeris, C. E. (1964). *Life Sci.* **3**, 391–394.
MacDonald, R. A. (1961). *Archs. intern. Med.* **107**, 335–343.
Mercer, E. H. (1962). *Br. med. Bull.* **18**, 187–192.
Mittermayer, C., Braun, R. and Rusch, H. P. (1964). *Biochim. biophys. Acta* **91**, 399–405.
Monod, J. and Jacob, F. (1961). *Cold Spring. Harb. Symp. quant. Biol.* **26**, 389–401.
Montagna, W. and Hamilton, J. B. (1949). *J. exp. Zool.* **110**, 379–396.
Osgood, E. E. (1959). *In* "The kinetics of cellular proliferation" (Stohlman, F. ed.). Grune and Stratton, New York.
Prop, F. J. A. (1964). Personal communication.
Rigas, D. A. and Johnson, E. A. (1964). *Ann. N.Y. Acad. Sci.* **113**, 800–818.
Saetren, H. (1956). *Expl. Cell Res.* **11**, 229–232.

Saetren, H. (1963). *Acta chem. scand.* **17**, 889.
Scott, E. J. van and Ekel, T. M. (1963). *Archs. Derm.* 88, 373–381.
Scott, R. B. and Bell, E. (1964). *Science* **145**, 711–713.
Sekeris, C. E. and Lang, N. (1964). *Life Sci.* 3, 169–174.
Takeda, Y., Ichihara, A., Tanioka, H. and Inoue, H. (1964). *J. biol. Chem.* **239**, 3590–3596.
Talwar, G. P., Segal, S. J., Evans, A. and Davidson, O. W. (1964). *Proc. natn. Acad. Sci. U.S.A.* **52**, 1059–1066.
Yaffe, D. and Feldman, M. (1964). *Devl. Biol.* **9**, 347–366.

Environmental Feedback Control of Cellular Growth

ANDRÉ D. GLINOS

Department of Cellular Physiology, Walter Reed Army Institute of Research
Walter Reed Army Medical Center, Washington, D.C.
and Department of Zoology, University of Maryland,
College Park, Maryland, U.S.A.

FOR a number of years my associates and I, first at Johns Hopkins University and later at the Walter Reed Army Institute of Research and the University of Maryland have focused our research efforts on the subject of this symposium, because of its great theoretical interest and because controlled cellular growth and proliferation is the determining factor in all healing and repair processes following injury. The first model system used in these studies was the regenerating liver of the rat and our findings together with their subsequent confirmation by other independent investigators are summarized in Table I. Based on the first three observations listed in this table, it seemed reasonable to conclude that humoral factors are involved in the control of liver regeneration and that their physiological mode of action consists of growth inhibition rather than stimulation (Glinos and Gey 1952, Glinos 1956, 1958a). On the basis of the fourth finding it was concluded that the decrease of these factors in the environment of the cells of the liver leads to activation of RNA and protein synthesis which in turn induces DNA synthesis and mitosis (Glinos 1956, Glinos and Moore 1956, Glinos 1958a,b, Glinos 1959). By postulating that the site of synthesis of these inhibitory factors is the liver itself, a feedback control emerges (Glinos 1958a) which is in good agreement with both the Weiss (1947, 1955) theories of autoregulation of growth and with recently developed concepts regarding the sequence: derepression of mRNA directed ribosomal synthesis of necessary enzymes→DNA synthesis→cell division (reviewed by Baserga 1965).

In addition to the type of experiments described in Table I, the finding that liver autografts respond to partial hepatectomy with changes in cytoplasmic basophilia, DNA synthesis and mitosis (Virolainen 1964, 1965, Leong *et al.* 1964) provides additional confirmation for the existence of humoral factors in liver regeneration without, however, any indication of their physiological mode of action.

In contrast to the consistency and reproducibility of the experiments hitherto described, parabiosis or sera injection experiments gave inconsistent results until recently. However, repeated review, discussion and investigation of these inconsistencies and their multiple causes (Glinos 1958b, 1960, Weinbren 1959, Bucher 1963, Leduc 1963, Echave Llanos 1963, McDonald *et al.* 1963) led to the recent very large series of observations on parabiotic animals which also confirmed the involvement of inhibitory humoral factors in the control of liver regeneration (McDonald *et al.* 1963).

TABLE I. Findings with regenerating liver.

1. Serum from partially hepatectomized rat inhibits the growth *in vitro* of liver explants and of some established cell lines considerably less than serum from normal rats (Glinos and Gey 1952, Glinos 1958, Moya 1963).
2. Dilution of the plasma *in vivo* by plasmapheresis induces mitosis in the intact rat liver (Glinos and Gey 1952, Glinos 1956, Naora and Naora 1957, Glinos 1958, Bucher 1958, Virolainen 1965).
3. Concentration of the plasma *in vivo* by fluid restriction inhibits mitosis in the regenerating liver (Glinos 1958, Hemingway 1961).
4. After partial hepatectomy or plasmapheresis the earliest detectable change in the liver concerns the distribution of the cytoplasmic basophilia (ribosomes) of the hepatocytes. This change evolves within the lobule according to a well-defined pattern which is also exhibited by the ensuing mitotic process (Glinos and Moore 1956, Glinos 1958, Oehlert *et al*, 1962, Bade 1964, Virolainen 1964).

The next question concerns the identity of these factors and based again on the findings of Table I, the suggestion was made that the proteins in the immediate environment of the regenerating cells i.e. in the extracellular fluid of the liver, may participate in regulating liver cell division (Glinos 1956, 1958a). To the extent that proteins were proposed as regulators of cell division, this hypothesis was found to be consistent with theoretical considerations derived from a formal treatment of template activity regulation (Kavanau 1964), with current concepts concerning mechanisms of DNA repression (Hamilton 1963) and with experiments on a cell free system derived from rat liver showing that normal liver microsomes contain more of a thermolabile inhibitor of protein synthesis than microsomes from regenerating liver (Hoagland *et al.* 1964). To the extent that these proteins were considered to be identical with the proteins of the circulating plasma, the hypothesis was found to be consistent with theoretical considerations on the function of the liver in maintaining plasma protein levels (Glinos 1956, 1958a, Naora and Naora 1957, 1964, Virolainen 1965) and on the possible participation of corticosteroids in controlling liver cell division (Hemingway 1961);

the hypothesis was not found to be consistent with theoretical considerations on hepatic hemodynamics and lymph formation (Brauer 1963) and with experiments on a cell free system derived from rat liver and showing that whole blood, dialyzed serum and purified rat serum albumin failed to inhibit protein synthesis (Hoagland *et al.* 1964).

These considerations indicate that while the growth of the liver after partial hepatectomy may, at least in part, be controlled by inhibitory proteins synthesized in the liver, these need not be identical with the bulk proteins of the circulating plasma, and that while depletion of the plasma proteins may induce liver cell proliferation, this need not be the main factor responsible for the growth of the liver after partial hepatectomy.

The identification of the effectors of humoral self-regulation in cell growth would be greatly facilitated by a model system where the immediate environment of the cells, as distinct from the circulating plasma, would be readily accessible and where other controlling and modifying factors (Abercrombie 1957), could be either eliminated or brought under strict experimental control. In this respect cell culture systems are vastly superior to regenerating liver and other tissues in situ. In such systems the immediate humoral environment of the cells is readily accessible and its modification by the metabolic activity of cell populations other than the one studied is eliminated. The operation of growth control mechanisms other than humoral, such as contact inhibition (Weiss 1950, Abercrombie 1961, Levine *et al.* 1965) can also be eliminated by using suspension rather than monolayer cultures. With the exception of medium renewal, which however is under strict experimental control, such systems are essentially closed and control of cellular growth can only be effected through environmental feedback, with the rate of metabolic reactions underlying growth and division depending on the composition of the medium while the latter in turn is affected by cellular metabolism. Figure 1 shows schematically how growth control through environmental feedback would be expected to result in a high cell density population exhibiting metabolic characteristics of maintenance and differentiated function. Removal of part of such a population would be followed by active growth and division of the remaining cells until the original population density would be restored. The requirement for suspension cultures in order to avoid contact inhibition can, of course, be satisfied only by using established heteroploid cell lines. Under these conditions, the occurrence of a reversible shift from growth and division to maintenance and differentiated function, would provide an exceptional insight into the humoral mechanisms involved in the self-regulation of cell growth.

Our investigations into whether such a shift does indeed occur, were carried out with L-929 cells, clone WRL10A (Glinos and Hargrove 1965) grown as stock suspension cultures in Eagle's spinner minimum essential medium (sMEM) supplemented with 10% horse serum. These cultures were maintained in logarithmic growth by periodic medium renewal and dilution so that the population was maintained between 3 and 10×10^5 cells/ml. Doubling time was 22 to 28 hours and plating efficiency, determined on trypsinized single cell suspensions to avoid any errors due to clumping, 50 to 60%.

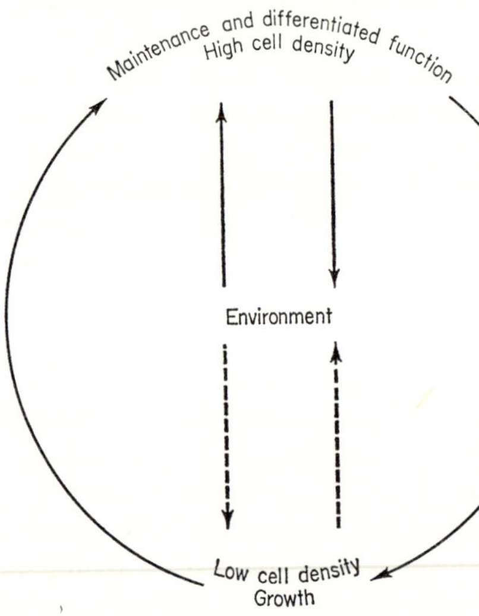

FIG. 1. Environmental feedback control of cellular growth in a closed system. The composition of the humoral environment of the cells affects their metabolism and is in turn affected by it. The solid and broken arrows indicate that this relationship would change as a function of cell density and would determine whether maintenance and differentiated function or growth would prevail. See also text.

Experimental cultures were set up in the same (sMEM) medium. Initial total suspension volume and population density were 200 ml and 3·5 to $4 \cdot 5 \times 10^5$ cells/ml respectively; temperature of incubation was 35°C and the pH was controlled by using 5% CO_2 either in a commercial incubator or in the incubation chamber previously described (Greer and Glinos 1964). In the first series of experiments, the cultures were grown for the indicated periods without medium renewal; in the second, the medium was renewed daily without dilution of the cell population, by centrifuging the entire culture at 1000 rev/min for 25 minutes and re-suspending the cells in fresh medium. In order to obtain proper dispersion of the cells, sampling began 2 hours after setting up or medium renewal, and continued at desired intervals. Cells were counted with a hemocytometer. In case more than 2 clumps per field were seen or more than 5 cells per clump were present, the counts were repeated on a single cell suspension obtained by trypsinization. Relative viability was evaluated on the basis of the ratio of the plating efficiency of trypsinized single cell suspensions from experimental cultures to the plating efficiency of suspensions from logarithmic stock cultures. Cell size was determined electronically with the Coulter Counter using a 100 μ diameter aperture tube calibrated with pollen grains of known size; medium pH, glucose and lactate, and cellular DNA, RNA and protein were determined using standard procedures. Oxygen utilization by the cells and oxygen dissolved in the medium were measured as previously described (Greer and Glinos 1964). Cellular ATP levels were measured by a modification of the luciferase reaction (Strehler 1963). Lactate dehydrogenase activity (LDH) and soluble protein were determined both in the medium and the 10,000 g supernatant of sonicated washed cells using the Bergmeyer *et al.* (1963), and Lowry *et al.* (1951) procedures respectively and the results were expressed as LDH specific activity in Wrobleski-La Due (1955) units per μg of protein.

Cultures with initial cell densities of 3·5 to 4·5 $\times 10^5$ cells/ml grown for 48 hours without medium renewal, reached the plateau phase at cell density levels of 10 to 20 $\times 10^5$ cells/ml, with relative viability and cell density declining rapidly thereafter (Fig. 2A lower curve, 2B, 2C, cell population curves). The final phase of these cultures, therefore, appears to be a starvation plateau of short duration.

In contrast, when carefully manipulated, cultures with daily medium renewal, reached density levels of 65 to 75 $\times 10^5$ cells/ml within 2 to 3 weeks (Fig. 2A upper curve). These levels were quite stable with daily fluctuations due to medium renewal or sampling manipulations less than 5% and no decrease in viability (Fig. 2D upper curve). These high density cultures, therefore, represent a stationary phase of long duration.

Since growth inhibitition and cell death may result from a low pH, frequent pH determinations were carried out during the growth of these cultures in the CO_2 incubator. An average value of pH 7·23 was found, which did not change significantly during the progression to either the starvation plateau or the stationary phase.

Rates of glucose utilization and lactate production and utilization in cultures progressing from logarithmic growth to starvation plateau were variable (Fig. 2B, 2C). There was no correlation between glucose utilization rates and the rates of logarithmic growth or the final cell densities achieved. Also, there was no correlation between the presence or the absence of glucose in the medium and the time of the termination of the plateau phase through loss of viability. Rates of glucose utilization in stationary cultures were also variable and in general lower than during logarithmic growth (Fig. 2D). Very low levels and finally depletion of glucose occurred always between 12 and 24 hours after medium renewal. During this period of glucose depletion, approximately one half of the lactate which had accumulated in the medium earlier, was also utilized.

The decrease of cell size and of cellular protein, RNA and DNA reported by earlier investigators (Kuchler and Merchant 1956, Saltzman 1959, Eidam and Merchant 1965) were also noted here as cultures progressed from

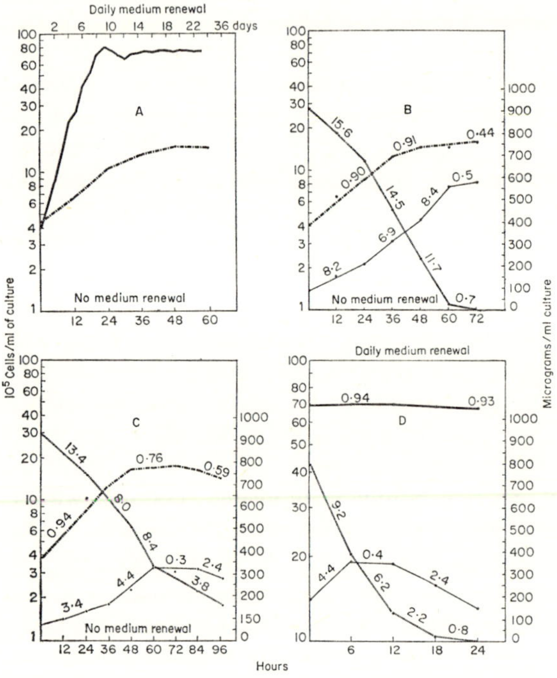

FIG. 2. Cell population kinetics in cultures with (——) and without (-·-·-·-·-) medium renewal (A) and concentrations of medium glucose (··············) and lactate (——) during growth without medium renewal of cultures with high (B) and low (C) glucose utilization rates and during a typical interval between successive medium renewals of high cell density stationary cultures (D). The curves in A represent typical single experiments; in B, C, D averages of three or more replicate experiments. In order to obtain proper dispersion of the cells the O time sample in A, B, C, was obtained two hours after setting up of the cultures and in D two hours after medium renewal. The numbers accompanying the cell population curves indicate relative viabilities as defined in the text, those accompanying the glucose and lactate curves, average rates of utilization or production over the corresponding curve segments in $\mu g \times 10^{-6}$/cell/h.

logarithmic growth to starvation plateau (Table II). Values for cell size, DNA, RNA and protein in high density stationary cultures remained within the limits characteristic for logarithmic growth.

Cultures progressing from logarithmic growth to starvation plateau exhibited minor fluctuations of the LDH specific activity (Table II) but due to variation from experiment to experiment no distinct pattern similar to the one reported by De Luca and Nitowsky (1964) could be established. In high density stationary cultures, LDH specific activity remained constant between media changes, at a level approximately ten times higher than during logarithmic growth. This increase of LDH activity was not due to decreased cell leakage, because unused culture medium and media from logarithmically growing, starvation plateau and stationary cultures had virtually identical activities averaging 40 units/ml of medium. Evidence that the observed

TABLE II. Constitution and LDH activity of cells in the log growth, starvation plateau and stationary phases. Logarithmic growth and starvation plateau data refer to the population curve shown in Fig. 2B; stationary phase data refer to the population curve shown in Fig. 2D.

	Log growth and starvation plateau				Stationary phase			
Hours	0	24	48	72	6	12	18	24
Cell volume (μ^3/cell)	1414	1356	1228	1152	1380	1496	1514	1459
DNA ($\mu g \times 10^{-6}$/cell)	24	24	23	16	25	22	21	28
RNA ($\mu g \times 10^{-6}$/cell)	41	34	38	31	37	36	30	33
Protein ($\mu g \times 10^{-6}$/cell)	336	274	324	250	307	365	287	363
LDH (units/μg protein)	1·04	1·14	0·92	0·79	8·36	9·23	10·5	9·93

differences in LDH activity were not due to the accumulation of dissociable activators or inhibitors was provided by the finding that mixing and dilution experiments on supernatants from sonicated cells of high and low specific activity gave the predicted results.

Chromosome counts and karyotype analysis showed that, on the basis of the rigorous criteria recently developed, the chromosomal complement of cells undergoing their first division after dilution of a high density stationary culture was identical with the complement of cells kept in continuous logarithmic growth described in detail elsewhere (Glinos and Hargrove 1965). This finding and the fact that a culture with typical logarithmic phase characteristics is invariably obtained by dilution of a stationary culture, indicate that the possibility of selection is remote and that most likely, logarithmic growth, starvation plateau and stationary phase represent different functional states of the same cell population.

The plateau phase of cultures without medium renewal appears as a transient state reflecting the decrease of cellular metabolism preceding cell death. Even with glucose present, cell proliferation ceased and relative viability invariably declined below 0·90 between 48 and 72 hours (Fig. 2C), thus indicating exhaustion of other limiting nutrients besides glucose.

The average cell density of stationary cultures was 7×10^6 cells/ml, approximately seven times higher than the geometric mean cell density of cultures grown without medium renewal from 0 to 72 hours. Exhaustion of glucose and other limiting nutrients, resulting in structural changes of the cells and loss of viability, would therefore be expected to occur within 10 hours after medium renewal. The fact that these high density cultures were maintained for prolonged periods on a daily medium renewal schedule (Fig. 2A upper curve) without changes in cell size, cellular DNA, RNA, protein (Table II) and viability (Fig. 2D upper curve), indicates that nutrient utilization rates were lower and that adaptive metabolic changes had taken place. Thus, because its availability per cell was lower, glucose was utilized at lower rates in these high density stationary cultures. This, however, did not prevent the virtual exhaustion of glucose from the medium at 12 to 18 hours, which in turn led to extensive utilization of lactate (Fig. 2D). During the maintenance of these cultures, lactate utilization was a recurrent event characteristic of the second 12-hour interval between medium renewals. Figures 2B and 2C, on the other hand, show that in cultures progressing from the logarithmic to the starvation plateau phase, lactate utilization occurred only as a terminal event concomitant with loss of cell viability. This indicates that in cultures kept in logarithmic growth by periodic medium renewal and dilution, utilization of the accumulated lactate does not occur. The striking increase of the LDH activity in the high density stationary cultures (Table II) would seem therefore to represent an adaptive response related to the recurrent utilization of lactate necessary for the maintenance of these cell populations.

The shift of cellular metabolism from growth to maintenance is further indicated by a consideration of glucose utilization in connection with ATP production and cell division (Table III). In starvation plateau cultures a relatively high mitotic index, in the absence of a net increase of the cell population, indicates replacement of nonviable cells (cf. Figs. 2B, 2C, relative viability data). The persistence of some mitotic activity in the stationary phase indicates primarily replacement of cells lost from the population through sampling and during medium renewal. Preliminary experiments with colchicine and with ^{3}H-thymidine indicate that the duration of mitosis in the stationary phase is increased and that the actual rate of cell renewal is even lower than shown by the mitotic index in Table III. The concurrent high levels of glycolysis, ATP and cellular proliferation during logarithmic growth must be viewed against the background of recent experiments (Goldberg *et al.* 1965) suggesting that glycolysis provides ATP for the

TABLE III. Interrelations between energy metabolism and growth in logarithmic, starvation plateau and stationary phase cells

Growth phase	Interval after medium renewal h.	Geom. mean cell number $\times 10^5$/ml	Glucose utilization micromoles $\times 10^{-7}$/cell Glyco-lytic [a]	Nonglyco-lytic [b]	Total [c]	ATP level micromoles $\times 10^{-9}$/cell	Relative cell number [d]	Mitotic index —
Log	0–24	6·55	15·3	5·9	21·2	7·4	2·50	1·42
Starvation plateau	48–72	17·00	0·0	4·8	4·8	3·7	1·02	0·73
Stationary	0–24	68·93	1·1	6·5	7·6	2·5	0·97	0·16

[a] Glucose converted to lactate.
[b] Glucose uptake)—(glucose converted to lactate); or (glucose uptake)+(glucose equivalent of lactate uptake).
[c] Sum (a+b)
[d] Ratio of the cell density at the end to the cell density at the beginning of each interval.

continuous phosphorylation of the available glucose and not for growth. Experiments on this point are now in progress in our laboratory.

The essential finding shown in Table III is that non-glycolytic glucose utilization is identical in logarithmic and stationary cultures while ATP and cell division in the latter are even lower than during starvation plateau.

A third indication of a shift in the emphasis of cellular metabolism from growth to maintenance is provided by the results of radiobiological experiments (Glinos and North 1963). These experiments showed that the minimum radiation dose necessary for inactivating cells with respect to their ability for indefinite division and clone formation (D_0 or D_{37} dose, cf. Puck and Marcus 1956) was the same regardless of whether the cells originated in the logarithmic or the stationary phase. When inactivated with a dose equal to D_0 or greater, cells of either origin completed a limited number of post-irradiation divisions inversely related to the dose. Given the same inactivation dose, cells originating in the stationary phase completed fewer divisions than cells originating in the logarithmic phase and this difference became more pronounced with increasing dose (Fig. 3).

Inactivated cells are able to complete a limited number of divisions, most likely because of the presence of a nongenetic constituent capable of initiating cell division (Glinos and North 1963, Elkind *et al.* 1963). At dose D_0 genetic damage would preclude further synthesis of this constituent; its concentration at the time of irradiation, therefore, would determine the number of divisions possible before its dilution below a minimum essential level. At doses above D_0 destruction or structural alteration of this constituent itself would limit further the number of possible post-irradiation divisions. On this basis, it can be seen that the difference between logarithmic and stationary phase origin cells (Fig. 3) would be due to a lower concentration of this constituent in the latter, a further indication of the shift of their metabolism from growth to maintenance. By extrapolation from the then available data on cellular protein and RNA during starvation plateau to the stationary phase, a parallel was drawn between a low concentration of this constituent(s) and a low protein and RNA content (Glinos and North 1963). Although these two molecular species are most likely to contain constituents initiating cell division, the data on the stationary phase now available (Table II) indicate that lower concentrations of these constituents need not be reflected in the total protein and RNA content of the cell. This is in good agreement with current concepts regarding specific messengers and enzymes where low relative concentrations would preclude their influencing total RNA and protein levels.

As illustrated in Fig. 1, in addition to maintenance, high density stationary cell populations are expected to exhibit differentiated function. L cells were derived from mouse fibroblasts and recent work on collagen biosynthesis in our laboratory (Maxwell and Glinos, 1966) demonstrated that it does indeed take place in stationary L cell suspension cultures. Additional work in progress on the differentiation of stationary phase cells includes DNA-RNA hybridization aiming at the determination of the number of DNA sites actively transcribing during the logarithmic, starvation plateau,

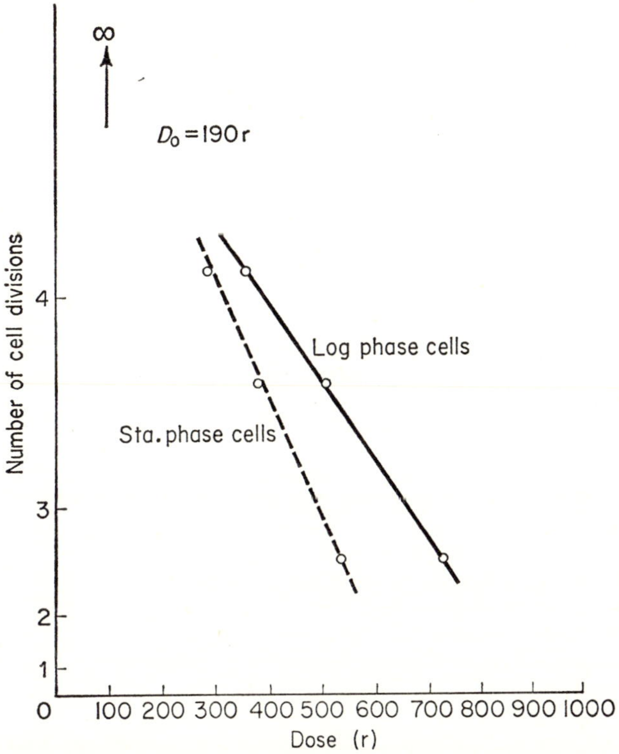

FIG. 3. Relation between dose and post-irradiation division in cells of logarithmic and stationary phase origin. Cells of either origin lose their ability to divide indefinitely at 190r (Do) and complete only a limited number of divisions inversely related to the dose. Given the same dose, stationary phase origin cells complete fewer divisions than cells originating in the logarithmic phase and this difference is more pronounced with increasing dose. (Redrawn from Glinos and North 1963.)

stationary and lag phases; also, studies on RNA messengers, polyribosome formation and protein synthesis in normal and irradiated cells during these different phases.

Pending the outcome of these experiments it would seem reasonable to state that the evidence obtained to date indicates that, in the cell culture system described, a shift of the emphasis of cellular metabolism from growth to maintenance and differentiated function does occur. The system is devoid of contact inhibition and would therefore appear to be well suited for studies of humoral environmental feed-back mechanisms including the possible elaboration by cells of inhibitory substances necessary for the self-regulation of cellular growth.

In terminating I would like to emphasize that although I was the only one able to attend this symposium in person, I spoke as a representative of a team composed of Drs. Maxwell, Bartos and Brown and Messrs. North, Werrlein and Vail, my close associates and co-workers at the Walter Reed Army Institute of Research and the University of Maryland.

REFERENCES

Abercrombie, M. (1957). *Symp. Soc. exp. Biol.* **XI**, 235.

Abercrombie, M. (1961). *Expl. Cell Res.* Suppl. 8, 188.

Bade, E. G. (1964). *Wirch. Arch.* 337, 503.

Baserga, R. (1965). *Cancer Res.* 25, 581.

Bergmeyer, H. U., Bernt, E. and Hess, B. (1963). *In* "Methods of Enzymatic Analysis" (H. U. Bergmeyer ed.), p. 736. Academic Press. New York.

Brauer, R. W. (1963). *Physiol. Rev.* 43, 115.

Bucher, N. L. R. (1958). *In* "Liver Function" (R. W. Brauer ed.), p. 432, Am. Inst. Biol. Sci., Washington.

Bucher, N. L. R. (1963). *Intern. Rev. Cytol.* 15, 245.

De Luca, C. and Nitowsky, H. M. (1964). *Bioch. Bioph. Acta* 89, 208.

Echave Llanos, J. M. (1963). *Revta. Soc. argent. Biol.* 39, 256.

Elkind, M. M., Han, A. and Volz, K. (1963). *J. natn. Cancer Inst.* 30, 705.

Eidam, C. R. and Merchant, D. J. (1965). *Expl. Cell Res.* 37, 132.

Glinos, A. D. (1956). *Science* 123, 673.

Glinos, A. D. (1958a). *In* "The Chemical Basis of Development" (W. D. McElroy and B. Glass eds.), pp. 813 and 841. Johns Hopkins Press. Baltimore.

Glinos, A. D. (1958b). *In* "Liver Function" (R. W. Brauer ed.), pp. 425 and 433. Am. Inst. Biol. Sci. Washington.

Glinos, A. D. (1959). *In* "Problemes d'Ultrastructures et de Fonctions Nucleaires" (J. A. Thomas ed.), p. 197. Masson et Cie. Paris.

Glinos, A. D. (1960). *Ann. N. Y. Acad. Sci.* 90, 592.

Glinos, A. D. and Gey, G. O. (1952). *Proc. Soc. exp. Biol. Med.* 80, 421.

Glinos, A. D. and Hargrove, D. D. (1965). *Expl. Cell Res.* 39, 249.

Glinos, A. D. and Moore, M. O. (1956). *J. Histochem. Cytochem.* 4, 434.

Glinos, A. D. and North, H. H. (1963). *Trans. N.Y. Acad. Sci.* **II**, 26, 145.

Goldberg, E. B., Nitowsky, H. M. and Colowick, S. P. (1965). *J. biol. Chem.* 240, 2791.

Greer, R. B. and Glinos, A. D. (1964). *Fedn. Proc. Fedn. Am. Socs. exp. Biol.* 23, 574.

Hamilton, L. D. (1963). *In* "Aspects of Protein Synthesis" (G. N. Ramachadran ed.), p. 161. Academic Press. London.

Hemingway, J. T. (1961). *Nature* 191, 706.

Hoagland, M. B., Scornik, O. A. and Pfefferkorn, L. C. (1964). *Proc. natn. Acad. Sci.* 51, 1184.

Leduc, E. (1963). *In* "The Liver" (C. Rouiller, ed.), p. 63. Academic Press. New York.

Kavanau, J. L. (1964). *In* "Bone Biodynamics" (H. M. Frost ed.), p. 353. Little, Brown and Co. Boston.
Kuchler, R. J. and Merchant, D. J. (1956). *Proc. Soc. exp. Biol. Med.* **92**, 803.
Leong, G. F., Grisham, J. W., Hole, B. V. and Albright, M. L. (1964). *Cancer Res.* **24**, 1496.
Levine, E. M., Becker, Y., Boone, C. W., and Eagle, H. (1965). *Proc. natn. Acad. Sci.* (US) **53**, 350.
Lowry, O. H., Rosebrough, M. H., Farr, A. L. and Randall, R. J. (1951). *J. biol. Chem.* **193**, 265.
Maxwell, R. E. and Glinos, A. D. (1966). *Fed. Proc. Fedn. Am. Socs. exp. Biol.* **25**, 297.
McDonald, R. A., Rogers, A. E. and Pechet, G. S. (1963). *Ann. N.Y. Acad. Sci.* **11**, 70.
Moya, F. J. (1963). *Expl. Cell Res.* **31**, 457.
Naora, H. and Naora, H. (1957). *Gann* **48**, 255.
Naora, H. and Naora, H. (1964). *Expl. Cell Res.* **33**, 151.
Oelert, W., Hammerling, W., und Buchner, F. (1962). *Beitr. z. Path. Anat. u.z. Allg. Path.* **126**, 91.
Puck, T. T. and Marcus, P. I. (1956). *J. exp. Med.* **103**, 653.
Saltzman, N. P. (1959). *Bioch. Bioph. Acta* **31**, 158.
Strehler, B. L. (1963). *In* "Methods of Enzymatic Analysis (H. U. Bergmeyer ed.), p. 559. Academic Press. New York.
Virolainen, M. (1964). *Expl. Cell Res.* **33**, 588.
Virolainen, M. This Symposium.
Weinbren, K. (1959). *Gastroenterology* **37**, 657.
Weiss, P. (1947). *Yale J. Biol. Med.* **19**, 235.
Weiss, P. (1950). *Q. Rev. Biol.* **25**, 177
Weiss, P. (1955). *In* "Biological Specificity and Growth" (E. G. Butler ed.), p. 195. Princeton University Press. Princeton.
Wrobleski, F. and La Due, J. S. (1955). *Proc. Soc. exp. Biol. Med.* **90**, 210

DISCUSSION SUMMARY

Chairman: Professor H. Teir

Professor Goss's systematic and lucid contribution to the strategy of growth resulted in a lengthy and interesting discussion. Professor Grundmann drew attention to hypertrophy of the cardiac muscle. It was known from Lindsbach's studies that the increase in weight of the cardiac muscle solely through hypertrophy was limited to 500 g. As there were hearts weighing 800–900 g an increase in the number of the fibrils must be expected.

Dr. Alexander pointed out that Professor Goss defined post-mitotic differentiated cells as those which cannot divide and queried the statement that neurons were post-mitotic cells. In Professor Goss's theory of neurons we did not know what constituted structural proteins. Certain derivatives of neurons, such as rods and cones, may fulfil this role. Professor Weiss suggested that the neuron reproduced daily its whole substance and all proteins. There were no structural proteins in the sense of static lasting fixtures.

Professor Goss stated that the inner lens fibre was the only tissue in the body which did not undergo a turnover when referring to this at the molecular level. There is a turnover in all other tissues; even the myofibrils of the muscle fibre had a half-life of approximately 30 days in the rat, so that nothing was static at this level. Professor Weiss pointed out that there was no turnover in the axon at all, except low molecular glucose utilization.

Dr. Lajtha asked Professor Goss to explain the qualitative difference between expanding and renewing cell populations. The answer was that there were two major distinctions: in the expanding tissues the mitotic activity was randomly distributed and not confined to a growth zone; in renewing tissues there was always a special segregation between the differentiated population and the germinating subpopulation. The other distinction was that in expanding populations the specific products of the cell by which it exerts its functions were exported from the cell, which remained compatible with mitosis. In the case of renewing tissues these products were retained intracellularly, and the differentiated cells were incapable of division.

Dr. Glinos's interesting paper was followed by a lively discussion. Professor Wrba asked whether the L-cell was a good system for the study of humoral factors as it was not possible to study organ-specificity. Professor Echave Llanos inquired further how this method could be used to explain Hemmingway's study in England whereby he noticed that adrenalectomy suppressed the response of the liver cells to plasmapheresis. In Dr. Glinos's opinion, Hemmingway's experiments did not show that the response to plasmapheresis was non-existent. Dr. Alexander asked Dr. Glinos why he thought that his *in vitro* model told anything about control in normal tissue. Cells were essentially malignant and if they were injected into animals which did not reject them on immunological grounds, they would grow without control as cancers and kill the animals. Dr. Alexander added that Fisher and Yale, who also grew cells in suspension culture were able to maintain them at a stationary level of about 10 million cells per ml. Yet, these same cells when injected intraperitoneally in mice will grow in ascites in a density of up to something like 10 million cells per ml. which will then kill the animal. Dr. Glinos replied that he was not familiar with Fisher's study but 10 million cells per ml *in vitro* meant that it was a renewal system. Energy was being produced but it was not being used for growth, so in this respect his system might be different.

Professor Wrba and Dr. Glinos discussed whether these cells could be called malignant. In Dr. Koss's opinion, the usage of the term malignant as applied to the cell strain was wrong, but perhaps there would be some agreement if one stated that any line of cells that grows indefinitely was abnormal. Hence any conclusions from the study of such a line of cells could only be limited, and may not have any bearing on normal tissue. Dr. Glinos said in reply that this was a heteroploid line; this was the best way to be more specific about the abnormality. He also said that any conclusions from cell systems *in vitro* have to be carefully extrapolated to the body and stated that his own system is only an approximation which had been chosen because it could be readily controlled. Professor Claude added that it may be exaggerating to say that cells have no bearing on normal cells, because cells are cells.

The paper by Professor Bullough and Dr. Laurence was followed by a long discussion. It was opened by Professor Bolden, who asked if a cell could re-enter the mitotic cycle from the dichophase and, if so, by what mechanism? In Professor Bullough's opinion the cells could return. Any cell which was still under gene control could alter its gene instructions. This is the really critical point, so that if a cell is issuing one set of instructions at the gene level it can also issue another. This must be an effector system and the chalone must play either a direct or an indirect part in this mechanism.

Professor Goss asked how soon after treatment with epidermal chalone was mitosis inhibited. Did it also inhibit DNA replication? Professor Bullough answered that if the epidermal chalone was applied by injection *in vivo* or if it was added to a culture medium *in vitro* the mitotic inhibition began immediately. Regarding the second question, Professor Bullough stated that the epidermal chalone would be expected to operate at the two short periods which occur before and after DNA synthesis. He also said that they knew in fact that the chalone had no effect during the mitosis proper, and that they had good indirect evidence that it had no effect during DNA synthesis.

Dr. Rytömaa raised the question of the exact chemical structure of the epidermal chalone. Dr. Hondius Boldingh answered that most purified fractions seem to be homogeneous when ultracentrifuged. This fraction contains some 65% amino acids and some 20% sugar materials, the rest being water. It is therefore assumed at the moment that the substance will be like a protein. Professor Goss said that this makes it very similar to erythropoeitin. This was true also in Professor Bullough's opinion, but he stressed that the two substances obviously acted in a completely different way.

The discussion continued by questions of Professor Weiss, Iversen, and Grundmann regarding the correlation between cell death (life expectancy) and replacement. Professor Iversen wondered whether it would be more correct to say "the shorter the life span, the higher the mitotic rate", rather than *vice versa*. Professor Bullough replied that although he had no idea which is first in this system, the point was that it was a beautifully balanced mechanism which made a single, integral whole. Professor Goss then brought out the fact that "the smaller the animal, the faster the turnover of red cells". Professor Bullough answered that there was a curiosity in blood cells; these systems were likely to be the most complex of all in body when it came to the mechanism of control. He emphasized, however, that he did not think that blood cells were going to break any fundamental laws.

Possible pharmacological properties of the chalone which switched off mitosis *in vivo* interested Dr. Alexander. One would not expect a physiological growth-controlling effector to work instantly, whereas something with pharmacological or vasal constricting properties might well do so. Does this material have any pharmacological action?

Professor Bullough said that there were two answers to these questions. Firstly, there were no pharmacological side effects. Secondly, the substance probably should act instantaneously. In Dr. Alexander's meaning there could still be a pharmacological action, and this is why it is rather important to establish that it has no such effects. For example, cystein produces an immediate mitotic inhibition within about 2 minutes of injection. Nobody suggested it was a chalone reaction; it was a complex pharmacological mechanism which was missed for 12 years because it was not easy to spot.

The long discussion ended with Professor Bullough's statement that it would be fantastic to find a pharmacological effect on mitosis which is tissue-specific, particularly when the substance was extracted from that tissue and the effective dose was incredibly low.

Section II

General Aspects (B)

Some Theoretical Aspects of Intratissular Regulations of Cell Multiplication

G. D. TUMANISHVILI

Institute of Experimental Morphology, Academy of Sciences of the Georgian SSR Tbilissi, U.S.S.R.

THE methods of dividing tissues and organs into embryonic and "adult" categories are not entirely satisfactory and inaccuracies may occur. Undoubtedly, differentiated embryonic tissues have more features in common with the homologous "adult" than any other form of tissues. The change of volume caused by growth, indicates sharp differences in the various tissues. For embryonic tissues $\Delta V > 0$ (where V is the volume of the tissue) and $\Delta V = 0$ is characteristic of the so-called "adult" tissues, although cell reproduction takes place in the latter case as well. Here the number of newly formed cells is equal to that of dead ones and compensates the loss. But the growth with an increase of volume takes place, not only in embryonic tissues, but at compensatory growth and organ regeneration as well. Hence all the tissues with respect to their growth can be divided into two groups corresponding to $\Delta V > 0$ and $\Delta V = 0$. The third group (nervous tissue) cannot grow practically and require special consideration, which I prefer to leave until more information on the causes of growth inability for these tissues becomes available.

If $\Delta V > 0$ and an increase in the volume takes place because of an increase of the number of cells, then ΔN and ΔV are greater than zero (N is the number of cell nuclei). Later we shall see that it is easier to deal with the number of nuclei than with the number of cells. We cannot doubt that there is a characteristic value in the ratio N/V for each tissue, and stage of ontogeny. The concentration of nuclei (C_N) shows a certain constancy. So, for instance, the value of C_N for the chick embryo liver remains constant from the eleventh to the fifteenth day. Usually we determine C_N by counting the nuclei in the field of vision. The number of nuclei in that field, being inconstant, are directly dependent on C_N and are quite suitable for corresponding calculations. The C_N of the chick embryo liver, obtained in this way were equal to 96 ± 2 from the eleventh to the fifteenth day of incubation. This was established for a large number of embryos. A very small deviation from the mean value shows that the mentioned quantity is constant for four days longer. It is clear *a priori* that C_N is more constant in organs of adult animals, which do not increase their volume. The number of cells N are constant, as $+\Delta N = -\Delta N$ the volume V remains constant and hence the ratio $C_N = N/V$ is constant. The value of C_N alters because of the change in condition of the growth system, such as that of the tissue. C_N depends, for instance, on the age of an animal. It is well known that the part of a rat or mouse liver, left over after partial hepatectomy, following compensatory growth, contains less nuclei per unit volume than intact ones. Then, the C_N began to increase, having remained on a constant level for a long time, but not reaching the value of C_N, which is characteristic of the intact liver. This means that the system has reached a certain level of stabilization, but its state is different from intact liver tissue.

It is quite clear that the quantity C_N reflects automatically a certain aspect of nuclear-cytoplasmic ratios in the tissue, but not in each cell separately. As distances between nuclei are sufficiently large, the C_N in tissues, especially epithelial, does not depend on their diameter but on their number and on the volume of cytoplasm as well in the given volume. Thus we can only judge the number of nuclei in a given volume by the quantity C_N, but we cannot say anything on the concentration of the nuclear matter, depending directly, not only on the number of nuclei, but on their volume as well. It is clear that the quantity C_N is much more suitable for expressing a certain extent of nuclear cytoplasmic ratios in a unique tissue system, than the concentration of cells.

Taking into account the above mentioned considerations we assume that the quantity C_N in the tissue is some parameter reflecting the change of the tissue state. Hence, during the periods when C_N is constant, tissues are in a relatively stable, stationary state. Each temporary deviation of C_N from its characteristic value means that the system has left its stationary state and is in the state of relative instability.

If the general form of the expression of C_N constancy for all systems is given, the cell components of which can multiply, we obtain

$$C_N = \frac{N}{V} = \text{const.} \tag{1}$$

but for the maintenance of the C_N constancy, the following condition should bc valid

$$\frac{\Delta N}{N} = \frac{\Delta V}{V} \tag{2}$$

Equations (1) and (2) characterize the stationary state of the tissue.

In our laboratory we mainly studied, tissues whose volume increased because of the cell multiplication ($\Delta V > 0$). They were the eleven-day-old chick embryo liver, a damaged part of the frog liver and a part of the rat liver remaining after the hepatectomy. In the first of these cases we increased the mitotic activity by means of an injection of 0·03 ml of the hen liver extract into embryos, thus breaking the stationary state of the embryo liver recipient. In all the other cases the interference itself was the condition breaking the equilibrium and the extracts were used as an additional influence for more complete study of the investigated phenomena.

During the work it was possible to observe that the C_N in the tissue depended on the mitotic activity. When the mitotic activity was increased over its normal value, C_N increased too; and when the mitotic activity

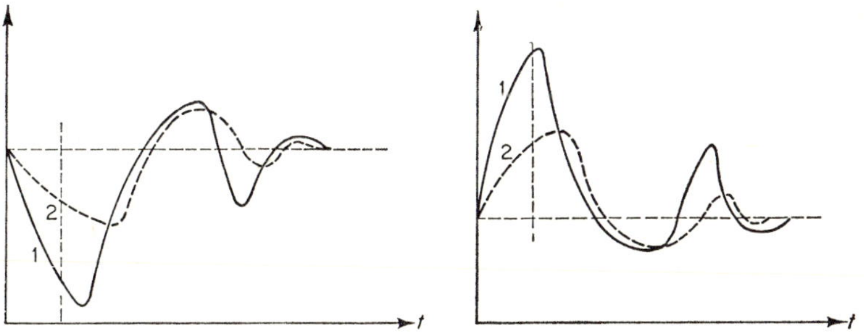

FIG. 1. Theoretical curves showing the relations between the changes of the mitotic activity (1) and the concentration of nuclei (2) in the tissue. A—The initial deviation of the mitotic activity towards its increase. B—The initial deviation of the mitotic activity towards its decrease.

decreased, C_N decreased. Therefore, there is no complete correspondence between the rate of nucleus cell multiplication and an increase in the volume of cytoplasm during the mitosis. The progress of the latter is slower than the former. It is clear from equations (1) and (2) that the change of C_N in such a case will directly depend on ΔN. But if the tissue is not injured the characteristic value of C_N returns rather quickly to its initial value. That can be realized either by an acceleration of the cytoplasm growth or by a decrease in the rate of N increase. It should be specially emphasized that the relationship of C_N and the value of the mitotic activity change in a characteristic way with the deviation of C_N from its characteristic value. Now at the change of C_N the height of the mitotic activity changes in the opposite way, I tried to show this in Figs. 1 and 2 in the form of theoretical curves.

When the primary deviation of C_N from its characteristic value increase is shown in Fig. 1 and when ΔC_N has a negative sign shown in Fig. 2. The real curves are not of a similar regular shape and they experience different variations. It may seem, if the period of the study is not chosen successfully enough (for example, the section up to the vertical line), that the C_N and the mitotic activity change in parallel. But in the following phase of the shown process the situation is sharply changed. The increase of the C_N is accompanied by a decrease of the mitotic activity and that in turn leads to a decrease of C_N. Such a situation may appear when there is a contour with a feedback. The latter can be represented in the following way

$$M \xrightarrow[+]{(+\delta N)} C_N$$
$$\underset{-}{\curvearrowleft}$$

where M is the mitotic activity, C_N in this contour is a peculiar monitor, i.e. it is a regulated parameter. But if this is the case, then the cells composing the tissue do not act as autonomous components, but interact, continuously

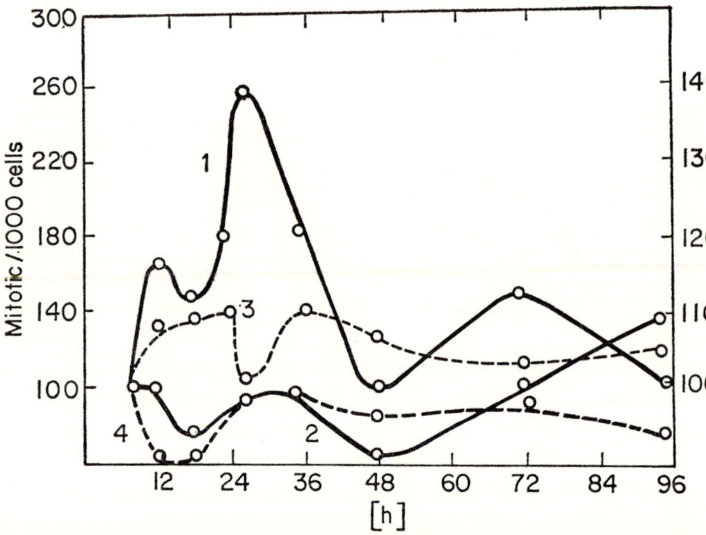

FIG. 2. The change of the mitotic activity and the concentration of nuclei in the eleven-day-old chick embryo liver after an injection of the cytoplasmic fraction (curves 1 and 3) and nuclear fraction (curves 2 and 4) of the hen liver. The solid line shows the changes of the mitotic activity; the dotted line shows the changes of the concentration of nuclei. The numbers are expressed in per cent of the corresponding values obtained for the control embryos.

exchanging information. We think it is possible to consider that certain substances, excreted by cells, can be considered as the means of the signal transmission. Some other scientists, such as Professor Teir and his co-workers came to similar point of view. As an excess of nuclei inhibits mitoses and an excess of cytoplasm causes an increase of the mitotic activity, one must assume that different parts of cells excrete different substances.

Cytoplasm should contain the substance promoting cell multiplication and nuclei the substance inhibiting it.

To verify these assumptions we made experiments with the eleven-day-old chick embryos. The nuclear and cytoplasmic fractions of the hen liver were injected separately into them (separation according to Hogeboom *et al.*, 1952). As shown in Fig. 3, the mitotic activity of the embryo liver injected with the cytoplasmic fraction increases, while the injection of the nuclear fraction causes a decrease of the mitotic activity. But some increase of the relative weight of the liver takes place at the injection of the hen nuclear fraction as well (Fig. 4). It should be emphasized especially that an increase of the relative weight of the chick embryo liver, caused by an injection of the nuclear fraction is accompanied by a decrease of the mitotic activity and a decrease of the C_N in the tissue. At the same time an injection of the cytoplasmic hen liver fraction is accompanied by an increase of the mitotic activity and C_N (compare Fig. 3 and Fig. 4). Thus the relative weight of the embryo organ increases under the action of the homologous cytoplasmic fraction because of promotion of cell reproduction and an increase of their number. The action of the nuclear fraction causes some smaller increase in the relative weight of the organ as well, but due to the promotion of the cytoplasm growth. It becomes clear that an excess of cell nuclei in the tissue causes not only the mitotic activity depression, but an increase in the total volume of the cytoplasm as well. The twofold action of nuclei promotes the fastest recovery of the characteristic value of C_N and the self-regulating contour takes the following form

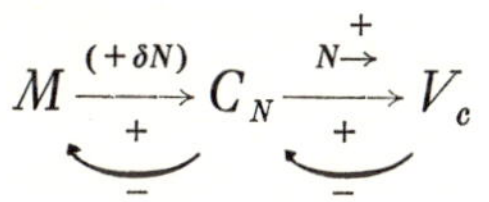

where V_c is the volume of the cytoplasm.

Further experiments have shown some other aspects of nuclear fraction action.

It has found, for instance, that the nuclear fraction of the eleven-day-old chick embryo has no growth inhibiting action with respect to the chick embryo liver of the same age. But if by means of a preliminary injection of the hen liver fraction to increase the mitotic activity of the eleven-day-old chick embryo, then as already mentioned, the concentration of nuclei in the tissue of the embryo recipient liver also increases appreciably by the 36th hour after the injection. If the nuclei is isolated from the embryo liver at this moment, it is found that they have a considerable growth inhibiting activity. Moreover the ability of the cytoplasm to promote cell reproduction in the tissue of the chick embryo liver is rather decreased with the increase of C_N in it.

Now we can construct some empirical system which can form the basis of the quantitative conclusions.

$$\text{I.}\quad \left.\begin{array}{l} N\xrightarrow{-}M \\ N\xrightarrow{+}V_c \\ N\xrightarrow{-}St \\ N\xrightarrow{+}In \end{array}\right\}\xrightarrow{-}C_N$$

$$\text{II.}\quad \left.V_c\xrightarrow[+]{(+\delta St)}M\right\}\xrightarrow{+}C_N$$

where
N is the number of nuclei
M is the mitotic activity
St is the factor stimulating the cell reproduction
V_c is the volume of the cytoplasm
In is the factor inhibiting the cell reproduction

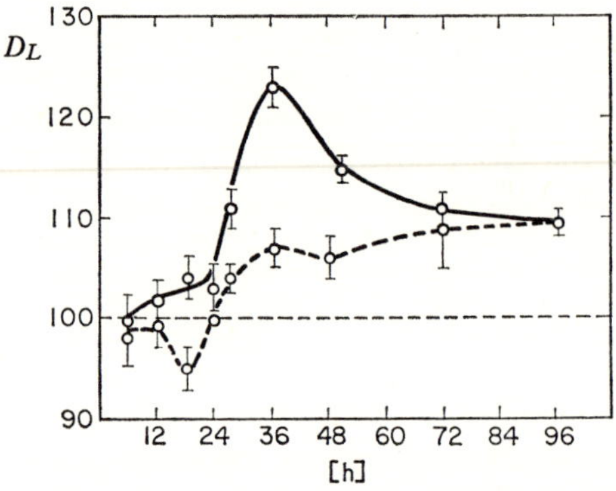

FIG. 3. The change of the relative weight of the liver (D_L) of an eleven-day-old chick embryo after an injection of the cytoplamic (a solid line) and nuclear (a dotted line) fractions of the hen liver.

It can seem that the constructed system contradicts the equation for the concentration of nuclei ($C_N = N/V$). However the system expresses the work of the regulatory contour with the negative feedback and the mentioned contradiction is false. Having assumed that the system is valid, we must conclude that the contribution made by the number of nuclei into the regulation of the parameter C_N is much more than the contribution made by the cytoplasm volume. Taking into account the fact that three from the four links of the first group are directly associated with the mitotic activity, we obtain

$$\left.\begin{array}{l} N \xrightarrow{-} \\[2ex] N \xrightarrow[-]{\overset{-}{\to St}} \\[2ex] N \xrightarrow[-]{\overset{+}{\to In}} \end{array}\right\} M$$

while in the second group only $V_c \xrightarrow[+]{\overset{+}{\to St}} M$. Hence the change of the number of nuclei in the taken volume, i.e. the concentration of nuclei, independently of the distribution of nuclei in the volume and of the medium filling the volume, acts much more on the rate of the cell reproduction than the change of the ratio of the total volume of the nuclear matter to the total volume of the matter of the cytoplasm.

The data taken as the basis of the considerations given in this paper were obtained in the course of study of the liver tissue growth. Nevertheless the extrapolation of our ideas does not encounter any theoretical difficulties. Proceeding from this theory it is easy, for instance, to explain the fact described by Bullough and Laurence (1960). These authors, having removed a part of the mouse ear skin on one side of the ear cartilage, obtained an increase of the mitotic activity in the epithelium on the other side of the ear cartilage. The results of the experiment allowed the authors to think that cells realize only an inhibiting control with respect to each other. Such a conclusion, however, may not be so. The point is that many authors (including those of our laboratory) are aware of the existence of some substances in tissues promoting the mitotic activity in mostly homologous tissues. Similar substances are found in the skin as well. This theory removes the contradiction. The removal of some amount of the promoting mitosis substance together with the cytoplasm is completely compensated by an increase of its formation because of the removal of nuclei. But as we have seen cell nuclei not only depress the mitotic activity, but also, according to the data available, promote the formation of the factor inhibiting the mitotic activity. It is clear that the total effect of the removal of the part of nuclei will be seen in an increase of the mitotic activity in the taken tissue.

I have implied in my considerations that certain substances are used as acting factors. In the most general case we can operate with the number of signals, or, even better, with the amount of information. That will not change anything in the theory. But it seems to me that there are grounds to believe that intercellular signalling is realized just by means of cell secretion (or by their different parts) of special substances.

We managed to establish in our laboratory that the synthesis of DNA in the cell nuclei also depends on the concentration of nuclei and hence on the intercellular interactions. At the same time it was found that *S* phase of the cell cycle is not the link directly influenced by actions regulating the mitotic activity. So, for instance, the mitoses are apparently inhibited when the premitotic synthesis of DNA has taken place, i.e. in G_2 phase of the cycle.

As the discussion of these problems would take too much time, I am obliged only to mention them.

The offered theory gives the possibility of approaching a number of problems from a different point of view and to make predictions, which are often justified.

The described regulation of the rate of the cell multiplication can be considered only as one of the links of the complex regulation chain directed to the conservation of correct size relations between organs.

In conclusion I should like to mention the co-workers of our laboratory N. V. Salamatina, D. D. Tabidze, K. K. Gunia and others who have made and are making a great contribution in the study of the problems which I have tried to elucidate.

REFERENCES

Bullough, W. S. and Laurence, E. B. (1960). *Proc. R. Soc.* **B**, 151, 517–536.

Hogeboom, G. H., Schneider, W. C. and Stribich, M. I. (1952). *J. biol. Chem.* 196, 111–120.

Autoregulation of Growth by Tissue Breakdown Products*

H. TEIR, A. LAHTIHARJU, A. ALHO and K.-J. FORSELL

II Department of Pathology, University of Helsinki, Finland

I. Introduction

Most theories relating to growth control start from the assumption that cells produce substances or factors which regulate their numerical proliferation (Robertson 1923, Turck 1933, Lee and Hanson 1947, Weiss 1947, 1955, Teir 1951, 1952a, Bullough and Rytömaa 1965). The active principle would thus be a product associated with the normal function of the cell.

The first observation at our laboratory in favour of potential tissue-specific growth stimulation was made with the outer orbital gland of the rat. After necrosis and inflammation had been induced experimentally in the gland, the mitotic count was elevated in the contralateral gland (Teir 1951). For some years after this, observations were made of the stimulative effect on the mitotic count of homogenates of the external orbital gland (Teir 1952a, 1952b), skin (Teir *et al.* 1951), the liver (Teir and Ravanti 1953, Blomqvist 1957), the stomach and the intestine (Sundell and Teir 1954) and the thymus (Telkkä and Teir 1955) after intraperitoneal injection of the homogenates of these organs into intact rats. These results were reviewed by Paschkis (1958).

The observations concerning the effect of corticosteroids on mitosis (Bullough 1952, Teir and Isotalo 1953) led us to study the relationship between the local growth factors and general hormones. As some correlations were demonstrated, we ascertained that the growth of tissue is regulated by a hierarchy of growth stimulators, organ- or tissue-specific agents and general (hormonal) factors (Teir 1952a). Our attempts to establish the chemical properties of these growth factors were not very fruitful (Teir 1952b).

In connection with a working series on hepatectomy, a part of the resected liver was left in the free abdominal cavity. The regeneration seen was more pronounced than usual. Ligation of the lobe and the subsequent ischemia and necrosis had a similar effect (Teir and Lahtiharju 1961). This question

* Supported by Sigrid Jusélius Foundation and Damon Runyon Memorial Fund.

was then studied in greater detail by Lahtiharju (1961) who noted distinctly how autolysis and necrosis both *in vitro* and *in vivo* stimulate post-hepatectomy regeneration. These results prompted us to continue the studies, using different organs of the rat as the investigation object.

The concept of the effect exerted on the regeneration phenomena by damaged or necrotized tissue is old. It is known, for instance, that the Egyptian Imhotep believed around 4000 B.C. that injury, traumatic or otherwise produce its effects by "not something entering from the outside", but by "the intrusion of something which the flesh engenders" (quoted by Breasted 1930). Virchow (1858) propounded that a formative substance is liberated in connection with tissue damage and that it stimulates the cells in the proximity to divide and proliferate. But it was nearly fifty years before Haberlandt, in 1921, showed clearly in his studies with cabbages that wound tissue really does participate in the proliferation of the cells in the immediate vicinity. Haberlandt was the first to speak of wound hormones. They have been studied subsequently in the experimental wound healing research field. So far, there is no unanimity concerning the existence and mode of effect of wound hormones of this type (Alho 1961, Teir and Nyström 1962).

The concept of necrohormones was introduced by Caspari (1927). He assumed that substances are released in necrosis which in a small concentration stimulate growth, and in a higher concentration inhibit it. Substances of this kind were studied intensively by Turck (1933) and Lee and Hanson (1947). Turck's point of departure was cytost which he regarded as an endocellular component which is capable of both stimulating and inhibiting the activity of cells close to it. Using the hanging-drop method, Turck found that in small concentrations cytost stimulated and in high concentrations inhibited growth (Turck 1921, 1933). It was emphasized by Lee and Hanson that growth factors produced by the autolyzing cells exert a stimulating or inhibiting effect on homologous species, whereas the factors secreted as a result of normal cell metabolism exert stimulating effects on any species.

Our increasing store of knowledge of the turnover of various tissues and organs has also given topicality to the idea of the physiologic death and elimination of cells. As the homeostasis of an adult organism presupposes that the number of cells is roughly the same, a number of cells corresponding to the production must be eliminated in one way or another. This happens through the necrobiosis of individual cells. An adult human loses considerable amounts of tissue daily in this way. Leblond and Walker (1956) estimated that *c.* $\frac{1}{2}$ lb of epithelial cells of the gastro-intestinal canal are eliminated in 24 hours. About 15×10^{10} granulocytes are produced daily. This same number of cells, about 75 ml in total, must also be eliminated. The amount of tissue to be eliminated in 24 hours from constantly renewing organs is indeed considerable, whereas it is relatively small in stable organs.

The working hypothesis has been that in connection with cellular necrobiosis substances are released which affect the organism as a whole via lymph or blood and especially influence the production of a homologous cell population. We have sought to approach this problem by studying experimentally *in vivo* and *in vitro* the effect of autolyzed tissue on cell production.

II. Outer Orbital Gland

This lacrimal gland, which in rodents is situated subcutaneously before the ear, is an interesting object for growth control studies. There are diploid cells at the moment of birth. During the growing period, 4-, 8-, and 16-ploids originate, the tetraploids being the most numerous (Teir 1944). Tissue homogenates make also tetra- and octoploid cells divide (Teir 1952a).

A. *Coagulation necrosis experiments*

The investigation comprised two experimental series with 2-month-old male rats, forty in one and twenty in the other. Their mean weight was 130 g.

The right outer orbital gland was exposed under superficial ether narcosis, whereupon a glass test tube containing 100°C water was pressed against its surface for 2 min. The colour of the gland surface turned yellowish grey during this procedure. The rats were decapitated after 1, 2, 4 and 7 days in the first and after 2, 6 and 10 days in the second experiment. A unilateral skin incision only was made on the control rats under ether narcosis.

An increase in the mitosis per 20,000 cells was used as the indicator of growth stimulation. It was observed when examining the control animals of both experiments, twenty-six in all, that mere skin incision failed to cause differences in the mitotic indices of the glands of the different sides.

In the first experiment (Fig. 1), mitotic stimulation was achieved after 4 days in the coagulated gland. The mean mitotic index was $10{\cdot}3 \pm 3{\cdot}0$; it was $1{\cdot}6 \pm 0{\cdot}6$ for the contralateral gland and $3{\cdot}2 \pm 0{\cdot}9$ for the control animals. The difference between the treated gland and the control glands was almost significant ($P < 0{\cdot}05$). There was mitotic stimulation in the coagulated glands also after 7 days. The mean mitotic index was $16{\cdot}0 \pm 5{\cdot}3$. The difference from the control value, $3{\cdot}2 \pm 0{\cdot}8$, was almost significant. The index for the contralateral glands of the experimental group was increased, $12{\cdot}0 \pm 5{\cdot}5$, but did not differ significantly from the control value.

In the second experiment signs of mitotic stimulation were seen only in the 10-day group. The mean mitotic index of the coagulated glands in this group was $8{\cdot}1 \pm 3{\cdot}4$, for the controls $1{\cdot}7 \pm 0{\cdot}5$. The difference was almost significant. No stimulation was established on the contralateral side.

B. *Ligation experiments*

There were two experimental series. In one of them the test animals were 24 six-week-old male mice, mean weight 23 g, and in the other 19 six-week-old male rats, mean weight 125 g.

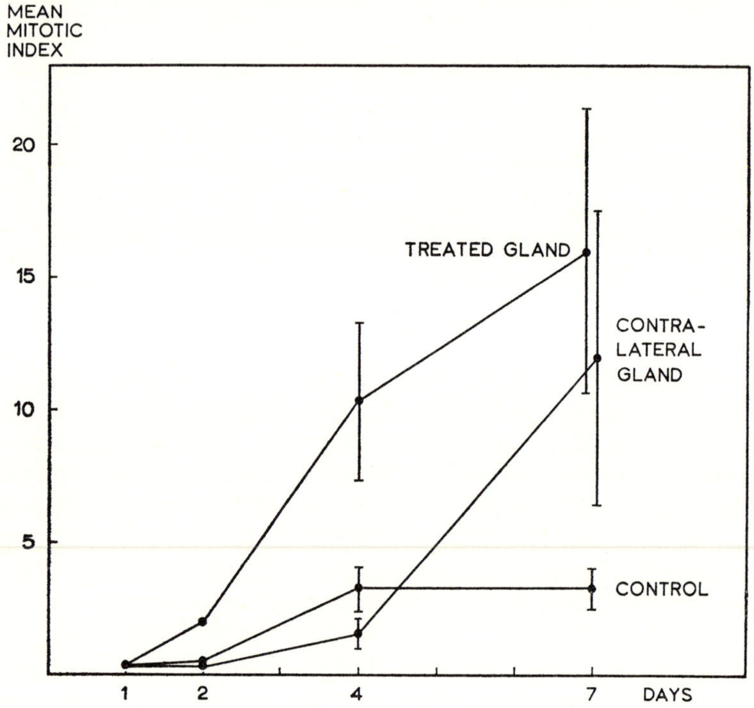

FIG. 1. Mitotic count 1, 2, 4 and 7 days after induction of coagulation necrosis in the outer orbital gland.

One outer orbital gland was exposed under ether anaesthesia and its circulation and gland duct were ligated with silk. A sham operation (skin incision) was performed on some of the controls and the others were not operated on at all. The mice were decapitated 1, 3, 6 and 9 days after the ligation, the rats after 1 and 3 days. Incorporation of tritiated thymidin was used as the indicator of growth stimulation. The mice were given 1 μC and the rats 0·75 μC of H^3-thymidine per gram of body weight 1 hour prior to decapitation. Incorporation was determined autoradiographically by counting the number of labelled cells per 10,000 cells in the control organs and in the outer orbital gland facing the ligated gland. In the control animals, there was no difference in the number of labelled cells between those subjected to a sham operation and the non-operated. Wound infection that had originated during the operation was seen in some of the ligated glands. Fairly even necrotization was established microscopically.

The mean thymidine index of the control group (40 animals in all) in the mouse experiment (Fig. 2) was $6{\cdot}0 \pm 0{\cdot}87$; the index of the contralateral organ was $15{\cdot}6 \pm 2{\cdot}89$ for the 1-day ligation group and $16{\cdot}4 \pm 2{\cdot}94$ for the

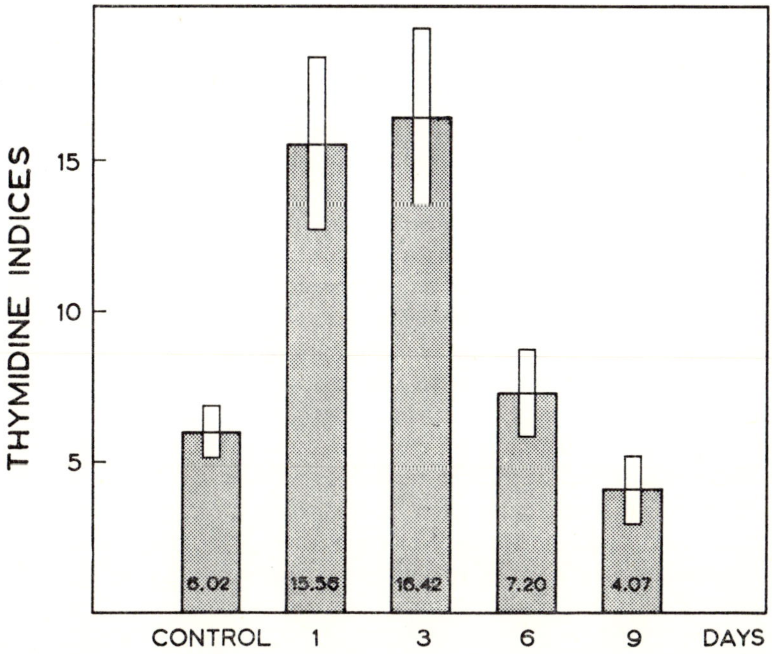

FIG. 2. Thymidine uptake in the contralateral outer orbital gland of mouse after ligature of the gland.

3-day group. The difference of both these groups from the controls was significant ($P < 0{\cdot}01$). No growth stimulation could be stated after 6 and 9 days.

The index for the experimental group in the rat experiment, $19{\cdot}8 \pm 6{\cdot}33$, did not differ from the control index after 1 day; the 3-day index, i.e. $28{\cdot}4 \pm 4{\cdot}66$, was higher than for the controls. Owing to the great deviation, the difference was not significant according to the t-test.

The experiments show that necrosis of the orbital gland affects the growth of homologous tissue. Necrosis stimulates the growth of the cells in the same organ and also of those in the contralateral gland. This is indicative of the presence of a circulating growth factor.

III. Liver

As post-hepatectomy liver regeneration has been studied so extensively, we decided to investigate in greater detail the role of autolysis and necrosis in this process.

A. *The influence of* in vitro *autolyzed liver tissue on post-hepatectomy liver regeneration*

Adult rats were used as the test animals in the first experimental series and, besides the mitotic index, the weight index was used as the indicator of regeneration (Lahtiharju 1961). Liver tissue removed at hepatectomy was incubated for 24 hours at 37°C, homogenized and injected into the abdominal cavity of the hepatectomized animals. The control animals, also partially hepatectomized, received physiological saline instead of the autolysate. Elevation of the mitotic and weight indices in the animals given autolysate was demonstrated in 10 out of 11 successive experiments, i.e. the results were fairly readily reproducible. The intensity of the effect was directly proportional to the dose up to the 2 g administered.

The rise in the mitotic index was elicited most clearly 24 hours after the injection, i.e. 48 hours after hepatectomy, whereas in the weight indices the effect was most distinct 3 days after the injection. A feature worthy of note was that the elevation of the mitotic indices was manifested as a sharp peak which probably lasted only a few hours, while the curve for the weight indices rose less steeply but lasted longer (Fig. 3). Although the effect was seen in the weight indices in one of the two experiments as late as 2 weeks post-operatively, it probably is not possible to achieve any lasting growth increase by a single injection, only a temporary acceleration of regeneration. Thus, a decrease to below the control level was seen in the mitotic and weight indices after the rise in the experiment illustrated in Fig. 3. The short duration of stimulation and the subsequent decline seen in the mitotic indices may explain in part the diverse information in the literature regarding the effect of tissue homogenates when a single time of sacrifice is used. Liver that had been autolyzed longer caused stimulation more rapidly than little autolyzed or fresh liver.

It was not possible to demonstrate species-specificity; calf liver caused an equally vigorous stimulation as the rat's own liver, but the effect was organ-specific in that kidney autolysate produced no stimulation. Boiling (15 min at 100°C) and lyophilization did not inactivate the autolysate.

In the next phase, the experimental animals were adult mice and the method employed was, similarly, injection of autolysate into the abdominal cavity. Here, however, the indicator of regeneration was the incorporation of tritiated thymidine into the tissues 24 hours after injection. The incorporation was determined autoradiographically (Lahtiharju and Teir 1964). A marked increase in the thymidine indices was seen in the parenchymal cells

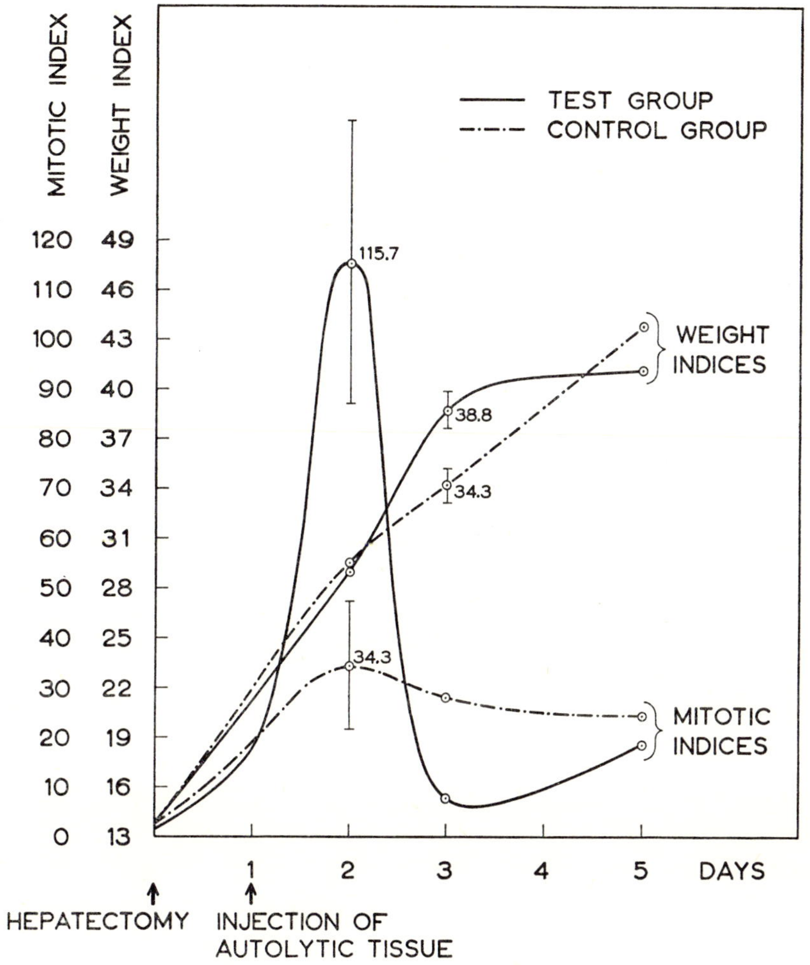

FIG. 3. Mitotic and weight indices of liver 1, 2, and 4 days after injection of autolytic tissue into hepatectomized rats.

of the liver but not in the mesenchymal cells, the epithelial cells of the stomach, the epidermis or the renal tubulus cells. The result suggests tissue-specific effect.

B. *The influence on normal liver of liver tissue autolyzed* in vitro *and injected into intact rats*

Only one experiment was performed. A significant increase was established 4 days after the injection. Necrotic foci were demonstrated in the livers histologically, and this may account for the increase (Lahtiharju 1961).

C. *The influence of* in vivo *necrotized liver tissue on post-hepatectomy liver regeneration*

The lobe of the liver resected at hepatectomy in the first experiments, or a part of it, was left in the abdominal cavity and the animals were decapitated 2 days after the operation (Lahtiharju 1961). The mitotic stimulation now varied according to the size of the dose, and sometimes there was no influence. A prominent feature in these experiments was that if a small dose caused mitotic stimulation, a large dose produced an increase in the weight indices. This led to the conclusion that the phase of mitotic stimulation was already over in the latter case; in other words, a large dose was quicker acting than a small dose. The results were similar when a ligated lobe was left in the abdominal cavity instead of a free lobe. Compared with the injection method, the reproducibility of the results was poorer with both these methods.

A new method in liver regeneration research was the production of necrosis by carbon tetrachloride (Lahtiharju 1964). The subsequent regeneration was studied by autoradiographic determination of the incorporation of tritiated thymidine in various tissues. An elevation of the thymidine indices in the parenchymal and mesenchymal cells of the liver was established 2 days after the carbon tetrachloride injection. No significant increase was demonstrable in the gastric epithelial cells and the renal tubulus cells.

Using various methods, liver experiments showed consistently that necrotic and autolytic liver tissue stimulate the proliferation of liver cells. An especially noteworthy point is that the effect was organ-specific in many of these studies.

IV. Granulocytes

A study of the extravasal distribution of granulocytes in the different organs of the rat directed our interest to granulocyte kinetics and especially to their elimination (Rytömaa 1960). It has been found that elimination occurs chiefly through the gastro-intestinal canal in rat (Teir *et al.* 1963, Teir and Rytömaa 1966), man (Rytömaa *et al.* (unpublished)) and fish (Teir *et al.*

(unpublished)). The majority of the granulocytes disintegrate in the mucosal lamina propria; a part of them undergoes epithelial diapedesis and moves to the intestinal lumen. A working hypothesis was formed in connection with this work series suggesting that substances which affect via the humoral route the granulocytopoiesis which occurs in the bone marrow, are liberated or activated from the disintegrating granulocytes. This question has been studied in *in vitro* experiments in which granulocytes in the process of disintegration were allowed to affect bone marrow cell suspensions (Teir and Rytömaa 1964, Rytömaa and Teir 1964).

In the granulocytopoiesis experiments rat bone marrow cell suspensions were incubated in the presence of cells taken from the abdominal cavity of rats 24 hours after the injection of polyvinyl pyrrolidone and after incubation of the cells for 6 hours at 37°C. The ascitic fluid contains many different disintegrated cell types, about 60 to 70% consisting of granulocytes. Later, a parabiotic chamber and the hanging drop technique were used. In these experiments, more than 100 in all, the total number of bone marrow cells was significantly increased in the presence of homologous disintegrated granulocytes. The relative and absolute numbers of polymorphonuclears were significantly increased in stimulated cultures studied in 38 series of experiments; disintegrated ascitic cells did not stimulate the proliferation of erythroid cells. Autoradiographic analyses of the incorporation of S^{35}-sulphate, C^{14}-formate and H^3-thymidine confirmed these observations and further demonstrated that the increase in marrow cell activities was really dependent on the proliferation of myeloid cells.

Rytömaa and Kiviniemi (1964) have found granulocytopoietic factors in the serum. Our earlier attempts to demonstrate corresponding factors in bone marrow homogenate did not give altogether definite results (Wasastjerna *et al.* 1958).

Although the kinetics of so-called blood cells is fairly complex compared with the kinetics of the parenchymal cells, chiefly because the cells are produced and eliminated in different places, the above experiments have shown that it is very possible that in normal conditions an equilibrium exists between the formation of granulocytes and the elimination through the gastrointestinal tract (Teir and Rytömaa 1966).

V. Erythrocytes

Early in this century several authors observed increased erythropoiesis after administration of haemolysates to bled animals, and this has been confirmed (Sanchez-Medal *et al.* 1963).

The influence of mechanically traumatized and *in vitro* autolyzed erythrocytes was studied by Forsell. An example from these series is given in

Fig. 4, showing the effect of intraperitoneally injected haemolysates of red cells on haemoglobin and erythrocytes.

The haemolysis was performed gradually, either mechanically or by incubating erythrocytes at 37°C. In both instances "washed" erythrocytes were suspended in saline. When about one-tenth of the red cells were haemolyzed, the supernatant was removed after centrifugation and used as test substance. This solution contained saline and red cell constituents, but no intact red cells. The greatest part of the red cell constituents in this haemolysate represented the most vulnerable, of "oldest" erythrocytes. The haemolysates were injected intraperitoneally into 5 to 8 month old rats. The animals were allowed to eat and drink freely. No deficient states could be detected.

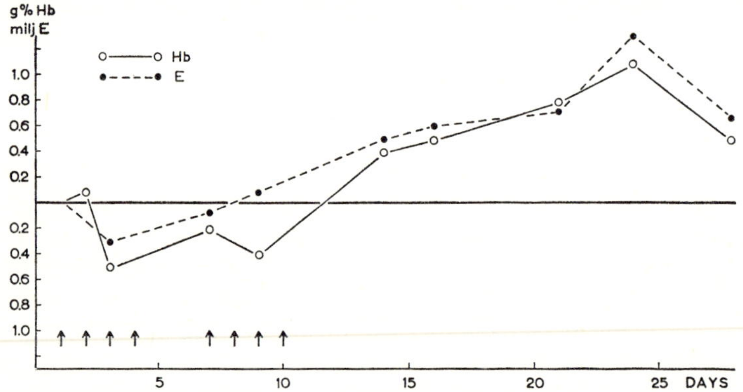

FIG. 4. Effect of intraperitoneally injected haemolysate of erythrocytes on erythropoiesis. Mean values of 25 rats.

When the injected volume of haemolysate corresponded to 2% of the animal's body weight, a clear effect on erythropoiesis was observed (Fig. 4). Injections of plasma and saline had no effect. Reticulocytes, venous haematocrit, bone marrow cell counts and Fe^{59}-turnover rate were also investigated and compare well with the changes in haemoglobin and the number of erythrocytes.

The results obtained so far concerning the effect of haemolysates show that it is possible to affect erythropoiesis and that a factor released in connection with the disintegration of erythrocytes may be involved here, too. This may be a factor which participates even in normal conditions in the mechanism regulating red cell production. Its relationship with erythropoietin remains an open question for the time being. It is probable that there are in play two different mechanisms influencing erythropoiesis.

VI. Discussion and Conclusions

The experiments we have conducted with different organs and cells have given fairly clear indications that autolytic and necrotic tissue influence mitosis and regeneration. The unevenness of the experimental results is probably due to the variation in the degree of autolysis and necrosis, although external conditions were kept as similar as possible.

Comparison of the new findings with earlier achievements showed that organ- and species-specificity have been further clarified. Liver experiments showed that this necrosis factor is organ-specific, but not species-specific. Granulocyte tests also suggested tissue-specificity.

Organ-specificity is the base for growth control theories which assume that there are local growth factors. Paschkis (1958) considered that the problem of organ-specificity requires further elucidation, primarily on the basis of his own experiments in which liver extract produced stimulation in other organs as well as in the liver (Paschkis *et al.* 1957). This could also be said today.

Comparison of the effect of fresh and autolytic tissue or tissue homogenate meets with difficulties since fresh tissue, tissue homogenate, cell suspension or their fractions also undergo autolysis after application.

Relatively little is known about the death of cells in normal multicellular organisms (Biggers 1964). Bessis's (1964) opinion that it is probable that substances are liberated during the death agony of the cell which attract neighbouring macrophages is interesting. He calls this necrotaxis. It was noted by Tsanev (1963) that irreversible mechanical injury leads to a splitting of the nucleoprotein complex, followed by a rapid degradation of RNA. The liberated breakdown products were assumed to be leukotactic substances which are released from injured tissue. Reversible injury, on the other hand, led in Tsanev's experiments to an activation of cellular multiplication preceded by a partial degradation of RNA and its subsequent resynthesis. He concluded from this that the reaction of nucleo-proteins to mechanical injury is the starting point of regeneration after injury.

The stimulation of cell production in the contralateral orbital gland in connection with ligation of the gland and also the ligation of a lobe of the liver can be compared with findings in the pancreas and kidney. Fitzgerald (1963) observed in the rat pancreas that ligation of the artery to the splenic segment produced a significant increase in H^3-thymidine labelled cells 3 to 5 days post-operatively. Ligation of the duct to the splenic segment caused a similar significant increased labelling in acinar cells. It was assumed by Fitzgerald that the removal of a normally occurring inhibitor of DNA synthesis is possible and that histone-DNA relationships are possibly involved.

Mitotic activity in the undamaged contralateral kidney of rat after 1 hour of ischemia and after unilateral nephrectomy and plain laparotomy was studied by Klinge (1964). A specific cell multiplication began on the second day of the experiments and continued for 8 days in three waves, each successive wave being weaker than its predecessor. Initially, the response to the injury after temporary ischemia was similar to that after unilateral nephrectomy; afterwards, however, the rate of mitosis sank to a lower level. Immediately after every operative procedure in which a laparotomy was performed, a mitotic peak was established on the first day in the contralateral kidneys and liver. Klinge assumed that these cells may have been in a post-synthetic phase and that nuclear divisions were caused by a still unknown humoral stimulus which leads to a non-specific mitotic reaction. He assumed, further, that organ-specificity exists only for a small part of this mitotic group.

Experiments in which an effect on the growth of the contralateral organ was established argue distinctly for a humoral influence. For the liver, this view has been supported by the studies of Glinos and Gey (1952), Wrba *et al.* (1960), Leevy *et al.* (1962) and Virolainen (1963). Our granulocyte and erythrocyte studies are also suggestive of this mechanism of effect.

A point to be taken into consideration in studying organ-specific growth factors is that numerous other factors also influence the biosynthesis and production of a cell population.

Our view of the relationship between general hormones and local organ-specific growth factors (Teir 1951, 1952a) has changed and widened subsequently, and now appears in the main as shown in Fig. 5. In addition to general hormones, the age, nutritional status and general metabolism, i.e. the homeostasis of the organism as a whole, are also in play. The next influence comes from the organ in which circulation and connective tissue are of interest. Mast cell degranulation in the tissue has been found to lower the mitotic count (Räsänen 1961). Heparin has been shown to be a strong inhibitor of mitosis (Räsänen 1963). This group of factors also includes the intracellular and extracellular (lymph, serum) specific growth-factor complex (template—anti-template, chalon—antichalon, growth stimulator—growth inhibitor, according to the customary terms used at different times) of the tissue type as a whole. Especially their amount and nature in the extracellular pole are of importance. The idea of factors liberated in connection with physiological cell death raises the question whether there are several factors which are close to one another and derive from the same cell population. They may be in relation also to differentiation, in organs in which differentiation occurs continuously, e.g. the epidermis, granulocytes and erythrocytes. Organogenetic studies of the last few years have also shown that there is a controlled autolysis during development (review by Zwilling 1964).

The innermost part of Fig. 5 consists of a cell unit with a cell in mitosis

(M), two daughter cells in the post-mitotic phase (G_1), one undergoing synthesis (S) and the other necrobiosis (NB). According to the hypothesis, S and NB are in mutual relationship. At the same time, a genetic response is involved. This unit constitutes the micro-environment of biosynthesis which is also influenced, but more weakly, by the factor groups mentioned above.

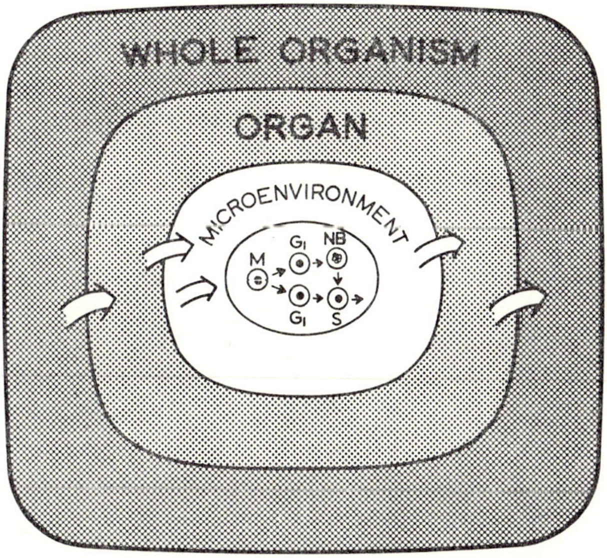

FIG. 5. The cellular growth control of a certain cell population is a complex system consisting in addition to the homeostatic factors for the organism as a whole also of unspecific (circulation, mast cells, etc.) and specific organ and micro-environmental growth factors.

A cell in mitosis (M) gives two cells in post-mitotic rest phase (G1). One of these two cells (NB) undergoes necrobiosis and is eliminated, which process (necrobiotic growth information) influences the synthesis in the other cell (S).

To date, we have relatively little experimental evidence of how these different factor groups affect the biosynthesis and mitosis of the cells. Räsänen (1965) recently performed an interesting experiment associated with this question. Rats were injected with 1 mg of dexamethasone three times a day for 2 days and on the third day at 8 a.m. The mitotic figures were now close to zero, as expected. At 12 noon Shay ligature was performed and at 4 p.m. the rats were killed. Erosions and loss of coherence between cells were seen histologically in the superficial parts of the mucosa. Unusually numerous mitoses were now seen in the renewal zone below it. There was, thus, a regeneration need which had overcome the dexamethasone-induced depression of DNA synthesis and of the mitoses. A single dose of 1·0 mg of cortisone causes a decrease in the DNA synthesis in the glandular stomach of rat, as has been observed autoradiographically with H^3-thymidine (Lahtiharju *et al.* 1964).

It may thus be assumed that, in physiological conditions as well, there is behind the normal kinetics and equilibrium of a cell population an interaction of many different factors associated with cell production.

The experimental conditions described here do not correspond to physiological conditions. We do not know how many cells there are at a given moment in a certain phase of the breakdown process. Moreover, the chemistry of autolysis, necrosis and necrobiosis is rather inadequately known so far. We sorely need additional information on this fundamental biologic event.

There is a continuous supply of informative feedback material in tissues which are constantly regenerated physiologically. For the moment, there seems to be a gap between the experimental results and the corresponding physiological phenomena. If we were able to arrange the experimental conditions in such a way that breakdown products and their mechanism of action would approximate more closely to the physiological conditions, we would be able better to understand the nature of the active agents.

VII. Summary

The working hypothesis was an assumption that the equilibrium between the production and elimination of the cell population of different organs is dependent, at least partially, on the necrobiosis of cells in connection with elimination.

Coagulation necrosis in the outer orbital gland of the rat was induced by heat treatment. An effect on the mitotic count was observed in the contralateral organ. In two other experiments the gland duct was ligated. An increased uptake of H^3-thymidine was observed autoradiographically in the contralateral gland.

Numerous studies on liver and granulocytes in which autolytic and necrotic tissue stimulated the growth of homologous tissue are reviewed. Haemolysates of erythrocytes were found to influence erythropoiesis.

The experimentally observed effect of autolysis and necrosis on the cellular growth of homologous tissue suggests the possibility that in normal conditions, too, factors which influence the physiological renewal of cells are released in connection with cellular necrobiosis.

Acknowledgement

We express our gratitude to Dr. Ahola and to Dr. Härkönen for assistance in the work "Coagulated necrosis experiments of orbital glands".

REFERENCES

Alho, A. (1961). *Acta path. microbiol. scand.* Suppl. 149.

Bessis, M. (1964). *In* "Cellular injury" (A. V. S. de Reuck and J. Knight eds.), pp. 287–328. Churchill. London.

Biggers, J. D. (1964). *In* "Cellular injury" (A. V. S. de Reuck and J. Knight eds.), pp. 329–351. Churchill. London.

Blomqvist, K. (1957). *Acta path. microbiol. scand.* Suppl. 121.

Breasted, J. H. (1930). *In* "The Edwin Smith Surgical Papyrus", vol. 1. University of Chicago Press. Chicago.

Bullough, W. S. (1952). *Biol. Rev.* 12, 133–168.

Bullough, W. S. and Rytömaa, T. (1965) *Nature* 205, 573–578.

Caspari, W. (1927). *Arch. klin. Chir.* 146, 711–736.

Fitzgerald, P. J. (1963). *Canad. med. Ass.* 88, 480–482.

Glinos, A. D. and Gey, G. O. (1952). *Proc. Soc. exp. Biol. N.Y.*. 80, 421–425.

Haberlandt, G. (1921). *S.-B. Preuss. Akad. Wiss.* 124, 221–234.

Klinge, O. (1964). *Beitr. path. Anat.* 130, 352–368.

Lahtiharju, A. (1961). *Acta path. microbiol. scand.* Suppl. 150.

Lahtiharju, A. (1964). *Virchows Arch.*, 377, 535–537.

Lahtiharju, A., Räsänen, T. and Teir, H. (1964). *Growth* 28, 221–224.

Lahtiharju, A. and Teir, H. (1964). *Exp. Cell Res.* 34, 205–207.

Leblond, C. P. and Walker, B. E. (1956). *Physiol. Rev.* 36, 266–275.

Lee, R. and Hanson, W. (1947). "Protomorphology." Lee Foundation for Nutritional Research. Milwaukee.

Leevy, C. M., George, W., Deysine, M. and Gnassi, A. M. (1962). *Exp. Molecular Path.* 1, 457–469.

Paschkis, K. E. (1958). *Cancer Res.* 18, 981–991.

Paschkis, K. E., Cantarow, A. and Goddard, J. W. (1957). *Fed. Proc.* 16, 98.

Robertson, T. B. (1923). "Chemical Basis of Growth and Senescence." Lippincott. Philadelphia.

Rytömaa, T. (1960). *Acta path. microbiol. scand.* Suppl. 140.

Rytömaa, T. and Kiviniemi, K. (1964). *In* "Proc. XIV Scand. Confr. on Path. and Microbiol" (L. Kreyberg, C. Lerche and O. H. Iversen eds.), pp. 167–170. Universitetsforlaget. Oslo.

Rytömaa, T. and Teir, H. (1964). *Sangre* (Barcelona) 9, 359–361.

Räsänen, T. (1961). *Gastroenterology* 40, 232–234.

Räsänen, T. (1963). *Acta physiol. scand.* 58, 201–210.

Räsänen, T. (1965). Personal communication.

Sanchez-Medal, L., Labardini, J. and Loria A. (1963). *Blood* 21, 586–593.

Sundell, B. and Teir, H. (1954). *Exp. Cell Res.* 6, 321–326.

Teir, H. (1944) *Acta path. microbiol. scand.* Suppl. 51.

Teir, H. (1951). *Soc. scient. fenn. comm. biol.* 13, 1–32.

Teir, H. (1952a). *Acta path. microbiol. scand.* 30, 158–183.

Teir, H. (1952b). *Growth* 16, 85–108.

Teir, H. and Isotalo, A. (1953). *Ann. Med. exp. Fenn.* 31, 171–180.

Teir, H., Kiljunen, A. and Putkonen, T. (1951). *Ann. Chir. Gyn. Fenn.* 40, 61–70.

Teir, H. and Lahtiharju, A. (1961). *Exp. Cell Res.* 24, 424–428.

Teir, H. and Nyström, B. (1962). *Arch. Path.* 74, 499–508.

Teir, H. and Ravanti, K. (1953). *Exp. Cell Res.* 5, 500–507.

Teir, H. and Rytömaa, T. (1964). *Sangre* (Barcelona) 9, 418–420.

Teir, H. and Rytömaa, T. (1966). *In* "Methods and Achievments in Experimental Pathology" (E. Bajusz and G. Jasmin eds.), Karger A. G., Basel.
Teir, H., Rytömaa, T., Cederberg, A. and Kiviniemi, K. (1963). *Acta path. microbiol. scand.* 59, 311–324.
Telkkä, A. and Teir, H. (1955). *Acta path. microbiol. scand.* 36, 323–326.
Tsanev, R. (1963). *In* "Symp. Biol. Hung" (G. Szántó ed.), vol. 3, pp. 55–73. Akadémiai Kiadó. Budapest.
Turck, F. B. (1921). *Med. Rec., N.Y.* **100**, 705–708.
Turck, F. B. (1933). "The Action of the Living Cell." Macmillan, New York.
Virchow, R. (1858). *Virchows Arch.* **14**, 1–63.
Virolainen, M. (1963). *Exp. Cell. Res.* 33, 588–591.
Wasastjerna, C., Teir, H. and Larmo, A. (1958). *Acta path. microbiol. scand.* 44, 80–87.
Weiss, P. (1947). *Yale J. Biol. Med.* **19**, 235–278.
Weiss, P. (1955). *In* "Biological specificity and growth" (E. G. Butler ed.), pp. 195–206.
Wrba, H., Ranz, H. and Ripoll-Gomez, M. (1960). *Naturwissenschaften* 47, 306.
Zwilling, E. (1964). *In* "Cellular injury" (A. V. S. de Reuck and J. Knight eds.), pp. 352–368. Churchill, London.

The Effect of Chalone on the Mitotic Rate and on the Mitotic Duration in Hairless Mouse Epidermis

O. H. IVERSEN and K. ELGJO

Institute for General and Experimental Pathology, University of Oslo, Oslo, Norway.

THE regulation of the mitotic rate in the epidermis has been the subject of much discussion. It can be concluded from a series of observations (see, for instance, Bullough and Laurence 1960, Oehlert and Block 1962, Iversen and Evensen 1962) that the regulation mechanism probably is of cybernetic nature based on the presence of a mitotic inhibitor substance, produced by the maturing cell in the epidermis.

Bullough, Hewett and Laurence (1964) have succeeded in extracting such an inhibitor from epidermal tissue. They have to a certain extent determined its nature and mode of action. The inhibitor has been called Chalone. According to Bullough the substance acts by forming an active complex with adrenaline.

Experimental work within epidermal carcinogenesis directed our interest towards the growth regulation mechanism of the mouse epidermis. After a single application of a carcinogen or an other skin irritating substance some cells are regularly killed. For the purpose of the present discussion we shall restrict ourselves to report the changes seen after a single application of one drop of a 1% solution of methylcholanthrene in benzene. The carcinogen has been applied with a special pair of forceps, and the effect always compared with a control area of a non-treated part of the epidermis of the same mouse.

A few hours after such an application the epidermis is thicker than normal. This is, however, caused solely by intra- and intercellular oedema, and not by an increased number of cells. A few days later the picture has changed. The epidermis is still thicker than normal, but now the number of epithelial cells have increased to about 140% of the normal. Thereafter this hyperplasia fades slowly off again. Figure 1 shows a quantitative analysis of this hyperplasia, with respect to the time after the application.

For a more functional study of this sequence of events we have applied different methods. The tetrazolium method is based on the reduction of a tetrazolium salt to formazane. As the tetrazolium salts are water-soluble and colourless, and the formazanes are strongly coloured and water-insoluble, it is possible to extract the formazane from the tissues and measure photometrically the amount of formazane per mg tissue. If this is performed after a single application of carcinogen, the results are as shown in Fig. 2, upper curve. The formazane production is increased with a maximum at 16 to 24 hours after the application. Then the amount of formazane falls again. We have interpreted the initial period of increased formazane production as a sign of cellular destruction, especially affecting mitochondrial function.

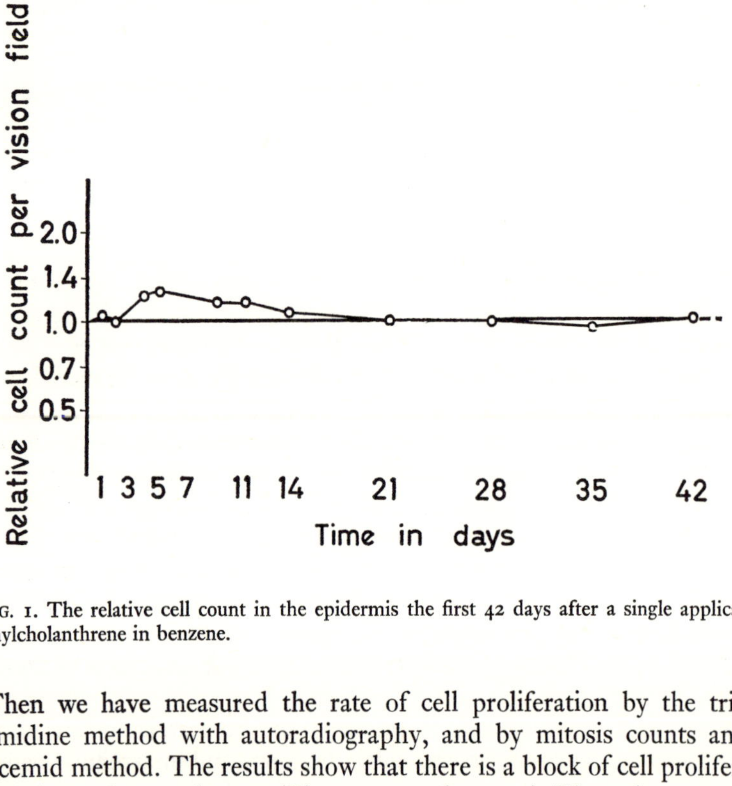

FIG. 1. The relative cell count in the epidermis the first 42 days after a single application of methylcholanthrene in benzene.

Then we have measured the rate of cell proliferation by the tritiated thymidine method with autoradiography, and by mitosis counts and the Colcemid method. The results show that there is a block of cell proliferation which lasts about 4 hours. (Fig. 2, second curve.) Then the rate of cell proliferation increases to about 3 times the normal value, whereafter it falls again. If the rate of cell proliferation increases to 3 times the normal value the first day after an application, without any preceding cell loss, the hyperplasia should develop as shown in curve m_1 in Fig. 2, third curve. The observed hyperplasia is labelled m_t. The area between this hypothetical hyperplasia (m_1) and the real hyperplasia (m_t) must be a measure of *cell loss*, the rate of which is calculated in the bottom curve in Fig. 2.

We have also measured the cell loss by means of a radioactive tracer as described by Skjaeggestad (1964), and proved that there is a heavy cell loss after one single application of skin irritants.

We interpret the observations in the following way: When cells are suddenly killed, and there is a chalone produced by the maturing cell in the

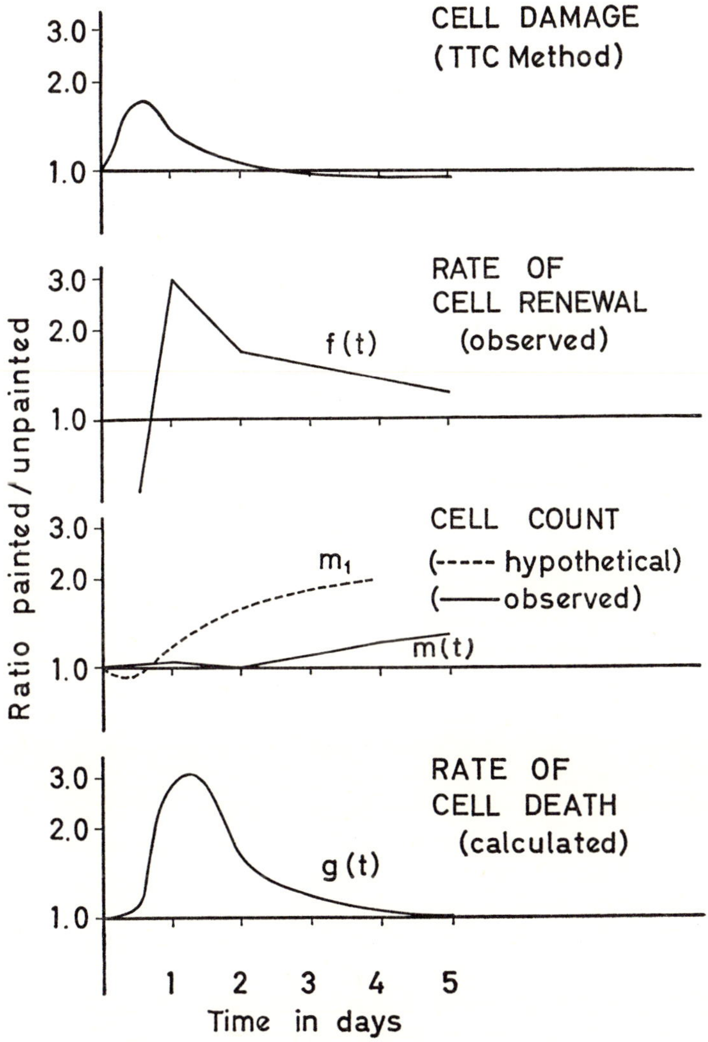

FIG. 2. Observations and calculations of the early effects of a single application of methylcholanthrene to mouse epidermis. (For a detailed explanation, see text.)

epidermis, the concentration of chalone falls suddenly, and the rate of cell proliferation increases correspondingly. In this way a population change is produced to an epidermis consisting mainly of newly formed cells. These need at least 3 days to be transformed to keratine, and in this way the transient hyperplasia develops.

We then went on to test if the cybernetic theory could explain the development of hyperplasia and the alterations in the rate of cell proliferation described above. We therefore constructed a mathematical model, which could be tested on an analogue computer (Iversen and Bjerknes, 1963). In this model the negative feed-back principle is the only principle regulating the thickness of the epidermis. We then simulated the effect of a sudden cell loss as provoked by the skin irritant methylcholanthrene. Figure 3 shows the results. It can be seen that both the changes in the rate of cell proliferation and the changes in the number of cells can be explained simply by assuming a negative feed-back principle based on the chalone effect.

As Bullough *et al.* (1964) have succeeded in extracting the chalone from epidermal tissue, we were interested in repeating this on the strain of hairless mice of this institute. Hairless mice of the strain hr/hr were used both for the production of epidermal chalone, and for the testing of its effect. The mice were about 12 weeks old. All experiments were started in the morning at about 9 o'clock to avoid, as far as possible, variations due to the circadian rhythm of the mitotic count. Equal numbers of females and males were used in each experiment.

The skins from 10 mice were utilized for the preparation of each batch of chalone. The animals were killed by neck extension, and the skins immediately removed. Epidermis was cut off by means of a Castroviejo keratotome. In this way it was possible to obtain flat pieces of skin containing the whole epidermis and a little of the upper part of corium. From each skin we could take 6 to 8 pieces of about 1 ×8 cm. These were immediately frozen down to —70°C by solid carbon dioxide.

When a sufficient amount of epidermis had been accumulated, the epidermal pieces were transferred to a porcelain mortar into which liquid air was added. This made the skin very brittle, thus making it possible to grind the epidermis to a fine white powder. Microscopical control revealed that very few unbroken cells remained after this procedure. The epidermal powder was then transferred to a test tube, and about 16 ml of saline was added. Further homogenization was performed partly by an Elvehjem Potter Homogenizator, later with an ultrasound homogenizator.

After the last homogenization procedure, the solution was centrifuged at 25,000 g for 15 min. The clear supernatant was pipetted off, and used as chalone solution. It was stored in the deepfreeze and used next day. Each animal was injected intraperitoneally with 2 ml of this solution.

The animals under study were killed at different time intervals after injection of the chalone. Skins were fixed in Bouin's solution and after paraffin embedding, sections from the dorsal skin were cut at 5 μ and stained

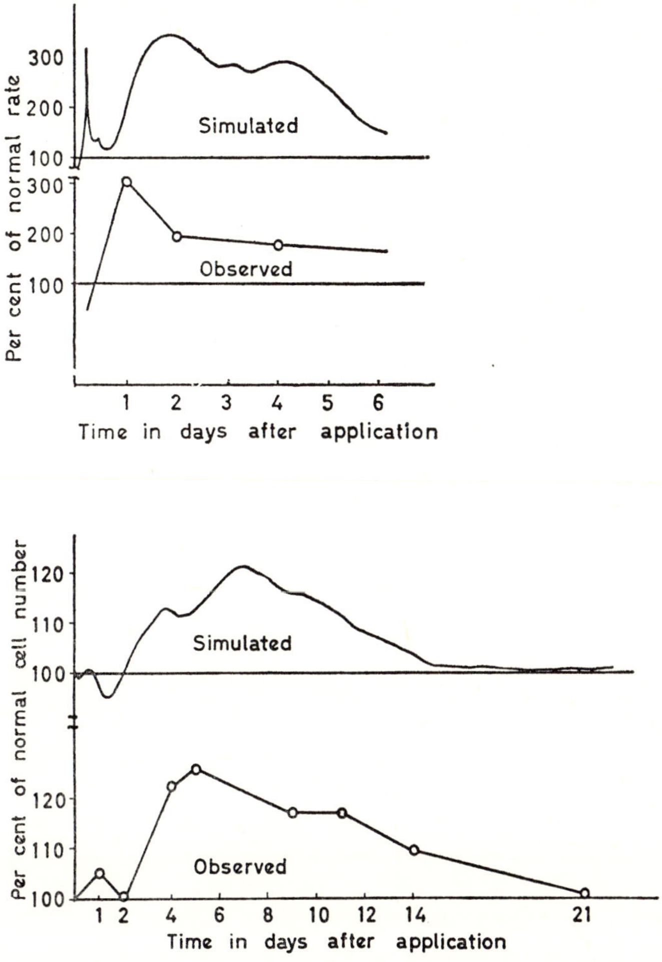

FIG. 3. A comparison of simulated and observed effects of a single application of methylcholanthrene to mouse epidermis. The upper curve concerns the rate of cell renewal. The bottom curve concerns the development of hyperplasia.

with Celestine blue. In each animal the number of mitoses was counted in 50 microscope vision fields using the objective 100, and ocular 8. Care was taken to count only interfollicular epidermis. A vision field of this type comprises of about 20 basal cells or, when no hyperplasia is present, about 30 nucleated cells. This means that 50 vision fields comprise about 1000 basal cells and about 1500 nucleated cells.

The mitotic rate was measured by the Colcemid method. Each animal was given 0·1 mg of Colcemid intraperitoneally. At regular time intervals during

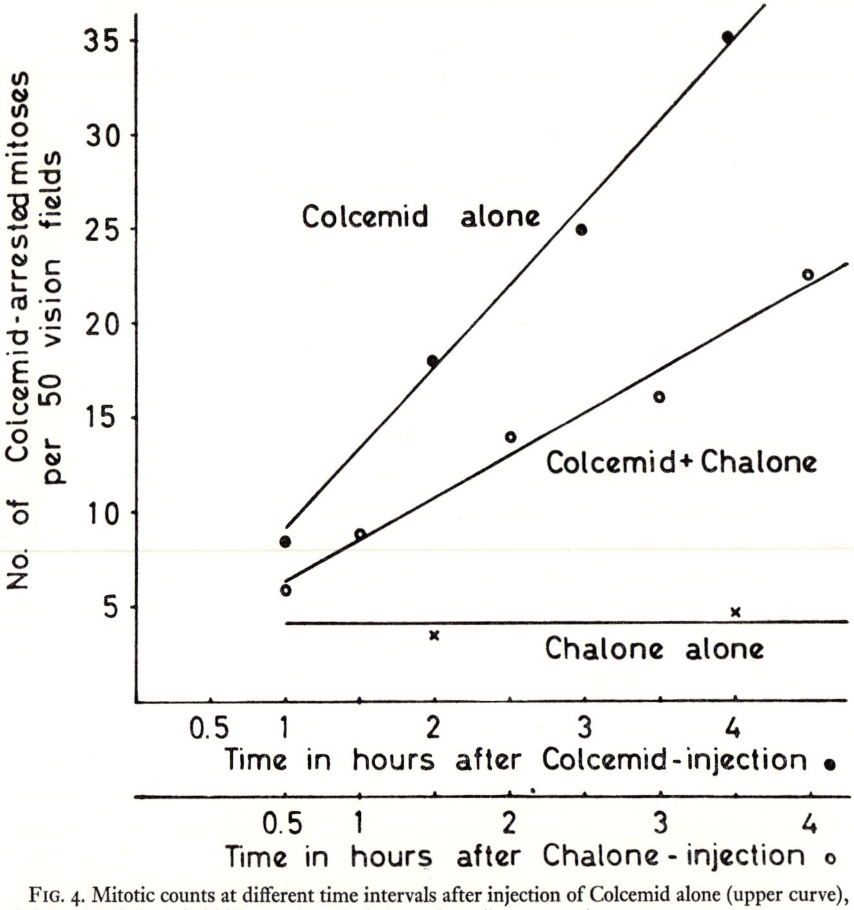

FIG. 4. Mitotic counts at different time intervals after injection of Colcemid alone (upper curve), Colcemid + chalone (middle curve), and chalone alone (lower curve).

the first 4 to 5 hours after the Colcemid injection, groups of mice were sacrificed and the mitotic counts determined. The line of best fit of the different counts was calculated by the least square method, and the tangent of the angle between this line and the abscissa was used as the measure of the mitotic rate.

The mitotic duration was calculated from the ratio between the mitotic count and the mitotic rate measured by the Colcemid method. The results are shown in Fig. 4. The normal mitotic rate in the epidermis of our hairless mice during the period from 9 a.m. to 1 p.m. is about 8·5 mitoses per hour per 50 vision fields. The normal mitotic count without Colcemid injection was measured at 9 a.m. in 80 animals. The average was found to be 8·5 mitoses per vision fields, and thus the mitotic duration was 1 hour.

The effect of chalone was measured in the following way: Mice were first injected with Colcemid and 30 min later with 2 ml of the chalone solution. Mitoses were then counted at different time intervals during the first 4 hours. The mitotic rate in the epidermis of our hairless mice under the influence of chalone was reduced to 4·5 mitoses per hour per 50 vision fields. Thus a reduction to about half the normal rate was achieved by the injection of chalone.

The mitotic count (with Colcemid injection) was also measured after chalone injection. The average mitotic count was found to be 4·2 per 50 vision fields. The mitotic duration can then be calculated to 0·9 hours.

As a control we injected mice with 2 ml saline only. The mitotic rate was then the same as in the normal group.

The conclusion is that under the experimental conditions described, an injection of 2 ml of the chalone solution provoked a significant and continuous depression of the mitotic activity that lasted for at least 4 hours. The mitotic count was lowered, but the mitotic duration was not significantly prolonged.

The production of chalone from epidermis is time-consuming, and requires a large amount of animals. As the chalone seems to be species unspecific, we searched for other possible sources of chalone. As it is easy to obtain skin of fish in Norway, we looked for possible chalone effect in extracts from skins of cod.

Skins from recently caught fishes were homogenized in a Sorvall Omnimixer, and then ground in a mortar with liquid air as described above. Saline was added, and the supernatant tested for chalone effect on mouse epidermis. The final solution contained less than 0·5 gram protein per 100 ml. With intraperitoneal doses as small as 0·1 ml of this solution, we have obtained a reduction of the mitotic rate in the epidermis to less than half the normal. We have separated the supernatant through a Sephadex column. Protein was found only in three of the fractions. As far as we have seen hitherto, the chalone effect is located to the protein fractions.

To test possible unspecific reactions from the injection of foreign proteins, we have also tested the effect of extracts of muscle tissue from fish and from ox. With heavy doses of such extracts, a reduction of the mitotic rate of the epidermis is seen, but the effect fades when less than 5 mg protein is given.

As we are engaged in further model studies using a digital computer, we were interested in determining the time sequences for the rate of cell proliferation which follows a sudden removal of epidermal cells. We have used the stripping method with Scotch tape on hairless mice, and stripped them on the back 4 times. The skin then had a slightly glistening and moist surface.

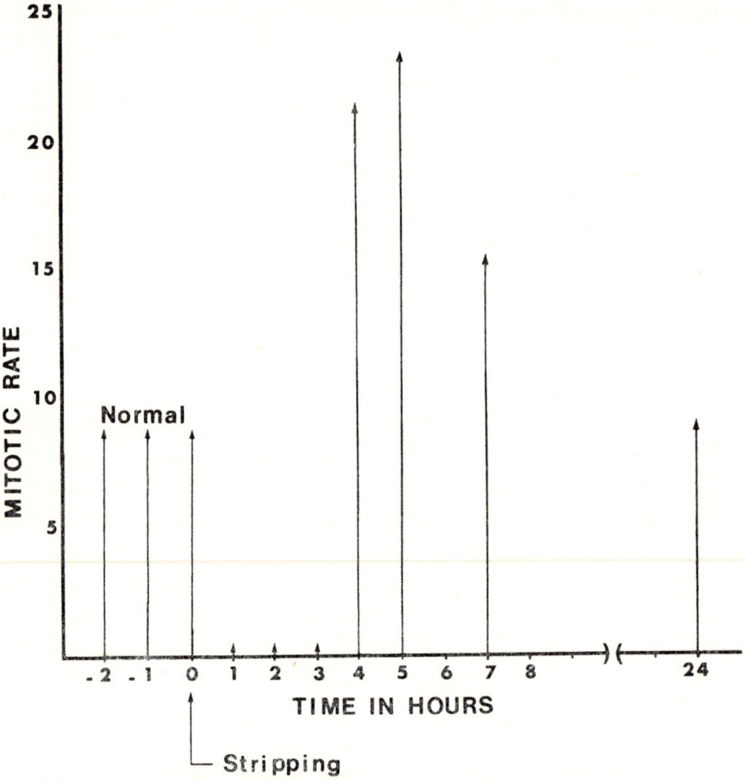

FIG. 5. The rate of cell proliferation two hours before and 24 hours after stripping with Scotch tape. The stripping provokes a 3 hours block of cell proliferation, and then the mitotic rate increases to more than twice the normal after 5 hours. After 24 hours the rate is normal again.

Figure 5 shows that after such stripping there is a 3-hour block of the cell proliferation. Then follows a steep increase which reaches its maximum 6 hours after the stripping. Subsequently the rate of cell proliferation falls to normal values again. The 3-hour block may come from a too heavy traumatic injury to the basal cells provoked by the four strippings. In the near future we are going to test the effect of one single stripping, which removes only a few of the cells. We have also performed these experiments with tritiated thymidine technique, but the sections are still in the refrigerator, and not yet developed.

If it is possible to measure chemically and quantitatively the amount of chalone per mg epidermal tissue, we should be in a quite new position in epidermal carcinogenesis studies. We shall know if the chalone is produced in normal amounts, in reduced amounts, or in increased amounts in epidermal carcinomas, and we can determine when in the process of chemical cancerization possible alterations occur.

REFERENCES

Bullough, W. S. and Laurence, E. B. (1960). *Proc. R. Soc. B.* 151, 517–536.
Bullough, W. S., Hewett, C. L. and Laurence, E. B. (1964). *Expl. Cell Res.* 36, 192–200.
Iversen, O. H. and Bjerknes, R. (1963). *Acta path. microbiol. scand.* Suppl. 165.
Iversen, O. H. and Evensen, A. (1962). *Acta path. microbiol. scand.* Suppl. 156.
Oehlert, W. and Block, P. (1962). *Verh. dtsch. Ges. Pathol.* 46, 333–344.
Skjæggestad, Ø. *Acta path. microbiol. scand.* Suppl. 169.

DISCUSSION SUMMARY

Chairman Professor Paul Weiss

The discussion began with a question from Dr. Lajtha who asked Professor Teir whether the erythropoiesis-stimulating effect of red cell haemolysates had been controlled by simply giving iron to the animals. Dr. Lajtha pointed out that most laboratory rodents, including rabbits, are commonly deficient in iron; blood haemolysates could thus stimulate erythropoiesis by providing iron to sideropenic animals. Professor Teir replied that this possibility had not yet been properly controlled.

Another question was asked by Dr. Lajtha, relating to the mode of action of chalones. He noted that if chalone acts at the G_2 phase of cell cycle, a vast majority of the cells should contain 4n amount of DNA. Because this can quite easily be checked spectrophotometrically, Dr. Lajtha asked whether the results, if available, are compatible with the concept.

The chairman mentioned some results published by Dr. Overton that had some bearing on this problem. In the study of local stimulation of skin proliferation in amphibians by grafts of nervous tissue, Overton found that the stimulated cell population in completely starved larvae contained many cells which had grown to twice their resting size but failed to go through the actual act of mitosis. The chairman then emphasized that one had to be very careful to differentiate between those visible signals when the cell stops its metabolic activity, as opposed to the basic problem discussed here, namely, whether or not the cell metabolism is redirected. The chairman referred to some recent findings which indicated that certain mitochondrial and mitotic reactions might well be related processes in the cell. He concluded, by agreeing with the opinions of many previous speakers, that cell metabolism could be switched, "dichotomized" so to speak, either toward synthesizing DNA, so as to go through mitosis and produce another cell of the same kind, or (perhaps through the RNA mechanism) producing more of the specific products of the cell, thus passing through various stages of cyto-differentiation.

Coming back to the original question, Professor Bullough stated that they had checked the 4n and 2n cells in epidermis, and that there is no increase in 4n's. Professor Bullough further stressed that if one inhibits mitosis * for any length of time bearing on the chalone mechanism, there is no pile up of 4n cells, which again indicates that the blockage is prior to the DNA synthesis.

Professor Goss pointed out that there is a basic paradox in some of the studies suggesting that products of dying or destroyed cells are acting as a feedback to stimulate mitosis; in many tissues mitoses can also be stimulated by the simple

* In this book the word mitosis has been used in two meanings: it refers either to the actual act of mitosis ("karyokinesis") or to the whole generative cycle of a cell. It should be evident from the context, however, what is meant in each particular case. (Editor's note.)

removal of a proportion of the cell population, which precludes the feedback of the products of cell destruction. Professor Iversen replied to this that if one accepts the template-antitemplate theory of Weiss, it is possible to consider that some autolytic substance is binding the inhibitor (antitemplate); this would straighten out the paradox very beautifully.

The chairman added that some years ago when Ilse Fischer worked with him on these problems, they conducted some experiments in tissue culture and assumed that the so-called "growth-promoting" power of embryonic extract contained cell debris; that is necrotic cells of practically all the embryonic cells. So, it had to "desinhibit" all the tissues put into it. Actually Doljansky showed this effect for heart extract on heart fibroblasts. The chairman and Fischer compared extracts from *whole* embryos with those from which the particular organ to be tested had been omitted, ascribing the difference in growth to the differential between the two media—a "minus" experiment. They got some rather dramatic results: for example in experiments with explanted heart, extracts that did contain heart debris stimulated proliferation, whereas those extracts that did not, maintained the differentiated condition, as manifested by continuous pulsations, for many weeks. However, this difference showed up only if the embryo that had furnished the extract had been at least 10 days old—so that obviously the heart is affected between the 7th and 10th day embryo. The difference is unknown.

With respect to the problem introduced by Professor Goss, it was pointed out by Professor Teir that electron microscopic studies have indicated that certain degenerative changes seem to occur in the remaining liver during the first day after partial hepatectomy, that is, prior to the start of the actual regeneration. Professor Teir also suggested that lysosomes, the "suicide bags", may play an important role in these phenomena.

Referring to some fairly recent studies where it has been shown that both RNA and DNA, or their breakdown products, are capable of causing depression of genes, and also to other studies which have indicated that they are capable of stimulating certain syntheses as well, it was suggested by Dr. Wood that such effects are presumably due to the binding of nucleic acids with histones which have been shown to be inhibitors of cellular proliferation. Dr. Wood therefore asked Professor Teir whether it would be conceivable that his extracts contain nucleic acids, or their breakdown products, which may be operating as gene depressors in the system studied. Professor Teir replied that it is difficult to define the stage of breakdown of the active material, because catabolic processes evidently go on also after the injection of the material.

Professor Koss directed the discussion to the possible relationship between carcinogenesis and chalone action. He wondered whether it would be possible to tie together the extremely protracting process of carcinogenesis and the short-acting effect of chalones. Professor Iversen replied that these two processes could be linked in theory, but not at present in experiment. Professor Bullough, mentioned that they have very recently tested the epidermal chalone on an epidermal mouse tumour. They found that in certain circumstances—though the answer was not very simple—the chalone could inhibit mitotic activity in the tumour. Professor Echave Llanos added that they had also made some observations which indicated that at least a particular mammary tumour in CBH mice appeared to be sensitive to the chalonal type of control, as judged from the existing circadian rhythm of mitoses.

Dr. Alexander put forward some criticism regarding the specificity of the chalone effects. He pointed out that mouse skin has been shown quite clearly to be hypoxic, and that any vasal constricting effect was going to be more pronounced on skin than on other organs. Because anoxia, like certain changes in pH as quoted earlier by the chairman, would certainly stop mitoses, Dr. Alexander wondered whether this might have a bearing on the effects of the epidermal chalone. Professor Bullough replied to this that vasal constriction is utterly without significance *in vitro*.

The discussion was brought to an end by the chairman's summarizing comment in which he re-iterated once again that "growth" is an extremely complex network of processes, with many component pathways, inter-related and interdependent, and no claim to any monopolistic explanation by one or other exclusive principle, ending in some sort of religious warfare between the divergent opinions, has much of a chance of leading to true understanding of the growth process. The time has not yet come to formulate a comprehensive theory of growth control.

Section III

Haematopoiesis

Proliferation Kinetics of Steady State Cell Populations

L. G. LAJTHA

Paterson Laboratories
Christie Hospital & Holt Radium Institute
Manchester, 20

I. Introduction

In the adult organism in most species, the various cell populations constituting that organism are in steady state. This means that the numbers of cells within the population remain constant, apart from relatively minor oscillations, and cell proliferation and cell loss are balanced. This constancy of numbers has certain time limits, since, in reality there is no real constancy of cell numbers: they increase fast during early post-natal life, then this increase slows down and gives place to a slow loss during senescence. The slopes of this increase or decrease in the adult, however, are small enough that for relatively long periods—years in the case of man—a steady state is a close approximation of the situation.

A cell population is defined merely as a number of cells under discussion. It may be a homogeneous population of one type of cells only, or it may contain various subpopulations, e.g. the bone marrow cell population containing the erythroid, myeloid and other subpopulations. Alternatively, the erythroid cells may be taken as the population in question, and the early and late erythroid cells as the subpopulations. For quantitative studies it is frequently necessary to subdivide the cells into "compartments" in which *all* cells are identical in *all* respects (for the purposes of the experiment). This is different from a subpopulation in which some cells may be in early or late G_1, early or late S, or other parts of the cell cycle.

Some of the general principles of cell population kinetics have been discussed in detail elsewhere (Gilbert and Lajtha 1965, Lajtha and Gilbert 1966); this discussion will be limited to the steady state cell populations only.

There are two main types of steady state cell populations. The first is characterized by the fact that for the maintenance of cell numbers it depends on a precursor cell population. This means that the cells in this population are in "transit" coming from the precursor population, and leaving eventually to differentiate and/or to die. Such transit populations are not self-maintaining. There may or may not be cell division in such transit populations, but even with cell division the numerical maintenance depends on the precursor population. An example for a nondividing transit population may be the reticulocytes—coming from the nucleated red cell precursors and leaving to become mature erythrocytes; and an example for a dividing transit population may be the early nucleated red cell precursors, coming from the bone marrow stem cells, and leaving after several divisions, for the late nondividing normoblasts.

The second main type of steady state population is characterized by the fact that it is self-maintaining. It is therefore a "stem cell" type which reproduces itself at a sufficient rate to cope with the physiological rate of cell removal from the population. The two best examples for such stem cells would be: (a) the intestinal crypt cells which maintain their numbers in spite of cell loss into and from the villus, and (b) the bone marrow stem cells which maintain their numbers in spite of the continuous removal of cells for differentiation into the erythroid, myeloid or thrombocytic series.

In the following, some properties of these two steady state populations: the transit and the stem type, will be discussed.

II. Transit Populations

The basic characteristic of a transit population is that it is not self-maintaining, it is originating from a precursor cell population. The name transit signifies that these cells have a temporary existence: they either proceed on to a pathway of differentiation to enter another cell population, or they die.

Of nondividing transit populations, there can be two kinds. There is one in which the cells spend a finite time in the population, e.g. reticulocytes in the marrow, coming from normoblasts, maturing for about 40 hours in the marrow, and then released into the blood.

FIG. 1.

Figure 1 represents a "pipe-line" of cells of various ages. The length of the pipe, the transit time, is about 40 hours. The cells at the beginning of the pipe are "young", at the end of it are "old". The pipe, in fact, can be represented as a series of "compartments" of cells of equal ages. The other nondividing transit population is the granulocytes in the blood. They come from the bone marrow and are destroyed in the tissues. Unlike the reticulocytes, the blood granulocytes have a random half-life in the blood, in man this is about 7 hours. A diagrammatic representation is therefore somewhat different from that of the pipe-line of the reticulocytes since the population can be considered as one compartment in which all the cells have an equal chance of being removed (Fig. 2).

FIG. 2.

The dividing transit populations are different only because the cells can perform a certain number of divisions during transit. However, notwithstanding these divisions, they are not self-maintaining; they come from a precursor cell population and either differentiate during their few cell cycles allotted, or they are removed after a few cell cycles without further differentiation. An example for the first is the bone marrow nucleated erythroid cells; for the latter possibly the cells in the surface of the villi in the intestine. A simple diagrammatic representation for both would be:

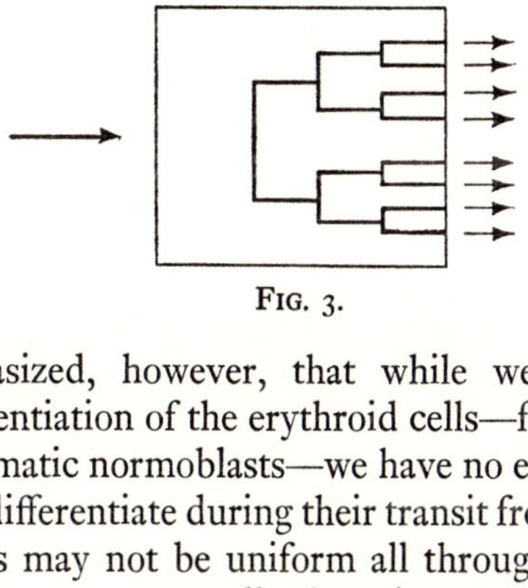

FIG. 3.

It must be emphasized, however, that while we have experimental evidence for the differentiation of the erythroid cells—from early pronormoblasts to late polychromatic normoblasts—we have no evidence to prove that the villus cells do not differentiate during their transit from the base to the tip.

The cell cycle times may not be uniform all throughout transit in these populations. In the bone marrow cells there is a tendency for longer cell cycles with increasing degrees of maturation, and eventually a stage is reached (late polychromatic normoblasts) which is not capable of further cell division. This stage may be treated as a separate nondividing transit population, but it can also be added on to the dividing transit population as an additional "storage time".

Although there are a number of dividing transit type of cell populations in the body, there are very few which can be studied in sufficient detail for quantitative kinetics. The erythroid cells in the marrow are eminently suitable because they are morphologically identifiable, they can be obtained in large numbers with ease, it is possible to make smear preparations from them which are suitable for quantitative autoradiographic analysis at the cellular level, and finally because they have a specific marker of the differentiation process: the synthesis and accumulation of haemoglobin.

In transit populations there is no need to seek for a population size control mechanism, their size is defined by the rate of "feed" into the population and by the transit time. This is usually a property of the cells as a result of the biochemical processes of differentiation. The erythroid cells may be considered as one in which the control is "built in" by the relationship between the rate of accumulation of haemoglobin (rate of maturation) and cell cycle time. The concentration of haemoglobin increases in spite of the successive cell divisions since the *amount* of haemoglobin more than doubles during one cell cycle. Eventually the concentration will reach a sufficient degree to interfere with the process of proliferation. At the normal rate of haemoglobin synthesis this concentration is achieved after five cell cycles. However, if the rate of haemoglobin synthesis increases, the same concentration may be achieved after four cell cycles, and if it slows down the critical concentration will be reached after six or more cell cycles.

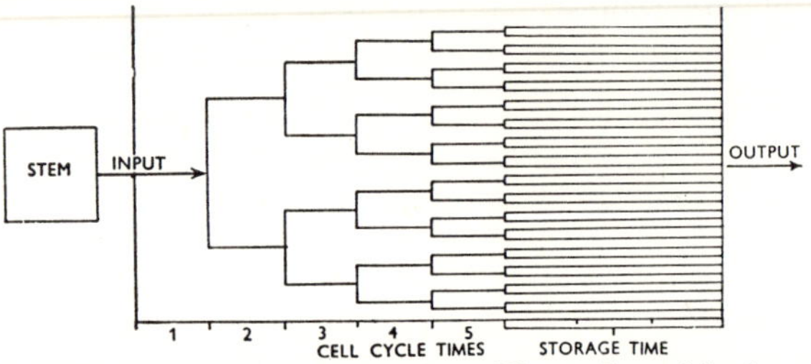

FIG. 4. Dividing Transit Population. Output Rate (O)$=2^n$ I, n=Number of divisions, I= Input rate; Transit time $(t)=n\tau+\sigma$, τ=Cell cycle time, σ—Storage time; Population (p)= I $[(2^n-1)\tau+2^n\sigma]$.

Figure 4 indicates that the population size depends partly on the number of cell cycles, and—if we discount the storage time—one extra cell cycle will double the population, one less will halve it. This way, the rate of haemoglobin synthesis, i.e. the rate of maturation can govern population size. This rate may or may not be externally controlled (i.e. externally to the cell population in question). While this mechanism can double or halve the cell population, these alterations in population size can be maintained only if the "feed in" from the precursor cell population (stem cells) is maintained. The real external control is by controlling the rate at which cells enter the transit population. In the case of bone marrow, this is achieved by the humoral factor erythropoietin which initiates differentiation in the stem cells. Any stem cell thus affected ceases to be a stem cell, it is freed from any feedback control applicable to stem cells. The result of uncontrolled growth would be malignancy, but by virtue of differentiation—i.e. continued haemoglobin synthesis triggered off by erythropoietin—the growth is limited to 4 to 5 cell cycles.

The general concept then is that the dividing transit population depends on a triggering stimulus which acts on its respective stem cells. This stimulus probably acts by masking or unmasking a gene. This process—in the case of erythroid cells at least—is irrevocable and results in specific messenger RNA production, which in turn produces the necessary enzyme (or enzymes) for the processes of differentiation. Growth is thus limited by the rate of the processes of differentiation, and the population is maintained by the rate at which stem cells are triggered (unmasked) to differentiate.

At present no humoral factor other than erythropoietin is known which can trigger off differentiation. However, there are good reasons to suspect the existence of other similar factors, acting on the bone marrow stem cells (granulopoietin, thrombopoietin). In other tissues the appropriate "poietin" may be produced locally, but there is so far no good experimental evidence for their presence.

III. Stem Cell Populations

The definition of the stem cell populations is that they are self-maintaining, i.e. they can and do maintain their own numbers without any "feed" into the population by some precursor cells. Removal of cells from stem cell populations may be continuous or discontinuous, and it may vary in rate. The cells "removed" may differentiate, i.e. enter a transit population, or die. The population size is controlled by some kind of feedback control. Unlike transit populations, the control of the stem cell population size must be a proper feedback control mechanism. The nature of this mechanism is not understood, but it is this mechanism which keeps the numerical balance between the various cell populations in the organism.

The simplest stem cell model, originally proposed for the bone marrow stem cells, implied an asymmetric cell division. It postulated that whenever a stem cell divides, one of the daughter cells turns out to be a differentiating cell, the other remains a stem cell. By simply varying the time interval between successive cell divisions, the model could produce an increased or decreased number of differentiating cells while maintaining the stem cell numbers. In this model the stimulus for differentiation is envisaged to act on the intermitotic cycle time of the stem cells.

The main criticism of this model is that it does not allow regeneration of the population size after a depletion by radiation or cytotoxic chemicals. If each cell division produces only one stem cell, then the only way such a population could respond to depletion would be a faster division rate. This could produce an increased number of differentiating cells, but could never restore the stem cell numbers to normal.

The experimental evidence has completely refuted this scheme and for this reason Osgood (1957) developed an ingenuous modification. He suggested that stem cells may perform two kinds of division, the $\alpha \rightarrow \alpha$,n and the $\alpha \rightarrow 2\alpha$. The result of the first is a stem cell (α) and a differentiating cell (n), that of the second two stem cells (2α). According to this concept, in normal steady state there would be predominantly $\alpha \rightarrow \alpha$, n divisions, during regeneration $\alpha \rightarrow 2\alpha$ divisions.

Although this scheme allows regeneration of the stem cells, it does not explain the mechanism for the asymmetic $\alpha \rightarrow \alpha$,n division, nor does it offer any mechanism to produce the rather basic change from an asymmetric to a symmetric division. Also, it implies that the stem cell is unipotential, unless a further unexplained mechanism is postulated to allow the differentiating cell to be erythroid or myeloid on demand.

For these reasons it seemed necessary to re-examine the available experimental evidence and to construct a stem cell model in its light.

As a simple example the phenomenon of liver regeneration was investigated. It is known that the mitotic rate and DNA specific radioactivity are both very low in the resting liver, and that after partial hepatectomy there will be a 12 to 18 hours lag before DNA specific activity increases. This will be followed by a similar increase in the mitotic index 3 to 4 hours later. Both these activities will peak sharply and decline steeply (Cater *et al.* 1956).

The time sequence of events is incompatible with the concept of a very long cell cycle being telescoped into a shorter one on regeneration. The explanation, however, which might be considered is that in the normal liver the vast majority of the cells are not in a state of cell cycle, but in a true resting stage. For this, following the "gap" terminology, the term "G_0" has been proposed (Lajtha 1963). From this G_0 state the cells are triggered off into cell cycle (into the G_1 phase). The rate of triggering is determined by a control mechanism governed by the rate of cell removal. Although the mean cell cycle time is probably variable, this variation is likely to be limited and contributes very little to the rate of population turnover.

The liver as a self-maintaining (stem) cell system may be illustrated by the diagram opposite.

The same principle has been proposed for the bone marrow stem cells and a kinetic model has been constructed accordingly (Lajtha *et al.* 1962,

Gilbert and Lajtha 1965). With recent developments in techniques it is possible now to study experimentally the kinetics of bone marrow stem cells in a quantitative fashion (McCulloch and Till 1962, Hodgson 1962, Cudkowicz *et al.* 1964, Porteous *et al.* 1963, Gurney *et al.* 1962) hence the situation is that quantitative kinetic models and their predictions can be subjected to experimental proof.

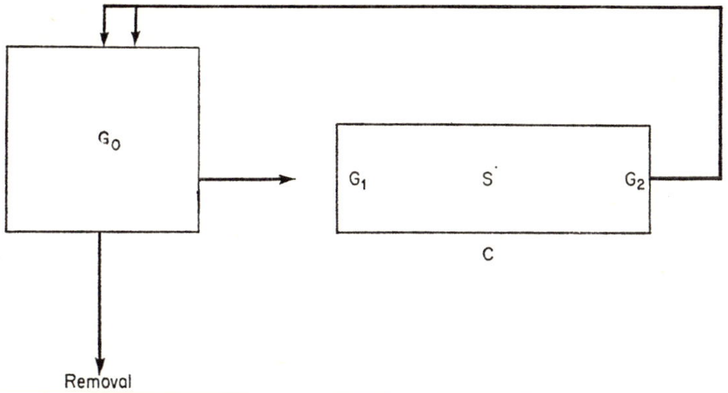

FIG. 5.

The available experimental evidence suggests the following:

(1) The bone marrow stem cells consist of two subpopulations, one in a G_0 state, the other in cell cycle (C). Cells completing the cycle enter the α_0 population from which each cell has an equal chance of being triggered into cell cycle.

(2) During regeneration the proportion of cells in G_0 decreases and that in cell cycle increases (the rate of triggering into cell cycle increases).

(3) Differentiation—induced by the humoral factor erythropoietin—is initiated in the G_0 cells ("available" for differentiation) but not, or very inefficiently, in the cells in cycle (cells in the G_2—S—G_2 cycle).

(4) For population size feedback control the cells in cycle are largely irrelevant, the population "size" is defined and controlled by the G_0 population.

(5) The relationship between rate of triggering into cell cycle and G_0 population size appears to follow a sigmoid curve.

Although we suspect that the population size feedback control is effected by a self-produced inhibitor, i.e. a substance which prevents the stem cells from entering into cell cycle, we have no experimental evidence for the biochemical nature of such a substance. However, an increasing number of computer predictions from the above stem cell model are being verified by experiment, therefore the model is possibly not very far from reality.

It seems likely that in any self-maintaining cell population—which on regeneration shows a significantly large increase in mitotic index or rate of DNA synthesis—we are dealing with a population consisting of the resting G_0 and the proliferating G_1—S—G_2 cycle subpopulations.

Probably most stem cell populations consist of the two subpopulations, G_0 and C. However, we have to consider a different type of stem cell population also, one in which there is no evidence for a G_0 subpopulation. There are two experimental systems, the bean root tip and the intestinal crypt which have been both extensively studied (Hall *et al.* 1962, Lamerton *et al.* 1963) but in which there is no evidence for a significant and functional G_0 subpopulation. In such systems all the cells in the population may be in cell cycle all the time (Fig. 6).

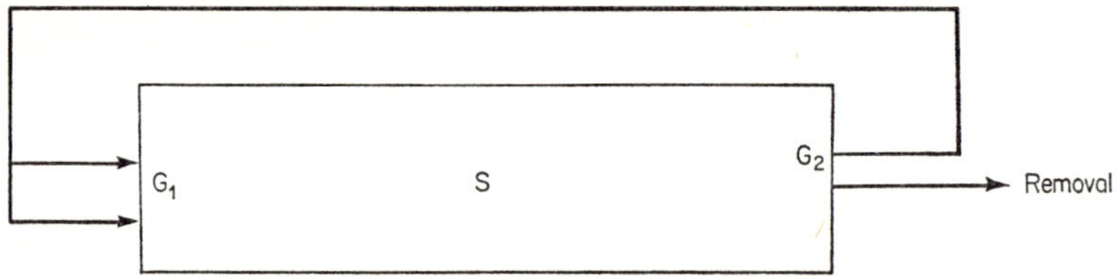

FIG. 6.

There are two characteristics of such stem cell populations. Firstly, in steady state 50% of the daughter cells leave the population (become elongating cells in the roof, or move up on to the villus in the gut); and secondly, in steady state the production rate is the maximal rate. Increase of cell production can be achieved only by shortening the cell cycle, a possibility apparently severely limited in these two systems.

Although 50% of the daughter cells leave the populations, this is not an inherent property of the daughter cells, but of population size, for on decreasing the population (by radiation or cytotoxic chemicals) the proportion of cells differentiating is decreased in a fashion which suggests a dependence on total population size.

This stem cell population is, to all intents and purposes, in a continuous growth which would be exponential were it not balanced by cell removal. This removal may resemble an "overflow" since cells reach the "outside" of a biochemically defined compartment.

As with the bone marrow stem cells, nothing is known of the biochemical factors which define this "overflow", one can only speak vaguely of the milieu in the elongating zone of the bean root, or "something up" in the villus wall, as determinants of differentiation, i.e. cell loss. The problem is obviously much more complicated than a simple overflow, since differentiation (cell removal) does not stop when the stem population is partially depleted, it only decreases in rate.

IV. Conclusion

The above examples show that there are several steady state cell populations which can be studied in a quantitative fashion. With the addition of some new techniques, and with a wider application of the already available methods we will be able to increase the number of cell populations under study, and also be able to study them in greater depth.

The outstanding problem is the mechanisms which governs the steady states. From models we can surmise possible regulating processes, but there is a profound lack of experimental evidence on their nature and of the substances involved. This may be due, in part, to the short life of such regulating substances—a short life-span of these is a prerequisite of a sensitive regulation—and partly to the relative crudity of biochemical separating methods available.

These difficulties, however, will be overcome in due course, and with a better understanding of the normal steady state regulation, we will advance a great step towards understanding one of the most basic problems of multicellular organization. The realization that malignant growth is characterized by the loss of the normal feedback controls of growth adds a great practical importance to this problem.

REFERENCES

Cater, D. B., Holmes, B. E. and Mee, L. K. (1956). *Acta radiol.* 46, 655–667.

Cudkowicz, G., Upton, A. C., Smith, L. H., Gosslee, D. G. and Hughes, W. L. (1964). *Ann. N.Y. Acad. Sci.* 114, 571–582.

Gilbert, C. W. and Lajtha, L. G. (1965). *In* "Cellular Radiation Biology XVIII Annual Symposium on Fundamental Cancer Research, 1964", pp. 474–497. Williams and Wilkins, Baltimore.

Gurney, C. W., Lajtha, L. G. and Oliver, R. (1962). *Br. J. Haemat.* 8, 461–466.

Hall, E. J., Lajtha, L. G., and Oliver, R. (1962). *Br. J. Radiol.* 35, 388–397.

Hodgson, G. S. (1962). *Acta physiol. latinoam.* 12, 365–369.

Lajtha, L. G. (1963). *J. Cell. comp. Physiol.* pp. 143–145.

Lajtha, L. G. and Gilbert, C. W. (1967). *Adv. biol. med. Phys.* (In press.)

Lajtha, L. G., Oliver, R. and Gurney, C. W. (1962). *Br. J. Haemat.* 8, 442–460.

Lamerton, L. F., Lord, B. I. and Quastler, H. (1963). *Radioakt. Iso. in Klin. Forsch.* 5, 493–501.

McCulloch, E. A. and Till, J. E. (1962). *Radiat. Res.* 16, 822–832.

Osgood, E. E. (1957). *J. natn. Cancer Inst.* 18, 155–166.

Porteous, D. D., Alexanian, R. and Lajtha, L. G. (1963). *Int. J. Radiat. Biol.* 7, 95–100.

Regulation System of Blood Cell Production*

T. RYTÖMAA and K. KIVINIEMI

II Department of Pathology, University of Helsinki, Helsinki, Finland

I. Introduction

It must be supposed on *a priori* grounds that in order to maintain a balance between supply and demand, specific mechanisms exist which transmit signals to the progenitor cells in the bone marrow. The aim of this article is to describe some features of these mechanisms in the light of new experimental findings.

The dynamics of cell proliferation may be more complex in the granulocytic (Cronkite and Fliedner 1964), erythrocytic and lymphocytic systems than in any other tissue. Among the complicating features is the fact that these systems must adjust their rate of cell production not only in terms of mass (cell numbers) but also in terms of functional demand.

The basic difference between the expressions "functional demand" and "functional load" deserves consideration in this context. As stated earlier by Weiss (1955) there are instances, such as after the removal of one orbital gland, where an excessive functional load placed upon the remaining gland cannot explain its subsequent compensatory growth (Teir 1951). This has led to the common postulate that maintenance of the total mass in equilibrium state represents the main principle in growth control (Weiss and Kavanau 1957, Osgood 1957, Iversen and Bjerknes 1963). It must be noted, however, that functional demand—from the viewpoint of the whole organism—represents *per se* a more important principle in blood cell regulation than the tissue mass (as in such examples as physiological "polycythaemia" at high altitudes, etc.). Moreover, this principle may not be rightfully neglected in other instances: in the case of the remaining orbital gland it is very difficult to understand why the gland could not respond to that part of the demand of the organism which was previously addressed to the excised gland. Thus it can be assumed that changes in tissue mass are realized by the cells (= negative feedback loop) as are also changes in the functional demand (= positive feedback loop).

* Supported by grants from the Damon Runyon Memorial Fund, New York (DRG-644) and Sigrid Jusélius Foundation, Helsinki.

Another point of interest, especially in the granulocytic system, is the theoretical possibility of controlling the number of mature cells (tissue mass and functional "power") by means of decreased death rate, i.e. by prolonged cell life. Evidence that this does not happen comes from the observation of the random loss of granulocytes (Cartwright *et al.* 1964): the cells die through hazardous environmental interaction and not due to intrinsic "ageing", which could be varied according to needs. Thus the life expectancy of granulocytes is unpredictable and must vary from time to time.

The aim of the present paper is to describe some results of investigations into the feedback mechanisms which control the rate of cell production in the erythrocytic and granulocytic systems. Many of the inferences based on these results have been presented earlier in a theoretical paper dealing with the control mechanisms on more general grounds (Bullough and Rytömaa 1965); and some of the preliminary results themselves have also been reported (Rytömaa and Teir 1964, Teir and Rytömaa 1964, Rytömaa and Kiviniemi 1964).

II. Earlier Studies

A. *Positive feedback factors*

1. Erythropoietin

The existence of a humoral factor, erythropoietin (EP), capable of stimulating red cell production has been firmly established (see Brecher and Stohlman 1959, Linman and Bethell 1960, Jacobson and Doyle, editors, 1962). Studies of the mode of action of EP have indicated that its primary action seems to be to promote the conversion of stem cells into erythrocytic progenitor cells (Alpen and Cranmore 1959, Lajtha and Oliver 1960, Erslev 1964). Induction of new erythrocyte producing units, erythrons, by cell conversion may be taken as an embryonic process, and EP should thus be regarded as an inducing substance (Bullough and Rytömaa 1965). This conclusion may appear as a matter of dispute because it presupposes that the formation of cells which are recognizable as belonging to the erythrocytic tissue is primarily due to the transformation of undifferentiated and hence pluripotential stem cells. It may be safer not to be too categorical with regard to the pluripotential nature of the stem cells although the available experimental evidence is clearly in favour of this thesis (Ford *et al.* 1956, Becker *et al.* 1963, Lajtha *et al.* 1964, Lajtha 1966).

A survey of the literature of the action of EP has also indicated that it does not change either the mitotic rate of the progenitor cells, or the maturation rate of the developing erythrocytes, or the rate of haemoglobin synthesis by more than approximately 10% (Alpen and Cranmore 1959, Lajtha and

Oliver 1960, Erslev 1964). Slight primary changes reported to occur in these parameters may well be due to impurities essentially present in all available EP preparations, which may also contain other specific "plasma erythropoietic factors" (Linman and Bethell 1960). Presence of such impurities is demonstrated by the finding of Fischer and Fruhman (1964) who noted that some EP-preparations caused significant increase in Fe^{59} uptake in the bone proper, but that this effect was not observed in a more purified preparation obtained from sheep plasma. A logical conclusion emerging from these and other results of a similar nature is that EP-preparations may act on the progenitor cells that have advanced beyond the stem cell stage due to "contaminants". As a matter of fact, Linman and Bethell (1960) have presented evidence which suggests that there appear to be at least two specific "plasma erythropoietic factors" which cause the observed reactions: the first factor is relatively thermolabile and ether-insoluble; the second is thermostable and soluble in ether. According to the terminology adopted here, the first factor would be identical to the "true" erythropoietin. The status of the second factor is not clear in view of the present study, but it may be a substance which would here be called "erythrocytic antichalone".

2. "*Granulopoietin*"

Experimental confirmation of the existence of humoral factors stimulating granulocyte production is not so firmly grounded as is the demonstration of erythropoietin. This somewhat curious situation may be mainly due to the complex physiology of the granulocytic cells—one of the major obstacles in the bio-assay techniques adopted.

Some of the data which has been submitted in support of the existence of "granulopoietins" has been affected by fallacies, such as the temptation to think that alterations in granulocyte production could be readily measured from the circulating granulocyte values. Changes in blood erythrocyte levels are known to reflect changes in the rate of red cell production under properly controlled assay conditions, but circulating granulocyte level represents an entirely fallacious index of the rate of granulocyte production in all assay conditions adopted so far. Some relation does evidently exist between these two phenomena, but rather in the opposite direction: anything which promotes removal of granulocytes from the blood or from the "marrow granulocytic reserve" (MGR) (Craddock *et al.* 1959) leads to a chain of events which in normal animals will ultimately cause an increased rate of production (Craddock 1962).

Numerous authors have published data which demonstrates the existence of one or more humoral factors that promote the delivery of granulocytes from the bone marrow to the blood (Menkin 1946, 1955, Steinberg and Martin 1950, Farr and LeQuire 1950, Wasastjerna *et al.* 1958, Fukuda *et al.*

1960, Gordon 1960, Gordon *et al.* 1964, Bierman 1964). The common feature of all these factors is their ability to produce a marked granulocytosis some 2 to 6 hours after administration; therefore the factors have been termed "leucocytosis-inducing factor" (LIF), "leucocytosis promoting factor", etc. Occasionally, however, such other names as "granulopoietin", "leukopoietin G", and "haemopoietin" have been applied which must not be confused with "true" granulopoietin or with the other direct production stimulators named according to the terminology adopted here. The work of Gordon and his associates (1960, 1964) demonstrates that LIF obtained from the plasma of rats after repeated leucocyte withdrawal (leucophoresis), evokes increased blood flow rate in perfused rat legs and femurs accompanied by increased discharge of leucocytes into the perfusate. It thus seems evident that LIF participates in the regulation of granulocyte physiology, but that it can only be a secondary controlling mechanism with regard to the stimulation of production of granulocytes.

It is almost trivial to state that the humoral control mechanism of granulocytopoiesis must be studied by such means as meet the many serious objections which have been raised against the earlier bio-assay techniques. The ostensible triviality of this statement fades when a search is in fact made for such means. Simple methods able to provide straightforward evidence and unambiguous inferences do not seem to exist.

B. *Negative feedback factors*

It is becoming increasingly obvious that all mitotic control mechanisms found in different tissues are merely variations on certain common themes (Bullough and Rytömaa 1965). Recently much new information has been obtained regarding mitotic control by humoral substances acting as inhibitors in a negative feedback mechanism. Such substances have been termed chalones (Bullough 1962). The first experimental verification of the "chalone" concept came from the studies of Bullough and Laurence (1964, 1966) who extracted one such substance from the epidermis and demonstrated its activity in depressing mitosis in normal epidermis, both *in vivo* and *in vitro*. These results have now been confirmed by others (Iversen and Elgjo 1966). This epidermal chalone has also been prepared in an almost pure form (Homan and Hondius Boldingh 1966); it appears to be a glycoprotein in nature.

With regard to the control of blood cell production, it has been obvious for some time that certain clinical and experimental observations appear incompatible with the idea that poietins act as the only control mechanism (Brecher and Stohlman 1959). Findings which lead Brecher and Stohlman to suggest the existence of a dual control for red cell production were based on the inconclusiveness of the evidence that hypoxia is the universal trigger for increased plasma levels of erythropoietin. For instance, in the box turtle, erythropoiesis is stimulated by bleeding but not by hypoxia (Altland and Parker 1955). Similarly, compensated haemolytic "anaemias" do not involve hypoxia and yet red cell production is still remarkably increased. It is also well known that erythrocyte production may be suppressed after an adequate erythrocyte transfusion, a fact which is clearly easier to explain in terms of chalone action than in terms of decreased erythropoietin level.

The existence of a granulocytic chalone has been indicated by some findings reported in the literature. Wetherby-Mein and Cottom (1956) noted that myeloid leukaemia was temporarily suppressed by transfusion of fresh blood but not of stored "bank" blood. Such a finding is compatible with a chalone-induced inhibition because only fresh blood could be assumed to contain enough chalone. Craddock (1960, 1962) has demonstrated that intact mature granulocytes produce, presumably secrete, a chalone-like substance which inhibits DNA synthesis in lymphoid cells. Certain leukaemic granulocytic cells do not produce this substance (Craddock 1962) which suggests that these results are not due to unspecific toxic influences. In the future it may appear unlikely that these results have been caused by a "true" granulocytic chalone because of the non-specific nature of the responding tissue. However these results could well fit the general concept if the inhibited lymphoid cells happen to be closely related to the stem cells, or if they act as "trephocytes" in the bone marrow (see e.g. Craddock *et al.* 1964, Fichtelius and Linna 1966, Kekomäki and Rytömaa 1966).

III. Bioassay Technique

Repeated bleedings are known to lead to the appearance of erythropoietin in the plasma or serum. There was thus reason to think that removal of large quantities of mature granulocytes from the animal's body (leucophoresis) would by analogy lead to the appearance of "granulopoietin(s)" in the serum. It was already known, however, that leucophoresis causes the release of "leucocytosis-inducing factor" in the blood (Gordon 1960). As mentioned earlier, this factor has been shown to promote the discharge of mature granulocytes from the bone marrow to blood (Steinberg and Martin 1950, Gordon *et al.* 1964, Bierman 1964) which in turn may initiate a chain of events which lead to increased rate of granulocyte production (Craddock 1962). It is thus clear that there was no point in attempting to demonstrate the existence of another humoral factor, direct production stimulator, by any means which failed to give due consideration to a co-existent action of the release factor.

Attempts to destroy or separate chemically the action of the LIF were unrealistic *per se*. Obviously, the complication could not be ignored, especially because of the unknown speed at which the signal, probably triggered by the release effect, would be transmitted to the progenitor cells. In order to settle this dilemma, use was made of test conditions in which the responding organ could rightfully be assumed to be resistant to the action of the release-factor without concomitant loss of responsiveness to the action of a test substance. Realization of this aim was basically very simple although the introduction of some new complications could not be avoided: use was made of *in vitro* assay conditions where the release phenomenon could no longer operate. The complication caused by LIF and other similar factors has thus been the basic reason why *in vitro* methods had to be adopted in this study.

A. *Sources of test material*

Leucophoresis was effected by injecting 15 ml of polyvinyl pyrrolidone (PVP, L. Light & Co., England)—3·5% solution in physiological saline—into the peritoneal cavity of rats. After 14 hours another 15 ml of PVP was injected and the ascites fluid, containing about 20,000 cells per mm^3 of which 60 to 70% were granulocytes, was removed by puncture 2 hours later. In most instances the PVP-injections were repeated 2 to 3 times before collecting the serum. Blood was withdrawn by cardiac puncture, serum was separated and stored at −20°C until used.

In addition to leucophoresis, a number of other treatments were applied to the rats in attempts to induce the appearance of other hypothetical factors in the serum. It may be mentioned here that most, if not all the treatments applied here are known to be followed by specific reactions in the bone marrow *in vivo*. Therefore one crucial assumption in the working hypothesis has been that the signals are transmitted to the bone marrow cells by humoral substances. Such an assumption is obviously plausible, especially since the signals must originate in the periphery (compare, however, the possible signal role played by a "leucocytosis-inducing factor" as described above).

The different types of sera assayed in this study are the following:

"Old" normal serum Blood was withdrawn by cardiac puncture from normal healthy-looking adult rats belonging to the same stock of animals as all treated rats (an "inbred" Sprague-Dawley strain, raised at Mankkaa, Orion Medical Factory, Helsinki). The serum was separated as usual and stored at −20°C for long periods of time (up to some months). Prolonged storage, combined with repeated freezing and thawing of the material, were believed to yield a reference serum which would be sufficiently free of chalone activity. Control serum used for reference purposes in the experiments was obtained from "old" normal serum by pooling 2 to 4 batches of sera, collected from different rats.

Erythropoietin serum. Serum was collected from animals after two or three daily bleedings by vena jugularis puncture. In each puncture 6 to 8 ml of blood was withdrawn from 350 to 400 g rats, corresponding to about one-third of the total blood volume. Serum was collected 5 hours after the last puncture and stored as usual.

Eosinapheresis serum. Eosinapheresis was performed by injecting 7 ml of a 0·1% asbestos suspension into the peritoneal cavity of rats. Another 7 ml of asbestos suspension was injected 14 hours later; and 2 hours after a third injection of 7 ml asbestos suspension, ascites fluid was withdrawn. Serum was collected 2 hours later. Asbestos has been reported to be the strongest eosinotactic matter (Speirs 1958); in this study the removed ascites fluid contained 10,000 to 20,000 cells per mm^3 of which about 60% were eosinophils.

Polycythaemia serum. Blood was collected from donor animals by cardiac puncture, red cells were separated by centrifugation, and injected into the peritoneal cavity of the recipient animals. This procedure was repeated daily for 3 days. The serum was collected 24 hours after the last injection. Recipient's red cell count was followed during the treatment; it revealed that at the time of serum collection the circulating red cell count increased from $6{\cdot}5 \times 10^6$ to $10\text{–}11 \times 10^6$ cells per mm^3, i.e. by some 60%. It was further calculated from the number of injected cells that 50% or more of them were recovered from the circulation.

Hyperoxia serum. Rats were kept at high oxygen tension, under normal barometric pressure. This was achieved by continuous insufflation of the "cages" (large glass jars) with oxygen. The oxygen flow was adjusted according to the condition of the rats: if too much oxygen was given they became ill. The red cell count was followed daily; serum was collected at the time when first signs of decreasing red cell values appeared. This occurred usually in about three days which indicates that erythropoiesis had been suppressed. It may be noted that the maximal drop in circulating erythrocytes in rats would be 5 to 6% after 3 days.

Chloroleukaemia serum. Serum was collected from rats suffering from transplanted chloroleukaemia (Shay *et al.* 1952). Animals with a generalized leukaemia (produced by intraperitoneal transplantation) and those with a solid tumour (subcutaneous transplantation) served as serum donors.

Serum fractions. Some test and control sera were fractionated by means of gel filtration with Sephadex types G-25, G-75, and G-200 (Pharmacia, Uppsala, Sweden). Serum fractions were eluted from the column in Hanks BSS as it was used as the basal solution in the culture medium. The effects of the serum fractions on bone marrow cells were tested with and without unfractionated control serum.

B. *Bone marrow cells*

Bone marrow cells were obtained from untreated adult male rats. The donor animals were killed by decapitation after light ether anaesthesia, and one femur was immediately removed and freed from surrounding tissues. Epiphyses were cut off and the marrow plug was blown out by forcing air through the marrow cavity. The cells were recovered in Hanks BSS and the plug was broken down by gentle agitation. An equal volume of cell suspension (avoiding the few remaining clumps) was added to the test and control media already containing all other components. After a little experience it was possible to estimate—and to know—the cellularity of the original suspension fairly accurately by naked eye within an error of less than ±50%. It was thus possible to proceed with the culture preparation without the need to perform cell counts. Speed in the last steps of culture preparation has been found to be crucial for the success of this technique. If too much time elapses between the killing of the rat and the start of the cultures, the bone marrow cells become "pseudo-synchronized": they pile up in the DNA-synthesis phase. Pseudo-synchronization creates several problems of its own which need not be discussed here.

C. *Culture technique, proliferation parameters and measurements*

Detailed description of the culture technique and the analytical procedures will appear elsewhere (Rytömaa and Kiviniemi, unpublished). A scheme for the originally adopted techniques has been published earlier (Rytömaa and Teir 1964), but the present assay system contains some modifications and alterations which have been considered important. In this article the adopted assay techniques will be described briefly.

Individual cultures were prepared according to a monolayer technique originated and developed by Earle and his co-workers (Merchant *et al.* 1960). In this system the cells settle from the suspension and become attached to the upper surface of a coverslip (10×20 mm) hanging from the top of a test tube. In each experiment equal numbers of test and control cultures were started simultaneously. Usually 4 to 6 replicate cultures were used in respect to each varied factor (e.g. type of serum, incubation time, time of isotope addition, etc.).

The basic culture medium was normally composed of serum (20%), cells (500 to 1000 per mm^3), and Hanks balanced salt solution (BSS). Low cell concentration was essential to obtain a useful cell density on the coverglass, but the given limits are not very rigid. In earlier experiments antibiotics (penicillin and streptomycin) were added to the culture medium. Although they were avoided later, no increase in the infection frequency occurred.

Air was always used as the gas phase for the bone marrow cells. Chloroleukaemia cells required better oxygen supply; this was provided "internally" by adding erythrocytes to the cultures.

Radio-actively labelled precursor substances have been used as the basic indicators of proliferation parameters. It was essential that more than one proliferation parameter was studied in each experiment; some of them were always obtained by using labelled substances. H^3 thymidine (DNA synthesis), C^{14} formate (DNA synthesis; RNA and protein synthesis), S^{35} sulphate (granulocytic progenitor cells), H^3 adenine (RNA synthesis), H^3 leucine (protein synthesis), and Fe^{59} citrate (haemoglobin synthesis; peroxidase synthesis), all obtained from The Radiochemical Centre, Amersham, England, have been used in the studies reported in this article. The amounts and specific activities of the labelled precursor substances have been varied from one experiment to another; normally, rather high concentrations were used. This was done on purpose because the possible dilution effect, due to variable pool sizes in the different types of sera, was considered to represent a much more important potential source of error than the "high" doses of radiation delivered to the cells. If the applied radioactivities are viewed from the point of view of label concentration in the culture medium (ranging from 0·1 μc/ml to 5·0 μc/ml), they do not differ from the commonly applied amounts in any significant degree, but if the available activities are expressed in terms of per cell label they appear high. Even in the latter case, however, any manifest radiation damage or chemical interaction must have been insignificant on the basis that a 100-fold variation in this figure failed to modify the results in any detectable way.

After harvesting of the coverglasses the cells were fixed *in situ*. All the treatments applied in this "pre-analytical" stage were principally quite similar to the generally known techniques used for blood smears for example, and need not be repeated here.

Total radioactivity incorporated in the cells was measured in an IDL liquid scintillation counter (Isotope Developments Ltd., England) by putting the coverglass into an angular position in a vial containing the scintillator liquid (an ordinary mixture of PPO and POPOP in dioxan). A constant position of the coverglass in respect to the phototubes was adopted throughout the study, although it was not found to be critical for the measurable activities. Counting efficiency was, however, reduced in the adopted system, but this was entirely due to the 2π geometry of the samples (calculations have shown that the efficiency was better than in autoradiography). One hundred seconds counting time was consistently used in the activity determinations; it was sufficiently long to give counts reproducible within $\pm 3\%$.

After the measurement of the total radioactivity each preparation was carefully washed in series solutions of dioxan and then water to remove all scintillator liquid from the samples. The coverglasses were then mounted "face up" on microscope slides and *autoradiograms* were prepared by applying Kodak AR-10 stripping film. After processing of the autoradiograms the cells were stained in Giemsa solution, pH 5·7 (Feinendegen and Bond 1962) or with haematoxylin. The quality of the autoradiograms was not significantly altered by their preceding exposure to the scintillator liquid.

Autoradiographic analyses were regularly made from code numbered preparations if performed by one person only. In some instances two persons analyzed the preparations independently: one used code numbering, the other one made the analyses "openly".

Total numbers of cells attached to the coverslips were determined from the autoradiograms by means of a counting device which is essentially similar to the commercially available Integrating Eyepiece I (Zeiss, Oberkochen).*

A set of 25 points was projected onto the microscopic field of the specimen and the number of "hits" was recorded. The ratio of the sum of the hits to the total number of points was measured; it represents the area share of the cells and is thus proportional to the total cellularity.

If the coverslip preparations were not used for autoradiography, the total amount of "nucleic acids" was measured spectrophotometrically (Beckman D.U.). For this purpose 2 to 4 replicate cultures were hydrolyzed in perchloric acid (hot 5% PCA) for 5 minutes. After neutralization and dilution of the hydrolysates, the UV absorbancies were measured spectrophotometrically. In some instances cells were hydrolyzed overnight in 0·5 N KOH at room temperature; UV absorbancies were measured as above.

Other analyses performed in this study include differential counts of the cells (unreliable except for morphologically distinct populations); rough estimation of overall mitotic index and form ("stretching") of cells on the glass; alterations in cell morphology, etc.

Statistical methods have been varied according to the problem, and to the type of data available. The analysis of variance (Simpson *et al.* 1960) has been adopted in most instances because more than one factor was deliberately varied in each experiment.

IV. Results

A. *Total radioactivity in the cells*

The time course of the total radioactivity in the cultured bone marrow is seen in Fig. 1. H^3 thymidine and C^{14} formate activities respectively have

* This device was introduced and set up by Dr. J. Puittinen, M.D., whose assistance is gratefully acknowledged.

shown very constant time courses between different experiments and between different sera, as seen in Figs. 1, 4, 5, and 6 (note the differences in the scales). The activity increases rapidly after labelling with H^3TdR and reaches 50% of the peak (or plateau) value in about 2 hours; mean incorporation time varied in different experiments from about 2 to 5 hours, with an average of 3·4 hours.

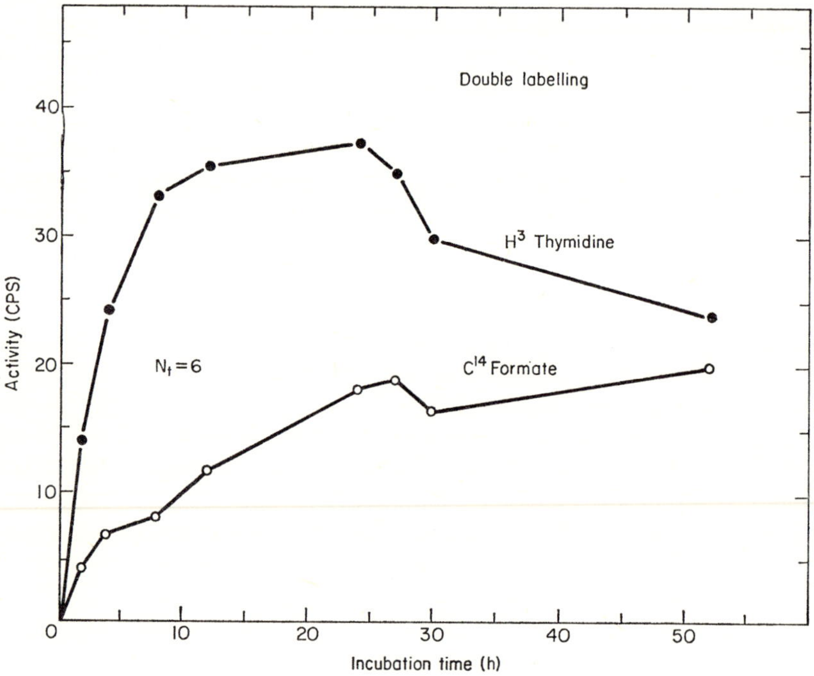

FIG. 1. Incorporation of H^3 thymidine and C^{14} formate in rat bone marrow cells *in vitro*. Activities represent the total amount of isotope which has accumulated in the cells of the original population, and in their descendants.
Double labelled cultures; liquid scintillation counter analysis. (N = 48).

The relative rate of increase in H^3TdR incorporation does not seem to be affected by the stimulating serum factors (leucophoresis serum) as indicated by the lack of significant serum-incubation time interaction in the analysis of variance. It thus follows that "leucophoresis serum" does not seem to influence the specific rate of DNA synthesis.

It is seen from Figs. 1, 4, and 5 that the total tritium activity in the H^3TdR labelled cells begins to decrease after 24 hours. This can be taken as the minimum life span of labelled cells under assay conditions, but it does not indicate whether these cells belong to granulocytic or erythrocytic systems.

The form of the activity curves allows another inference which may be of considerable interest for the present purpose. As seen in Figs. 1, 4, and 5 (best in Fig. 4 because of the convenient time scale), there exists an appreciable plateau in the curves. This suggests that significant overlapping of label influx and efflux does not occur; which in turn must indicate that premature cell death, if it occurs, cannot involve the majority of the labelled progenitor cells. Therefore, if the rate of cell production is controlled by premature cell death, the death must be limited to non-proliferating progenitor cells or, alternatively, to very early cell forms which constitute only a small fraction of all labelled cells. This has indeed been suggested by Stohlman (1959).

Lack of significant form differences between the TdR incorporation curves in the test and control cultures—"chloroleukaemia-serum" possibly excluded—suggested that it would be possible to estimate the "short-term" effect of a test substance in relation to the control effect, from any part of the curve not extending beyond 24 to 30 hours. As it has been possible to demonstrate that the majority of the TdR activity present in the cells at any one time, from between 1 and 30 hours, is a direct measure of the total DNA synthesis at zero time and shortly thereafter (based on results which are not given in this article), the suggestion made above could have been accepted without any further experiments. However, such experiments have been performed by different means; some of the results are presented here.

Ninety cultures were prepared and divided into five equal groups with respect to the number of cells added to them at zero time. Each group was further divided into three equal subgroups according to the incubation time to be used. The cell numbers were varied within a factor of 5 (from 400 to 2000 cells per mm^3) and the incubation times from 4 to 23 hours; H^3 thymidine concentration was kept constant (2 μc/ml) as usual. The results are shown in Fig. 2.

As indicated by the single regression line, different time-groups (subgroups) have been pooled. This was done to give a better visual picture of the fact that there really exists a strictly linear relationship between the number of DNA synthesizing cells present at zero time and the total TdR incorporation measured even as long as 23 hours later. It can be seen that the regression line passes through the origin (it is well within the confidence limits), and that a constant *concentration* of TdR yields activities which are directly proportional to the DNA synthesis. It may be noted here also, although not illustrated in Fig. 2, that the value of the regression coefficient is a function of incubation time; i.e. that the line has different slopes at different times.

It can also be seen in Fig. 2 that the combined effect of all known and unknown error factors as well as the "repeatability" of this effect in the adopted technique is expressed by $y \pm$S.E. Expressions of the standard errors of the means are more meaningful in this connection than expressions of standard deviations because means of replicate cultures (usually 4 to 5 in

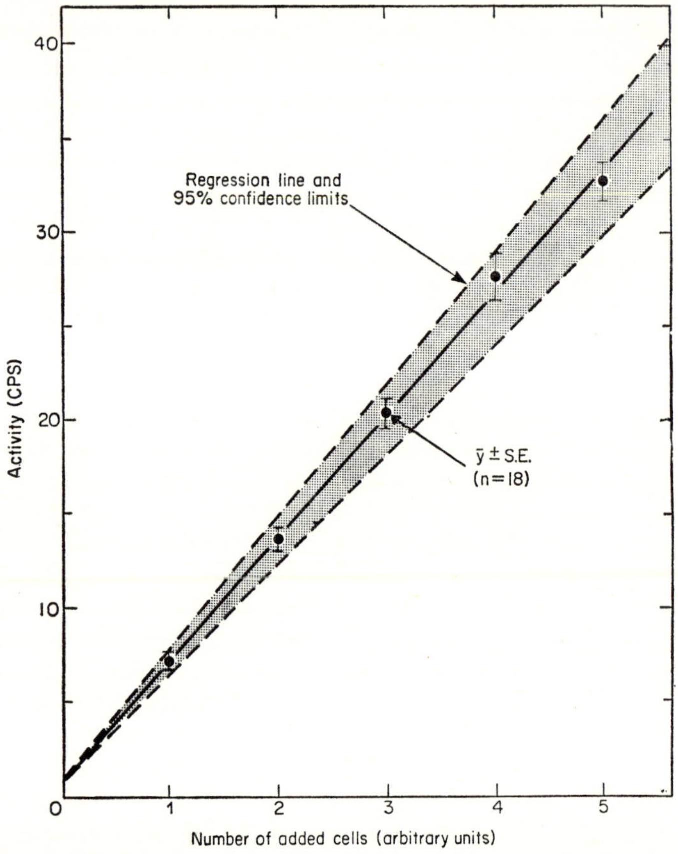

FIG. 2. Relationship between H^3 thymidine incorporation and total number of bone marrow cells present at zero time.

Activities have been measured by liquid scintillation counter after incubation times of 4, 8, and 23 hours, and pooled (4- and 23-hour values corrected to the 8-hour level). One unit of cells~ 160×10^3 nucleated bone marrow cells per culture~20×10^3 cells in DNA synthesis at zero time N = 90).

number) have generally been used as "single measurements" (cf. Table I). It can easily be verified from Fig. 2 that the average relative error of "single measurements", consisting of 4 to 5 individual cultures, has been about ±7%.

The significant difference between the forms of the curves of thymidine and formate incorporation (Fig. 1) deserves consideration. It is well known that these substances are involved in different metabolic pathways in the cells. Thymidine is incorporated into DNA in the last step of DNA synthesis (polymerization) whereas formate enters DNA as a member of the "one carbon pool". Formate *in vitro* labels mainly the methyl group of DNA-thymine (Totter 1954, Lajtha 1954, 1961, Smellie *et al.* 1958), and may therefore be taken as a rather specific indicator of *de novo* synthesis of DNA. Yet it can be shown that thymidine and formate incorporation curves do not match well even at the early times of incorporation, which must mean that there are other causes for the different labelling patterns.

It is known from previous work that only a small fraction of labelled thymidine is incorporated into the cells, since the available pool is rapidly catabolized (Rubini *et al.* 1960, Pelc and Appleton 1965). This is not the case with formate, which is catabolized slowly. Schroeder and Brecher (1960), on the basis of preliminary data, assumed that no new cells entered DNA synthesis after the start of culture *in vitro*; consequently, the differential availability of thymidine and formate could be neglected. However, it is now clear (unpublished data) that new cells do in fact continue to enter DNA synthesis for a long time after the start of the culture (up to 20 hours). It is thus apparent that formate incorporation may be affected by these cells, producing a continuous increase in the incorporation (Figs. 1 and 6).

It is well known that *in vivo* formate labels not only DNA but some other cell constituents as well (Lajtha 1961, Perretta and Pieber-Perretta 1962). The most important of these substances, from a quantitative viewpoint, are RNA and protein. To study the relative amount of formate that had been taken up by the cells in their DNA, their total activities were first measured as usual, and the preparations were then washed with successive solutions of dioxan and water to remove all scintillating liquid, dried, and hydrolyzed for 6·5 minutes in 1 N HCl at 60°C (Lajtha 1954) to remove all non-DNA label. After this the activities were remeasured. It was thought that this procedure would measure the minimum fraction of the incorporated formate which had been taken up in DNA, since it was more likely that the detachment of some cells from the glass would "overcompensate" the possibly incomplete removal of non-DNA label.

The results demonstrated (Fig. 3) that for at least 10 hours the majority of the formate activity was in DNA, but that between 10 and 23 hours an appreciable drop in this fraction occurred. The long interval does not allow unequivocal statements to be made about the time or type (slow or sudden) of the decrease. However, the presence of very few DNA synthesizing cells at 24 hours, and at later times, the suggestive shoulder in the formate curve coinciding well with the loss of H^3TdR labelled cells (Fig. 1), and the continuous increase in formate for at least 50 hours (Figs. 1 and 6), may all be taken as evidence favouring the interpretation that the proportion of formate which has been incorporated into the DNA may show a more sudden drop at about 20 hours than indicated in Fig. 3. In any case, it is quite clear that the increasing slope in formate activity after the first 24 hours is not a measure of the DNA synthesis rate.

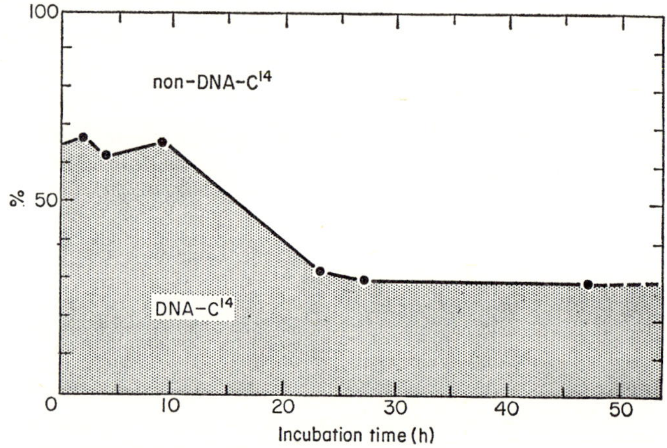

FIG. 3. Proportion of incorporated C^{14} formate present in DNA after different times of incubation.

Rat bone marrow cells *in vitro*; liquid scintillation counting before and after acid hydrolysis (see text). (N = 50.)

The general patterns of the TdR incorporation curve make it possible to compare test and control serum effects on the bone marrow cells (in relative terms) from any part of the curve before about 24 to 30 hours (cf. Fig. 2). This will also justify pooling of the different experiments without fear of systematic bias. Such a pooling is clearly necessary for obtaining evidence which would support or disprove generalized conclusions. In principle, the data at hand should be sufficient for this purpose because it consists of results from individual experiments repeated several times in a form where the non-essential details have not been kept constant. Furthermore, the whole material extends in time over a period of 2½ years, and can thus be subjected to statistical tests for cyclic, non-random influences. Such effects have not in fact been detected here (serial correlation).

A summary of the effects of some test sera on the total incorporation of H^3 thymidine in normal bone marrow cells is seen in Table I.

It may be noted that the mean percentages given in the table are possibly slightly biased because the material used for their calculation contained some obvious "failures"; i.e. it was not quite homogeneous. In some instances the cause of the failure was known with fair certainty; it was an "incompatibility" (not necessarily immunological) between the serum and the bone marrow cells which prevented adequate attachment to the glass surface, but this was not individually confirmed for each particular failure-experiment.

TABLE I. Relative effects of some unpurified "serum factors" on the total incorporation of H^3 thymidine per unit number of bone marrow cells *in vitro*. Liquid scintillation counting.

Type of serum according to treatment of rats	Number of repeated experiments	Total number of cultures (test+control)	Relative activity (% of control). Average[a]	Significance[b] 2α
Leucophoresis	25	664	+19·6	<0·001
Repeated bleedings	11	156	− 3·9	Insignificant
Polycythaemia (red cell transfusions)	2	54	−14·0	<0·01
High O_2-tension	4	102	−5·8	<0·05
No serum (Hanks BSS)	11	82	+4·8	Insignificant

[a] Calculated from the values obtained by comparison of means of replicate cultures in the test and control sera. Each group of replicate cultures consists of 3–5 individual cultures, both in the test and control group.

[b] Significances have been estimated with the sign test comparing the means of replicate cultures in pairs.

Rejection of some failure-experiments could also have been done on the basis of correct statistical criteria. Why, then, are the failures included in this material? The reason is that there were insufficient grounds for knowing that the rejection criteria were applied with equal rigour at both ends of the distribution. It was suspected in fact that the rejection was less rigorous with respect to too good results (slightly skewed distribution). Thus all rejections were omitted in the material summarized in Table I. Although the material in its present form is non-homogeneous, there is no reason to consider the results as being significantly biased from a biological point of view: the failure frequency was small enough (less than 0·10) to be without substantial effects, and furthermore, apparently randomly distributed between both ends of the distribution curve. Therefore it is quite clear that the *mean* values given in the table are relatively little affected by the failure-results, a fact which was actually verified by the closeness of the median and mean values as well as the different means calculated after applying the rejection procedure with varied efficiency. For these reasons only the means were calculated, with no attempt to estimate standard errors or deviations; the latter were ultimately biased much more than the means. Consequently, the statistical significance of the differences between the experimental and control cultures has been assessed by means of the "insensitive" sign test only.

It is seen from the results (Table I) that "leucophoresis-serum" has significantly increased thymidine incorporation in normal bone marrow cells, and that "polycythaemia-serum" has decreased it. It is also evident that EP-serum, obtained from rats made anaemic by repeated bleedings, has not exerted any detectable effect on the TdR incorporation. The lack of effect of erythropoietin on TdR incorporation *in vitro* has also been reported

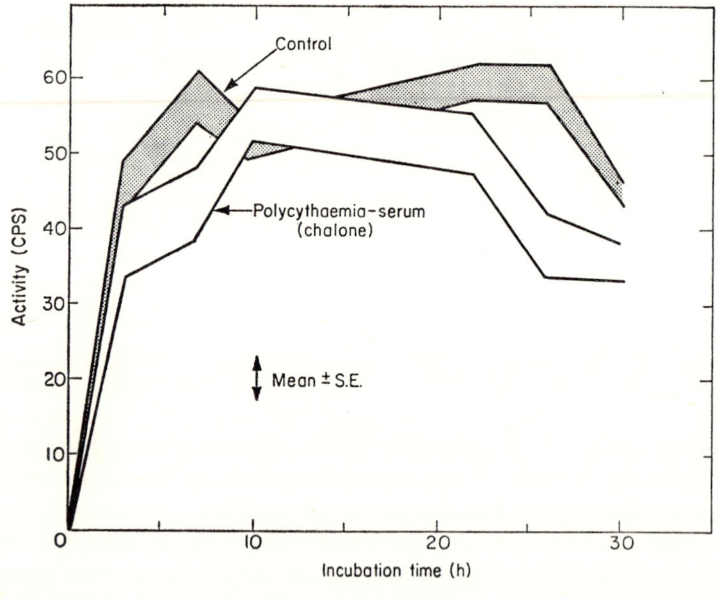

FIG. 4. Effect of "polycythaemia-serum" on H^3 thymidine incorporation in normal rat bone marrow cells *in vitro*.

Liquid scintillation counting; mean ± S.E. of 3 replicate cultures at each time point. The analysis of variance of the raw data is presented in Table II. ($N = 2\times 18 = 36$.)

by others (Erslev 1964). These findings permit some inferences which could be considered particularly important for the present purpose. The first of these concerns the modes of action of "polycythaemia-serum" which, because of the inactivity of EP-serum, cannot be explained in terms of erythropoietin lack. On the other hand, it is quite clear that the decrease in TdR incorporation produced by the "polycythaemia-serum" (Fig. 4 and Table II) could be readily explained in terms of chalone action since blood of the serum donors had a highly elevated red cell count. (From the viewpoint of commonly accepted negative feedback theories in biology the inhibitor concentration is proportional to the "tissue mass".)

TABLE II. Analysis of variance of the data seen in Fig. 4. Two-factor design, fixed variables.

Source of variance[a]	Sum of squares	Degrees of freedom	Mean square	F	Significance (P)
A-effect	920·2	1	920·2	24·80	<0·001
B-effect	1010·9	5	202·2	5·45	<0·0027
AB-interaction	389·2	5	77·8	2·10	Insignificant
Random deviations	889·3	24	37·1	—	—
Total	3209·6	35	(91·6)	—	—

[a] A-effect: Serum effect.
B-effect: Effect of incubation time.
AB-interaction: Difference in the form of the curves.
Random deviations: Combined effect of all error factors (known and unknown).

The second inference concerns the mode of action of "hyperoxia serum", or, more precisely, indicates the lack of knowledge of its modes of action. As seen in Table I, "hyperoxia-serum" decreased TdR incorporation significantly (according to the sign test 2P was <0·05, a probability value which has often been interpreted as meaning "almost significant"; the result is considered significant here because there is one apparent failure-experiment included in the material). This result is surprising and does not fit the "poietin-chalone-antichalone system" explanation: the inhibition cannot depend on poietin lack any more than in the case of "polycythaemia-serum", nor can it depend on antichalone lack (if it were present in the serum obtained from bled rats), nor on the presence of excessive chalone. Yet it is well known that erythropoiesis is suppressed *in vivo* at high environmental oxygen tension and it may be repeated here that erythropoiesis was suppressed also in those rats which served as serum donors in this study, judging from the decreasing erythrocyte counts in peripheral blood. Since the effect of "hyperoxia-serum" has not been studied with more specific techniques than overall TdR incorporation, except very superficially, attempts to "explain" this phenomenon are so far purely speculative. It is tempting to think however, that in these circumstances erythropoiesis could really be "death controlled" (Stohlman 1959).

It is also seen from the results presented in Table I that absence of any serum in the culture medium failed to cause detectable changes in TdR incorporation. Under these conditions there must have been a complete lack of environmental chalones which should, if the control serum still contained some, have resulted in stimulation. It may be argued that Hanks BSS is an inadequate culture medium and does not allow sufficient cell proliferation, and that the positive effect of chalone lack therefore balances out so that no change appears. This explanation does not seem plausible because (*a*) TdR will only label those cells which are in DNA synthesis close to the zero time; subsequent cessation in cell proliferation does not matter (the intracellular nutrient pool is presumed to be sufficient for the first cohort of cells); (*b*) cells do not die in Hanks BSS: labelled cells were not lost earlier than in the serum-containing media; and (*c*) the stimulating factor separated from the "leucophoresis-serum" with Sephadex gel filtration was able to increase TdR incorporation both in the presence and in the absence of any serum. The unaltered incorporation of TdR in Hanks BSS thus seems to demonstrate that control serum is devoid of detectable chalone activity and that it does not contain appreciable amounts of thymidylic acid or other "competitive inhibitors" which would dilute the labelled TdR.

Some other sera than those listed in Table I have also been studied by means of thymidine incorporation in the whole bone marrow population. It was found, for instance, that serum which had been obtained from rats after withdrawal of eosinophils ("eosinapheresis-serum") did not cause significant changes in the TdR incorporation. However, the presence of one, or possibly two failure-experiments in the small material at hand might be the reason for this negative result because there was a strong positive effect in the subsequent autoradiographic analysis of the failure-experiment.

Another serum, which is not listed in the table, was "chloroleukaemia-serum". It has been found to be the most potent inhibitor of TdR incorporation in normal cells—and in chloroleukaemia cells as well (Figs. 5 and 6).

The possibility that the effects were due to direct toxicity of the serum, particularly with regard to normal bone marrow cells, has not been ruled out conclusively. However, it has been possible to separate a chalone-like substance from the chloroleukaemia serum using Sephadex gel filtration. In Sephadex, it behaves similarly to the chalones separated from fresh serum. Both factors have shown an apparent tendency to be adsorbed to the

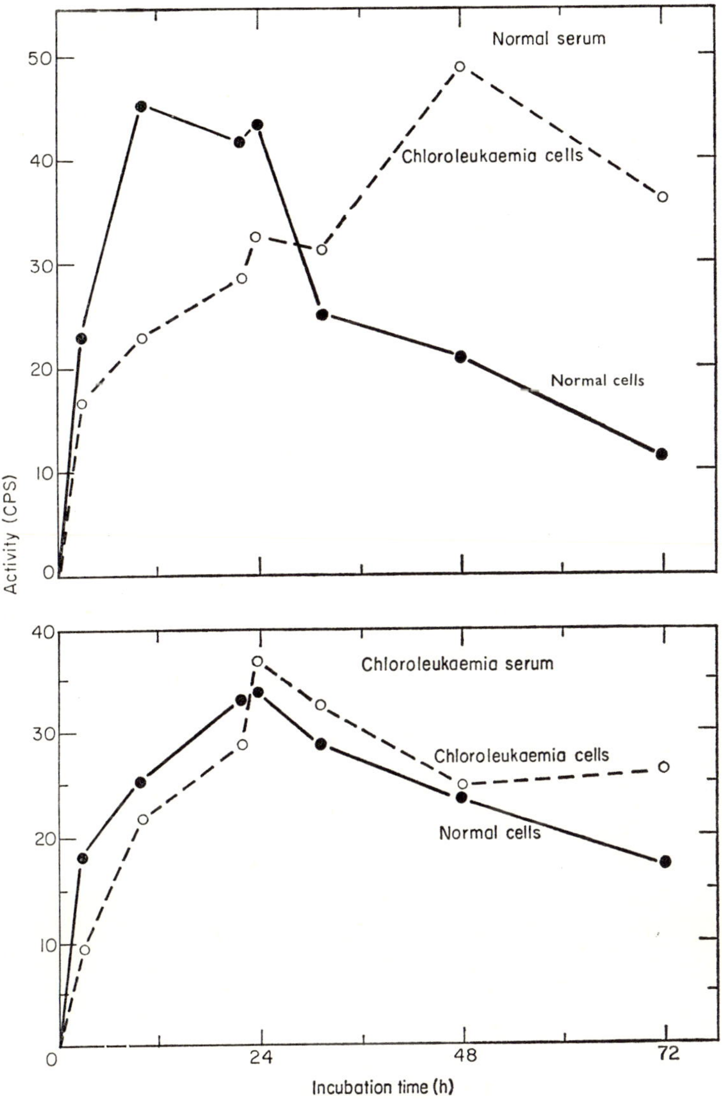

FIG. 5. Effect of "chloroleukaemia-serum" and normal serum on the incorporation of H^3 thymidine in normal bone marrow and chloroleukaemia cells *in vitro*.
Liquid scintillation counting. (N = 4×21 = 84.)

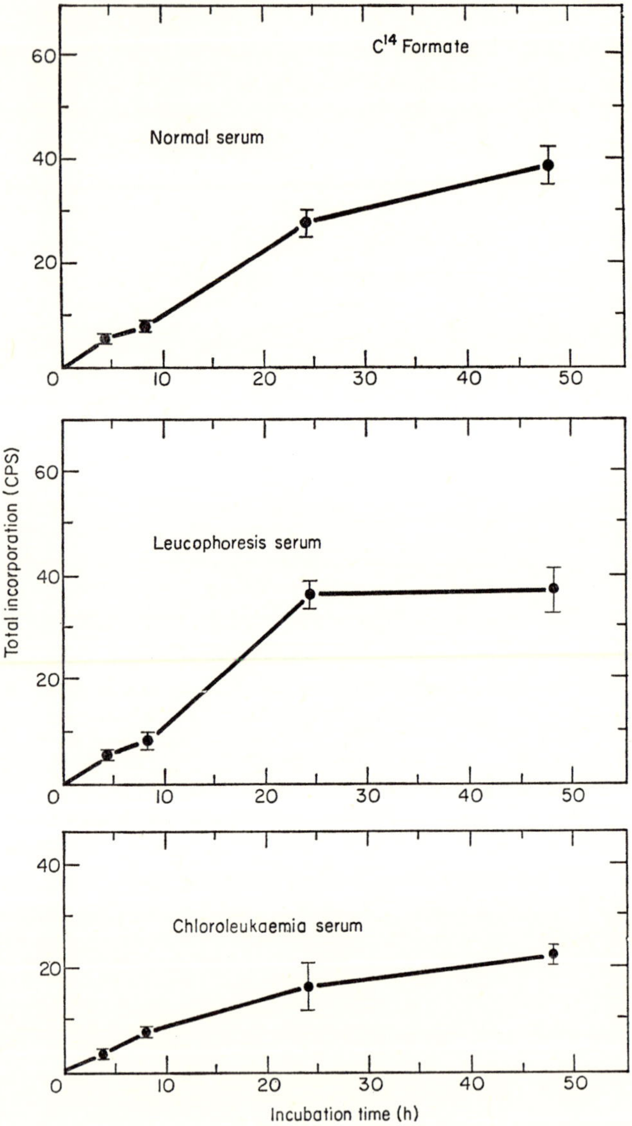

FIG. 6. Effect of "leucophoresis-serum", "chloroleukaemia-serum" and normal serum on the incorporation of C^{14} formate in normal bone marrow cells *in vitro*.
Liquid scintillation counting. ($N = 3 \times 20 = 60$.)

gel matrix: they are eluted from the column as a broad zone and occasionally have K_d-values above 1.

Most of the sera studied by means of thymidine incorporation in normal bone marrow cells have also been tested with other labelled precursor substances. This was done on the basis of one main argument: that the detected effects might have been peculiar to thymidine without being

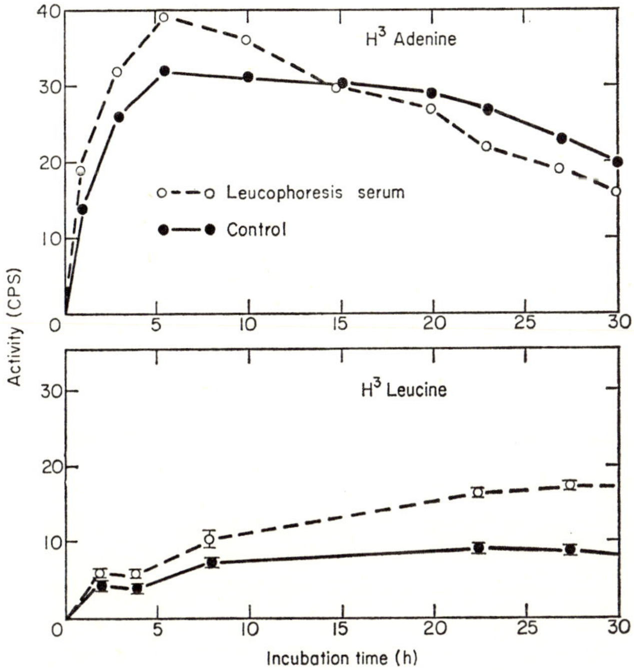

FIG. 7. Effect of "leucophoresis-serum" on RNA (H^3 adenine) and protein (H^3 leucine) synthesis in normal bone marrow cells *in vitro*.
Liquid scintillation counting. ($N_{RNA} = 2 \times 78 = 156$; and $N_{protein} = 2 \times 36 = 72$.)

primarily active at the cellular level; i.e. if the effects were due to specific changes in proliferation kinetics of the cells, such as triggering of resting progenitor cells in the generative cycle (Lajtha *et al.* 1964, Bullough and Rytömaa 1965), they should have been detected by the other parametric indicators as well.

Systematic studies have so far been applied to the "leucophoresis-serum" only, which was the original object of this study. Consequently, the rest of this report will deal almost entirely with this factor.

It can be seen in Fig. 6 that C^{14} formate, which is a fairly specific indicator of *de novo* synthesis of DNA, was utilized by the normal bone marrow cells more in "leucophoresis-serum" cultures and less in "chloroleukaemia-serum" cultures than in controls. Therefore the effects of these sera are not explicable in terms of "catalysis" of DNA polymerization as might be argued on the basis of thymidine data only. Basically similar results of C^{14} formate incorporation as described here have been reported by others (Gidáli and Fehér 1964). "Leucophoresis-serum" has also been tested for its ability to influence—directly or indirectly—RNA and protein synthesis. The results, based on the incorporation of tritium-labelled adenine and leucine, are shown in Fig. 7. All these results indicate that "leucophoresis-serum" must act at the cellular level because existence of a "universal" stimulus acting at some other level does not seem to us to be plausible.

B. *Autoradiographic studies*

It is quite clear that the results described have not provided relevant evidence from which to infer which type of cell progenitor is responding within the bone marrow. To obtain this crucial evidence autoradiographic analyses were performed, making use of the same preparations which had been previously analyzed for their total radioactivity. In the first step of the analytical procedure the relative number of labelled cells in the whole population was determined. The main reason for the application of this new "unspecific" procedure was to determine the type of analysis which could be expected to give the desired information. There were three problems. Firstly, had the increase in total incorporation been due to increase in the rate of DNA synthesis per cell (= shortening of the synthesis time)? Secondly, had the increase been due to increment in the number of DNA synthesizing cells (= triggering of the resting cells into DNA synthesis/shortening of the "dichophase"—or G_0—G_1 in the "gap" terminology)? Or thirdly, had the increase been due to both mechanisms?

Mental comparison of these three alternatives seemed to favour the idea that the second would more likely be the result of a specific action than the other two. Furthermore, it was also known that the specific rate of TdR incorporation was not significantly different in the test and control groups, which evidently did not support the first alternative. Analyses of the relative numbers of labelled cells in "leucophoresis-serum" and in controls indicated readily that there was a significant difference between the two groups (Fig. 8).

The overall labelling index in the control group was about 12 to 15% "at zero time" which is consistent with the *in vivo* labelling index in our rats and also with the *in vivo* data reported by others (Lamerton and Lord 1964). In the "leucophoresis-serum" the overall labelling index was higher, occasionally reaching values up to 20%. A more accurate estimate of the relative overall labelling values in the test and control groups can be derived from the time course of the labelling indices (Fig. 8). Assuming that in both groups cell cycle times remain unaltered, as is actually indicated by the available data, the slopes of the time course curves are directly proportional

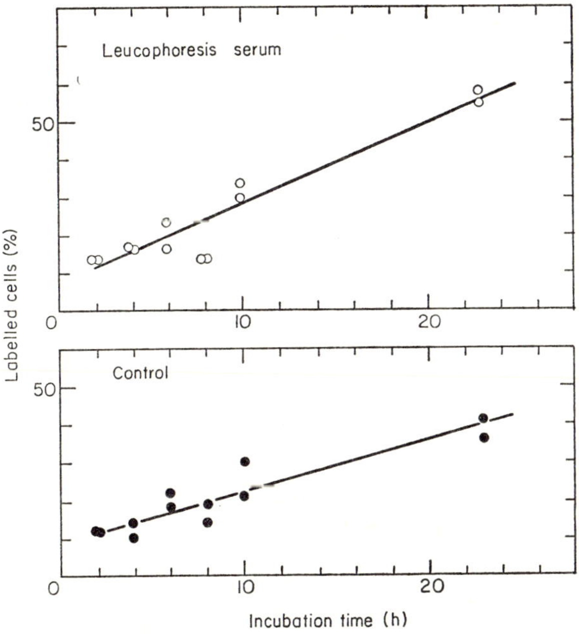

FIG. 8. Effect of "leucophoresis-serum" and normal serum on labelling indices in normal bone marrow cells *in vitro* (H^3 thymidine).

The two dots at each time point refer to two analyses of the same preparations. The first analysis was made by one of the present authors; the second analysis was made by three medical students, who were completely unfamiliar with the nature of the preparations.

(The results have been corrected for systematic errors before pooling.) "Least squares" regression lines through the points gave regression coefficients of 1·3% per hour in the control group, and 2·10% per hour in the "antichalone" group.

to the original labelling indices. The other assumption on which this statement is based is that the death rate of unlabelled cells does not differ significantly in the two groups. Although such a possibility can hardly be more than a theoretical argument, it was however studied here.

Total cellularity in the "leucophoresis-serum" and in control cultures was measured by the point counting method and also by the spectrophotometric determination of "nucleic acids" in the cell hydrolysates. Analysis of variance revealed that there was no significant "serum-effect" on the total cellularity or on the "nucleic acids" (there may, of course, be such effects in reality but they must ultimately be so small as to remain undetected). It was also observed that the total cellularity decreased in time and that the slope was significantly curvilinear ($P < 0 \cdot 01$ according to the analysis of variance).

The above results demonstrate that a significant increase in the *absolute* number of labelled cells had occurred in the "leucophoresis-serum". The available quantitative estimates of this increase showed that it had been the sole source for the increment in TdR incorporation (cf. e.g. Table I and Fig. 8). Similar estimates in the cases of chalone-like inhibition have demonstrated that reduction has apparently also occurred in the per cell labelling.

Overall changes in the absolute numbers of labelled cells, and the subsequent increase in DNA synthesis per unit population, did not indicate whether or not the "leucophoresis-serum" effect had been specific for granulocytic cells. To study this problem in autoradiography one is inclined to think that the analyses must be combined with complete differential cell count and subsequent determinations of labelling indices in each subpopulation. This is obviously true in many instances. One must realize, however, that the results are easily biased because cell morphology is impaired in autoradiography, and morphological criteria are not sufficient to classify all cells properly (see e.g. Boll 1963), especially in rat bone marrow (Endicott and Ott 1945). Therefore, search was made for shortcuts in this study in order to avoid a laborious method of analysis which might still yield more or less biased results.

When performing the first complete autoradiographic analyses—using code numbered preparations—it appeared that some of the slides contained an unusually high number of labelled "mature" granulocytes (metamyelocytes and more mature cell forms). The preparations were therefore carefully analyzed with respect to this morphologically rather distinct sub-population only, and supplemented with similar analyses of autoradiograms of some other experiments. The pooled results are seen in Fig. 9.

More than 1000 mature granulocytes were scored at each time point in both groups and therefore the counting error is not intolerably large although the frequency of labelled cells was generally rather low. The "impossible" finding that about 1% of mature granulocytes was labelled already at 2 and 4 hours cannot be due to chance labelling: the limit of labelled cells was set at 10 nuclear grains, which was at least 10 times higher than the average background level. A plausible explanation of this finding is that some of the morphological metamyelocytes have actually been myelocytes. This explanation is consistent with the findings of Boll (1963) and of Lamerton and Lord (1964).

It is seen from Fig. 9 that there was a very large difference between the slopes in the two groups. A linear regression through the points from 2 to 23 hours gave a value of 0·20% per hour for the regression coefficient in the control group and 0·54% per hour in the test group. This indicates that an increase of about 170% was caused by the "leucophoresis-serum". The analysis of variance, performed with the individual counts, showed that both the serum-effect and the time-effect were highly significant ($P<0{\cdot}001$) but

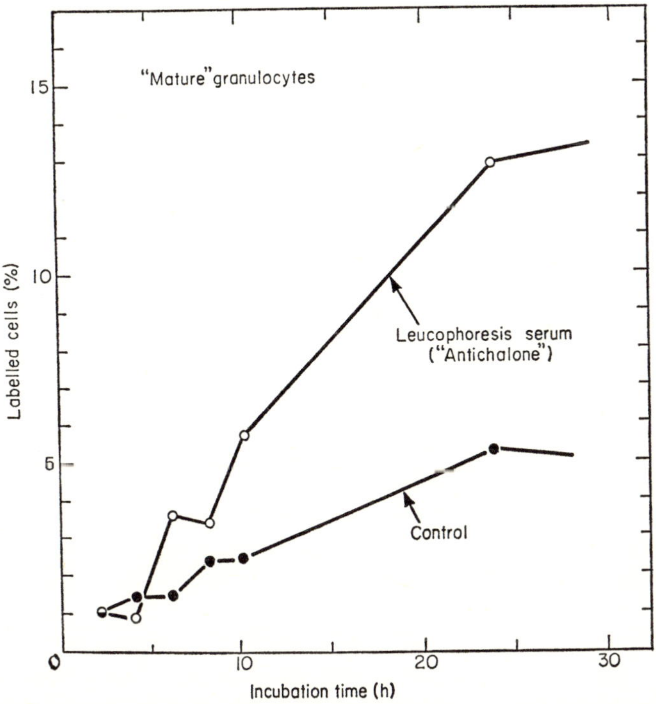

FIG. 9. Effect of "leucophoresis-serum" and normal serum on the accumulation of labelled "mature" granulocytes (metamyelocytes and more mature cell forms) in normal bone marrow cell cultures, labelled with H^3 thymidine.

Each point is a "weighted" mean of "blind" autoradiographic analysis of three experiments. ($N = 2 \times 30 = 60$)

that the interaction-effect was not significant at all. The biological interpretation of the last-mentioned result is that the relative rates of increase were not statistically different in the test and control cultures. This in turn indicates that labelled myelocytes mature at an equal rate in both groups. In this study there was no detectable difference between the relative numbers of mature cells in the test and control groups, which, on the basis of equal cellularities, means that the absolute numbers of mature cells must also have been "equal".

It is obvious that most, if not all, labelled mature cells must have originated from the very last myelocyte division because much more time would have been needed for maturation from any earlier progenitor stage. If it is assumed that in both groups all labelled myelocytes divide prior to, or at the moment of crossing the morphological metamyelocyte-myelocyte border, then there has been 170% more labelled myelocytes in the "leucophoresis-serum" than in the control cultures. This suggests strongly that in normal conditions fairly large numbers of myelocytes are not involved in the generative cycle but are in the "dichophase", in a stage of "undecision".

Myelocytes constitute in most mammals a considerably large proportion of all divisable cells in the bone marrow (Donohue *et al.* 1958, Killman *et al.* 1962, Patt and Maloney 1964, Warner and Athens 1964). Consequently, an increase of the order of 150 to 200% in their labelling index is entirely sufficient to explain all other quantitative reactions described in this study. The obvious conclusion thus arises that the "leucophoresis-serum" has exerted a highly specific effect on granulocytic cells at the myelocyte level of maturation. This conclusion is consistent with the findings reported previously (Rytömaa and Teir 1964, Rytömaa and Kiviniemi 1964), e.g. with a more marked increase in S^{35} sulphate incorporation than in TdR incorporation in those studies utilizing double labelling. Although the precise relationship between the results described in this article and those reported by Warner and Athens (1964), Cartwright *et al.* (1964), and Patt and Maloney (1964) is obscure at present, it may not be a coincidence that in all these studies myelocytes seem to play a central role.

Other findings relating to the mode of action of the "leucophoresis-serum", and studies with extracts of the feedback factors from the different sera obtained by Sephadex gel filtration, will be reported elsewhere.

C. *Tissue specificity*

It was demonstrated in the preceding chapter that "leucophoresis-serum" appeared to exert a highly specific stimulatory effect on the granulocytic cells. The lack of response in other cell systems in the bone marrow has also been demonstrated by more direct means: no difference was found between the "leucophoresis-serum" and control cultures in a population termed "small mononuclears". Because this category consists of erythroblasts and lymphoid cells, it was concluded that neither of these systems responded to the "leucophoresis-serum". It is well known that in rat bone marrow distinction between these cell types is difficult, and no attempt was made to analyze them separately in this study. This was thought justifiable because there was good reason to think that if "leucophoresis-serum" should turn out to be unspecific it would be likely to affect all cell systems. In the unlikely event of some other cell system being affected besides the granulocytic cells, the effect would still show up in the adopted analysis just as it did when the overall labelling index was determined. As already mentioned, no indications of unspecificity were found.

The mitotic independence of various tissues indicates that the humoral substances which control mitotic activity are generally tissue-specific. However, the mechanisms underlying such specificity need not be due to some steric (chemical) affinity of the substances with the target-tissues. The specificity might in some cases be due to much simpler mechanisms: the other potential target-tissues could, for instance, be beyond the effective range of such substances. Thus the epidermal chalone could be "chemically" a potent inhibitor of bone marrow cells but not normally reaching them. Because of the fundamental importance of tissue-specificity in a biological control mechanism (Bullough and Rytömaa 1965) and the scanty empiric evidence available at present, some preliminary result are reported here.

Four extracts were tested for their effects on H^3 thymidine incorporation in normal bone marrow cells as usual. These extracts were: (a) an extract from pig epidermis (epidermal chalone, kindly supplied by Homan and Hondius Boldingh, N.V. Organon, Holland); (b) "bone marrow chalone" (separated from rat serum in gel filtration, presumably containing both granulocytic and erythrocytic chalones); (c) a positive fraction separated from "leucophoresis-serum"; and (b) unfractionated whole rat serum. Half of the cultures contained adrenalin (0·001 μg/0·4 ml) because it had been found by Bullough and Laurence (1964) that the epidermal chalone was a more potent depressor of mitotic activity in epidermal cells in the presence of sufficient adrenalin. The applied concentrations of epidermal chalone and adrenalin were equivalent to the amounts which had been found to reduce the epidermal mitotic rate by about 50% (Bullough and Laurence 1964).

The results of the experiment and the respective analysis of variance are seen in Table III.

The mean radioactivities may be somewhat misleading because of the small numbers of samples used in each sub-group ($N = 3$). According to the conventional way of experimentation, these numbers are really "too" small. Despite this the analysis of variance was capable of demonstrating that at least one of the test substances had significantly influenced TdR incorporation ($P \approx 0{\cdot}03$); and that the effect of adrenalin, either alone ($F = 1{\cdot}5$) or in combination with any of the test substances ($F < 1$), was insignificant. Which group or groups of the test substances were out of line with the rest could only be resolved by applying individual t-tests to the pairs of groups. It appeared from these tests that "bone marrow chalone" was the sole source of the detected significance. It may seem from the mean activities in Table III that epidermal chalone, too, had "inhibited" TdR incorporation slightly; there is, however, no evidence in the results which would support this apparent "inhibition". The lack of effect of the "leucophoresis-serum" fraction—except in comparison to "bone marrow chalone"—may be due either to the fact that it was used in a much smaller concentration than the other substances (half the "bone marrow chalone" concentration), or that the "whole serum" contained some stimulator. This latter possibility is actually suggested by the fact that "bone marrow

TABLE III. Incorporation of H^3 thymidine in normal bone marrow cells *in vitro* under the influence of epidermal chalone, and some other test substances (see text); both with and without the presence of adrenalin. The upper table includes the mean radioactivities in the cells in each subgroup; the lower table gives the results obtained by the analysis of variance.

(Incubation time 23 hours; liquid scintillation counting.) (N = 24).

B A Test→ ↓ substance	Mean radioactivity in cells (CPS)				$MEAN_B$
	Epidermal chalone	"Bone marrow chalone"	Fraction of "leucophoresis-serum"	Whole serum	
With adrenalin	30·7	29·4	37·1	34·5	32·9
Without adrenalin	34·3	27·4	39·5	44·1	36·3
$MEAN_A$	32·5	28·4	38·4	39·3	(34·6)

Source of variance[a]	Sum of squares	Degrees freedom	Mean square	F	Significance (P)
A-effect	63·3	1	63·3	1·5	Insignificant
B-effect	489·2	3	163·1	4·0	≈0·03
AB-interaction	99·5	3	33·2	<1	Insignificant
Random deviations	654·0	16	40·9	—	—
Total	1306·0	23	(56·8)	—	—

[a] A-effect: Effect of adrenalin
B-effect: Effect of "feedback" factors
AB-interaction: Modification of the feedback factor effects by adrenalin.
Random deviations: Combined effect of all error factors.

chalone" decreased TdR incorporation by 28%, which was considerably more than usual (cf. Table I).

These preliminary results have thus indicated that neither adrenalin nor epidermal chalone influence bone marrow cell proliferation significantly. Consequently, the findings tend to favour the hypothesis that tissue-specificity of the feedback factors is based more on complicated mechanisms than on their effective ranges.

V. Conclusions

The results presented in this article demonstrate that significant and reproducible changes occur in the rate of cell production of normal bone marrow cells *in vitro* under the influence of certain humoral factors present in the rat serum. The test sera were obtained from the animals after different treatments, which were known to produce similar changes in the bone marrow *in vivo*. The most extensive and detailed studies have so far been performed on the effects of a factor present in the serum of rats subjected to repeated leucophoresis. This factor does not change significantly either the rate of DNA synthesis, or the rate of maturation of granulocytic progenitor cells, or the rate of cell production in non-granulocytic systems. Its primary action seems to be to promote the conversion of "resting" myelocytes to the generative cell cycle. Due to this difference from erythropoietin in its apparent mode of action, this factor has been tentatively termed "granulocytic antichalone" rather than "granulopoietin".

Erythropoietin, evidently present in the serum of rats subjected to repeated bleedings, has not been found to change significantly the proliferation parameters studied here. Erythropoiesis has been suppressed by the serum of rats obtained after hypertransfusion with red cells, and after treatment with high exogenous oxygen tension. In the first-mentioned case, the inhibition has evidently been due to excessive erythrocytic chalone, whereas the cause of inhibition in the latter case remains to be explained. It is tentatively suggested that erythropoiesis is here "death controlled", although direct data is lacking.

The most active inhibitor of the production of normal bone marrow cells *in vitro* has been a chalone-like substance present in the serum of chloroleukaemic rats. Whether this effect is partially or totally due to toxic interaction remains a matter of some dispute. The results have indicated that inhibition of cell production by chalones and chalone-like substances is due both to the reduction in the number of proliferating cells and to the decrease in the rate of DNA synthesis.

Some of the serum factors studied here have been partially purified by means of gel filtration.

Acknowledgement

We are grateful for the most valuable help of Mr. Robert Oxlade, B.A., Oxon, Guy's Hospital Medical School, London, in correcting the grammar and in preparation of the manuscript.

REFERENCES

Alpen, E. L. and Cranmore, D. (1959). *In* "The Kinetics of Cellular Proliferation" (F. Stohlman, Jr., ed.), p. 290. Grune and Stratton, New York.

Altland, P. D. and Parker, M. (1959). *Am. J. Physiol.* **180**, 421–427.

Becker, A. J., McCulloch, E. A. and Till, J. E. (1963). *Nature* **197**, 452.

Bierman, H. R. (1964). *Ann. N.Y. Acad. Sci.* **113**, 753–765.

Boll, I. T. M. (1963). *In* " Proc. 9th Congr. Europ. Soc. Haemat., Lisbon 1963" **II/2**, pp. 1518–1526. S. Karger, Basel/New York.

Brecher, G. and Stohlman, F., Jr. (1959). *In* "Progress in Hematology" (L. M. Tocantis ed.), vol. 2, pp. 110–132. Grune and Stratton, New York.

Bullough, W. S. (1962). *Biol. Rev.* **37**, 307–342.

Bullough, W. S. and Laurence, E. B. (1964). *Expl Cell Res.* **33**, 176–194.

Bullough, W. S. and Laurence, E. B. (1966). (This volume.)

Bullough, W. S. and Rytömaa, T. (1965). *Nature* **205**, 573–578.

Cartwright, G. E., Athens, J. W., Haab, O. P., Raab, S. O., Boggs, D. R. and Wintrobe, M. M. (1964). *Ann. N.Y. Acad. Sci.* **113**, 961-967.

Craddock, C. G. (1960). *In* "A Ciba Foundation Symposium on Haemopoiesis" (G. E. W. Wolstenholme and M. O'Connor eds.), pp. 237–261. Churchill, London.

Craddock, C. G. (1962). *In* "Progress in Hematology" (Tocantis, L. M. ed.), vol. 3, pp. 92–104. Grune and Stratton, New York.

Craddock, C. G., Perry, S. and Lawrence, J. S. (1959). *In* "The Kinetics of Cellular Proliferation" (F. Stohlman, Jr., ed.), pp. 242–259. Grune and Stratton, New York.

Craddock, C. G., Rytömaa, T. and Nakai, G. (1964). *In* "The Thymus in Immunobiology: Structure, Function, and Role in Disease" (R. A. Good and A. E. Gabrielsen eds.), Chap. 17, pp. 317–331, Hoeber Med. Div., Harper and Row, New York.

Cronkite, E. P. and Fliedner, T. M. (1964). *New Engl. J. Med.* **270**, 1347–1352 and 1403–1408.

Donohue, D. M., Reiff, R. H., Hanson, M. L., Betson, Y. and Finch, C. A. (1958). *J. clin. Invest.* **37**, 1571–1577.

Endicott, K. M. and Ott, M. (1945). *Anat. Rec.* **92**, 61.

Erslev, A. J. (1964). *In* "Abstracts Xth Congr. Internat. Soc. Haemat. Stockholm 1964", M:2. Ljunglöfs Litografiska AB, Stockholm.

Farr, R. S. and LeQuire, V. S. (1950). *Proc. Soc. exp. Biol. Med.* **75**, 661-666.

Feinendegen, L. E. and Bond, V. P. (1962). *Expl Cell. Res.* **27**, 474–484.

Fichtelius, K. E. and Linna, J. (1966). (This volume.)

Fischer, S. and Fruhman, J. (1964). *Nature*, **201**, 836-838.

Ford, C. E., Hamerton, J. L., Barnes, D. W. H. and Loutit, J. F. (1956). *Nature* **177**, 452.

Fukuda, T., Oide, H. and Miyasaka, A. (1960). *Nature* **188**, 860–861.

Gidáli, J. and Fehér, I. (1964). *Am. J. Physiol.* **206**, 585–588.

Gordon, A. S. (1960). *In* "A Ciba Foundation Symposium on Haemopoiesis" (G. E. W. Wolstenholme and M. O'Connor eds.), pp. 325–362. Churchill, London.

Gordon, A. S., Handler, E. S., Siegel, C. D., Dornfest, B. S. and LoBue, J. (1964). *Ann. N.Y. Acad. Sci.* **113**, 766–789.

Homan, J. D. H. and Hondius Boldingh, (1966). (N. V. Organon, Oss, Netherlands.) Personal communication and discussion comment by Dr. Hondius Boldingh in this symposium.

Iversen, O. H. and Bjerknes, R. (1963). *Acta path. microbiol. scand.* Suppl. **165**.

Iversen, O. H. and Elgjo, K. (1966). (This volume.)

Jacobson, L. O. and Doyle, M., editors (1962). "Erythropoiesis." Grune and Stratton, New York.

Kekomäki, M. and Rytömaa, T. (1966). (This volume.)

Killmann, S.-Å., Cronkite, E. P., Fliedner, T. M. and Bond, V. P. (1962). *Blood* **19**, 743–750.

Lajtha, L. G. (1954). *Nature* **174**, 1013–1014.

Lajtha, L. G. (1961). "The Use of Isotopes in Haematology." Blackwell Scientific Publications, Oxford.

Lajtha, L. G. (1966). (This volume.)

Lajtha, L. G., Gilbert, C. W., Proteous, D. D. and Alexanian, R. (1964). *Ann. N.Y. Acad. Sci.* **113**, 742–752.

Lajtha, L. G. and Oliver, R. (1960). *In* "A Ciba Foundation Symposium on Haemopoiesis" (G. E. W. Wostenholme and M. O'Connor eds.), pp. 289–314. Churchill, London.

Lamerton, L. F. and Lord, B. I. (1964). *In* "Control of Cell Division and the Induction of Cancer" (C. C. Congdon and P. Mori-Chavez eds.), National Cancer Institute Monograph **14**, pp. 185–197. U.S. Government Printing Office, Washington, D.C.

Linman, L. W. and Bethell, F. H. (1960). "Factors Controlling Erythropoiesis." Charles C. Thomas, Springfield.

Menkin, V. (1946). *Arch. Path.* **41**, 376–387.

Menkin, V. (1955). *Ann. N.Y. Acad. Sci.* **59**, 956–985.

Merchant, D. J., Kahn, R. H. and Murphy, W. H., Jr. (1960). "Handbook of Cell and Organ Culture." Burgess Publishing Co., Minneapolis.

Osgood, E. E. (1957). *J. natn. Cancer. Inst.* **18**, 155–166.

Patt, H. M. and Maloney, M. A. (1964). *Ann. N.Y. Acad. Sci.* **113**, 515–522.

Pelc, S. R. and Appleton, T. C. (1965). *Nature* **205**, 1287–1289.

Perretta, M. and Pieber-Perretta, M. (1962). *Biochim. biophys. Acta* **61**, 828–829.

Rubini, J. R., Cronkite, E. P., Bond, V. P. and Fliedner, T. M. (1960). *J. clin. Invest.* **39**, 909–918.

Rytömaa, T. and Kiviniemi, K. (1964). *In* "Proc. XIV Scand. Congr. Path. Microbiol. Oslo 1964", pp. 169–170. Universitetsforlaget, Oslo.

Rytömaa, T. and Teir, H. (1964). *Sangre* **9**, 359–361. (9th Congr. europ, Soc. Haemat., Lisbon 1963.)

Schroeder, L. and Brecher, G. (1960). Cited by F. Stohlman, Jr. *In* "A Ciba Foundation Symposium on Haemopoiesis" (G. E. W. Wolstenholme and M. O'Connor eds.), discussion, p. 272. Churchill, London.

Shay, H., Gruenstein, M., Harris, C. and Glazer, L. (1952). *Blood* **7**, 612–622.

Simpson, G. G., Roe, A. and Lewontin, R. C. (1960). "Quantitative Zoology" revised edition. Harcourt, Brace and Co. New York/Burlingame.

Smellie, R. M. S., Thompson, R. Y. and Davidson, J. N. (1958). *Biochim. biophys. Acta* **29**, 59.

Speirs, R. S. (1958). *Ann. N.Y. Acad. Sci.* **73**, 83.

Steinberg, B. and Martin, R. A. (1950). *Am. J. 2Physiol.* **161**,14–20.

Stohlman, F., Jr. (1959). *In* "The Kinetics of Cellular Proliferation" (F. Stohlman, Jr., ed.), pp. 318–324. Grune and Stratton, New York.

Teir, H. (1951). *Soc. Sci. Fenn. Comm. Biol.* 13, 1–32.

Teir, H. and Rytömaa, T. (1964). *Sangre* 9, 418–420. (9th Congr. europ. Soc. Haemat., Lisbon 1963).

Totter, J. R. (1954). *J. Am. chem. Soc.* 76, 2196.

Warner, H. R. and Athens, J. W. (1964). *Ann. N.Y. Acad. Sci.* 113, 523–536.

Wasastjerna, C., Teir, H. and Larmo, A. (1958). *Acta path. microbiol. scand.* 44, 80–87.

Weiss, P. (1955). *In* "Biological Specificity and Growth" (E. G. Butler, ed.), pp. 195–206. Princeton, New Jersey.

Weiss, P. and Kavanau, J. L. (1957). *J. gen. Physiol.* 41, 1–17.

Wetherby-Mein, G. and Cottom, D. G. (1956). *Br. J. Haemat.* 2, 25.

The Development of Responsiveness to Erythropoietic Stimuli during Ontogeny

EEVA JALAVISTO†

Department of Physiology, University of Helsinki, Helsinki, Finland

I. Introduction

THE production of erythrocytes is very delicately adjusted to meet the demand of tissues for oxygen. The control mechanism is probably complex and not at all well understood. Humoral factors, erythropoietins, play a decisive role, but the mechanism of their action is not known. Furthermore we may ask which are the conditions necessary for the erythron to be able to respond to the lack of oxygen and whether the factors produced as a response to increased demand of oxygen are strictly tissue specific or do they have some general growth promoting effects as postulated by Leaders *et al.* (1962) for the purified erythropoietin. A further question is whether responsiveness is bound to the presence of renal tissue in a developing organism as it is in adult animals.

The experiments to be presented using developing chick embryos and young chicken as the object of study cannot give any definite answers to these questions. The investigations so far carried out must be considered as pilot studies only. However, the fact that the typical erythrocytosis as a reaction to hypoxia develops fairly late in the embryonic phase and the fact that we may observe the total growth rate of the organism and (in the early chick embryo) the size of the blood forming vascular area by simple direct means, give us information difficult to obtain otherwise, particularly in the adult organisms. This information may indirectly aid in the elucidation of the mechanism of red cell production control in adult organisms. I hoped that an analysis of the physiological events taking place when reactivity to hypoxia appears would give an idea of what is necessary for a reaction to develop.

II. The Procedure

The experimental procedure was very simple and will not be described in detail here; it is sufficient to say that the eggs (White Leghorn) were incubated in containers which were ventilated with air, 15% or 40% oxygen-nitrogen gas mixtures. The incubation temperatures were in the experiments with

† Deceased 1966.

early embryos 38°C and 40°C, in the later experiments with older embryos 39°C.

Because cobalt is known to lead to the appearance of erythropoietins in adult animals, a number of experiments with injections of cobaltous chloride were performed as well. The doses varied from 2 to 5 μg of metallic cobalt per gram embryo weight and it was injected twice with 2-day interval on the chorio-allantoic membrane. The dose is about 3 to 5 times the dose effective in an adult hen or cock (1 mg/kg body weight). The haematological determinations were (1) in the early chick embryo: determination of total haemoglobin with a benzidin reaction in homogenates made of the embryo, the membranes and the vascular zone after appropriate dilution (Astola *et al.* 1959); (2) in the 9 to 19-day-old embryos the haemocrite and the haemoglobin concentration in blood were determined in samples taken from the vitelline vessels (Jalavisto *et al.* 1965). The same procedure was employed in the cobalt-injection experiments (Jalavisto *et al.*, unpublished). The samples were taken 1, 2 and 3 days after the second injection (and consequently 4, 6 and 7 days after the first injection). The same experiment was extended to chickens 2 days after hatching and on 1-month-old chickens; (3) in the early chick embryo the size of the vascular zone was measured and the area calculated by assuming that the vascular zone formed en ellipse (Astola *et al.* 1959).

III. Results

A. *Weight gain*

The effect of the various experimental conditions on the growth of the total embryo may first be mentioned. The embryos reared in the lower temperature and lower percentage of oxygen were as to be expected (Cruz and Romanoff 1944), lighter than those incubated in higher temperature and/or higher oxygen percentage. The combined effect on the early chick embryo is illustrated in Fig. 1. In the older embryos the same effect of decreasing oxygen content on embryo weight is also noted. The effect of cobalt on the development of the embryo weight is not as clearcut. The mean weight of embryos aged 10 to 15 days seems to be less in the cobalt-injected embryos but the difference is not statistically significant. After the fifteenth day the weight curves of the cobalt-injected and controls cross and in this phase the injected embryos are significantly heavier than the controls, as seen from Fig. 2. The 2-day-old chicken became also slightly heavier than the controls following injections of cobalt.

The consequence of these differences in weight of the embryos in the various experimental groups is that it is not feasible to compare the haematological values in embryos of the same age and that the comparison must be made between embryos of approximately the same weight.

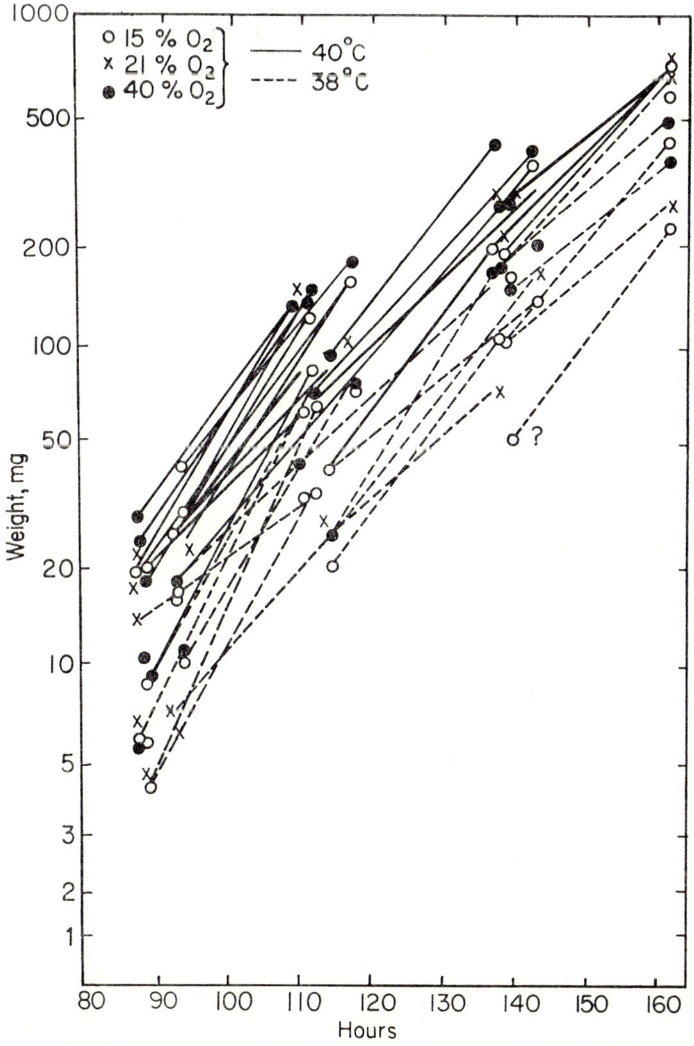

FIG. 1. Weight gain of chicken embryos under varying conditions of incubation. (Reproduced with permission from Astola *et al.* 1959.)

B. *Size of the vascular zone in the early chick embryo*

In the embryos less than 5 days of age the vascular zone can be measured. In the youngest embryos 64 and 67 hours old no clearcut difference in the size of the vascular zone can be noted between the 15 and 40% oxygen groups. Later the eggs incubated in 40% oxygen have in proportion to their weight a smaller vascular zone than those incubated in air and especially smaller than those reared in 15% oxygen.

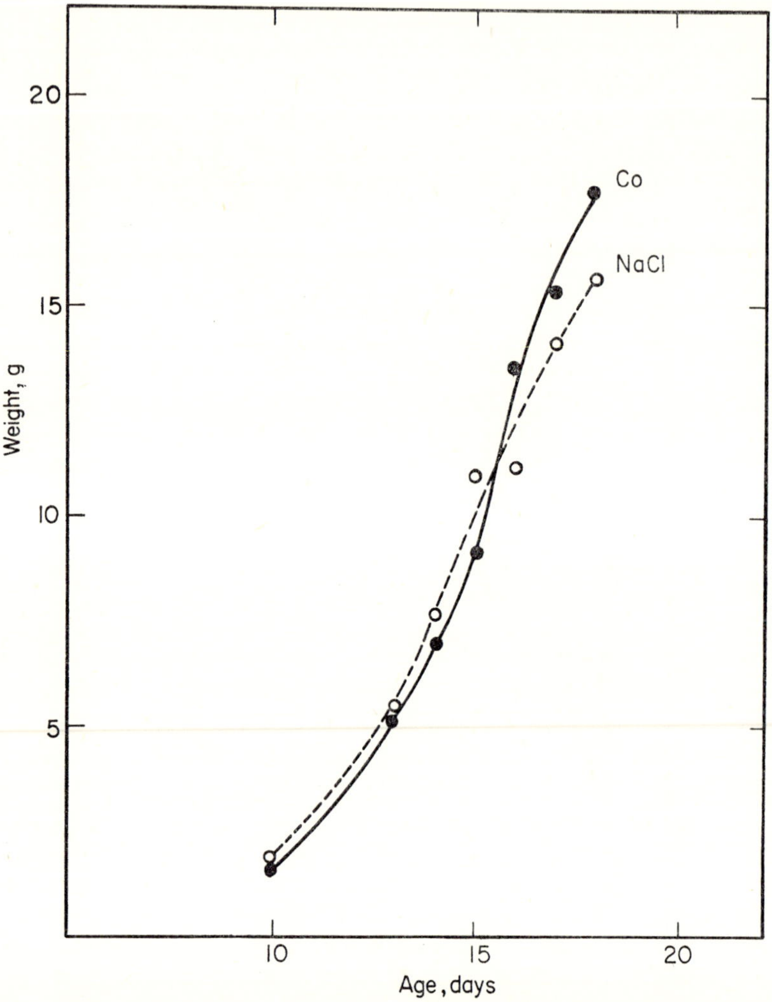

FIG. 2. Weight gain of chicken embryos injected with cobalt** (3–5 μg metallic cobalt per g embryo weight) solid circles and solid line. Empty circles and strippled line: controls.

C. *Haemoglobin*

The total amount of haemoglobin in 4 to 11 days old embryos is a function of weight irrespective of the oxygen content of ambient atmosphere. This is illustrated in Fig. 3. Thus in this phase the haemoglobin synthetizing cells do not react specifically on the differences in the supply of oxygen, i.e. they do not increase the production of haemoglobin beyond the increase in the total tissue mass. However, if the size of the vascular zone is related to the amount of haemoglobin a striking difference is seen: The haemoglobin

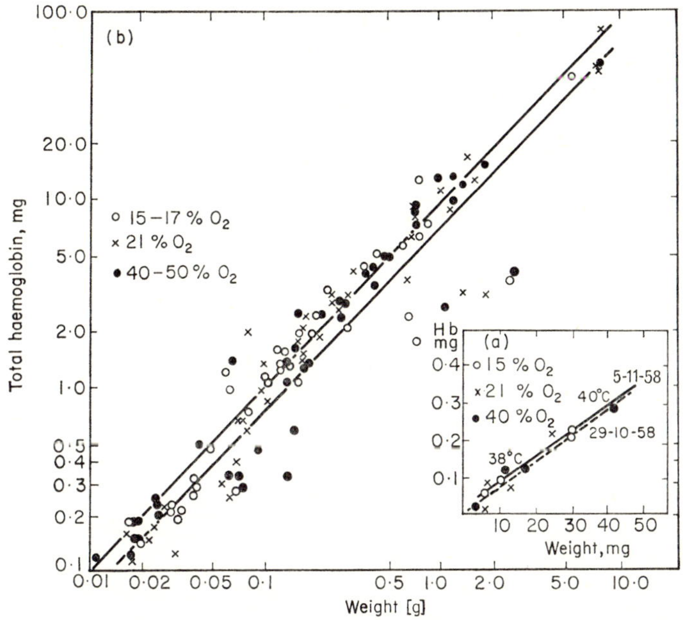

FIG. 3. Total haemoglobin as a function of embryonic weight. 4–11 day-old chicken embryos incubated under different oxygen tensions (Reproduced with permission from Astola *et al.* 1959.)

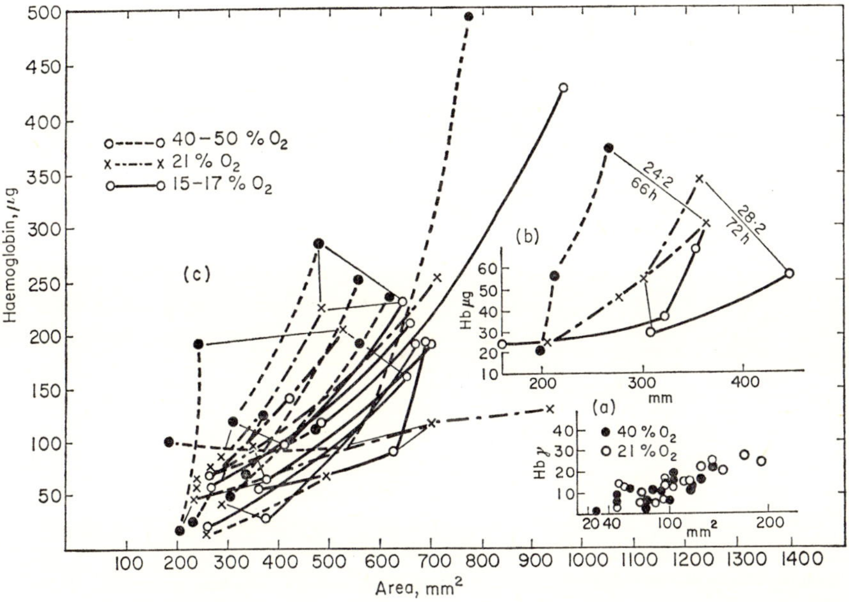

FIG. 4. Total haemoglobin (ordinate) of early chick embryo and area of the vascular zone (abscissa). (*a*) The determinations of haemoglobin, and of area are performed on the same, single embryo. (*b*) Pooled samples. Large scale. (*c*) Smaller scale, all experimental values except those shown in (*b*). (Reproduced with permission from Astola *et al.* 1959.)

content as a function of area of the vascular zone is definitely increased in the 40% oxygen-embryos—or to put it in another way—the vascular zone is reduced in relation to haemoglobin and weight of the embryo. Thus, when the eggs are incubated in 15% oxygen the vascular zone obviously reacts to the shortage of oxygen supply by increasing its size. Fig. 4 shows this effect of the oxygen on the vascular zone and total haemoglobin content of the embryo.

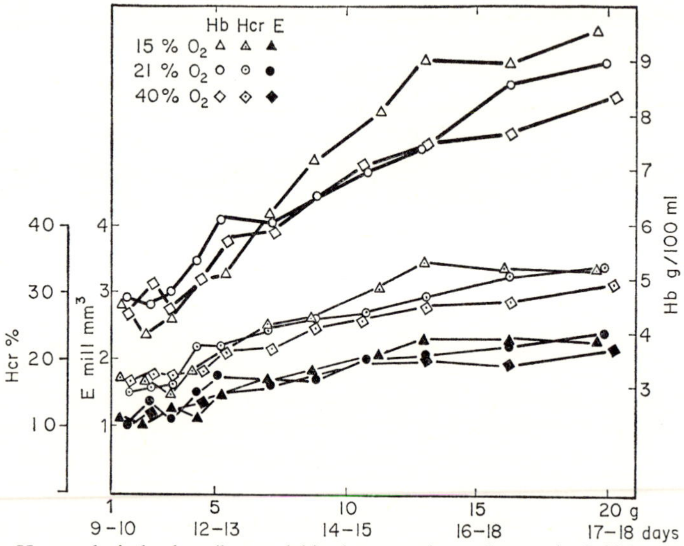

FIG. 5. Haematological values (haemoglobin, haematocrit, erythrocytes) of chick embryos 9–18 days old, incubated in 15, 21 and 40% O_2. (Reproduced with permission from Jalavisto *et al.* 1965.)

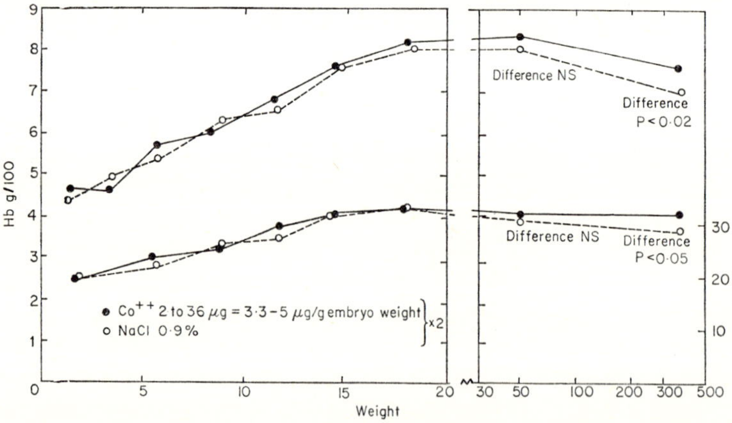

FIG. 6. Haemoglobin and haematocrit as a function of embryo and chicken weight in experiments with injections of cobalt. Left ordinate: haemoglobin g/100, right ordinate: haematocrit in %. Abscissa: weight in g. Solid circles: cobalt-injection experiments. Empty circles: control experiments with saline injections.

If the concentration of haemoblogin and the haematocrit are followed during the embryonic age 9 to 18 days there is a trend of continuous increase, which slows slightly down when the weight exceeds 13 g (Fig. 5). The effect of oxygen concentration of ambient atmosphere on haemoglobin is quite clearcut: in embryos lighter than 7 g there is no regular difference between the experimental groups. After the embryos have attained this weight and until the weight of some 13 g is reached the haemoglobin concentration increases more in the 15% O_2-group that in the 21 and 40% groups. In the heaviest embryos the haemoglobin content of the 15, 21 and 40% O_2 differs and decreases in this order of sequence.

Cobalt, although effective in the adult hen and in the 1-month-old cockerel has very little effect on the chick embryo haemoglobin level. Although in the heaviest embryos the average haemoglobin concentration exceeds numerically that of the controls, the difference is not statistically significant (Fig. 6). The same is true of the 2-day-old chicken as seen from Fig. 6. However, the dose might not be quite sufficient but it cannot be much increased before the mortality begins to rise.

IV. Discussion

The results presented above are not easy to interpret. What is quite clear is the absence of reaction to hypoxia of haemoglobin synthesis above the requirement imposed by the weight development of the embryo in the early and middle phase of incubation period although the hypoxia does somehow interfere with the "physiology" of the embryonic development. This interference is manifested in the smaller weight and, what is particularly interesting, in the relative increase of the vascular zone in response to hypoxia. Obviously in this phase the synthetic activities are hampered by lack of oxygen and the synthesis of haemoglobin is proportionally restricted. The question arises of the causal chain: is the growth rate limited because there is not enough haemoglobin to bind transport oxygen or is not the haemoglobin synthesis stimulated because the energetic needs are decreased by a smaller growth rate? Furthermore, are there any indications of circulating erythropoietins? Because lack of oxygen as such seems never to stimulate the growth of any structures the growth and adaptation of the size of the vascular zone to the varying oxygen concentration in ambient air seems to imply the existence of some humoral factor, but as yet I have no direct experiments to demonstrate the transmissibility of this effect.

If it is taken into account that the target cells of erythropoietin action in adult organisms are the stem cells, the effect on the vascular zone could be considered as analogous to that effect. But since the total haemoglobin is regulated only to keep pace with the general growth the feedback system responsible seems to be somehow related to some general growth factor which is consumed or bound by growing tissues. One could then assume that when growth is slow the amount left for action on the haemopoietic tissue would be increased and vice versa. However, this concept would be only applicable as long as the haemoglobin content is regulated to correspond to the general growth rate, i.e. in these experiments in the early phase until approximately the fourteenth day and in the experiments with cobalt during the entire (or nearly so) embryonic period because, as shown in Figs. 2 and 5, in the later phases the growth rates were slightly speeded up but the haemoglobin content as a function of embryo weight was not noticeably increased.

In the last phase of embryonic development when the erythron reacts to hypoxia by increasing the haemoglobin concentration in excess of the increase accountable for the increase in weight, some additional factor probably comes into play. Either the elaboration of a more specific erythropoietin begins or the feedback system changes. It was hoped that the determination of the phase in which this change takes place in the embryonic development could give an answer to these questions. However, unfortunately, the fourteenth to sixteenth days are critical in many ways. The erythropoiesis in the bone marrow begins to dominate. It begins there on the twelfth incubation day (Romanoff 1960). The hypophysis becomes functional approximately on the sixteenth day (Case 1952). The mesonephros increases its relative weight and its mass begins to overweight the mesonephros (Romanoff 1960). All these findings are unfortunately facts which could fit in *some* of the existing hypothesis about the erythropoietin production. According to Jacobson *et al.* (1957), Naets (1958) and Osnes (1958) the main site of erythropoietin production is the kidney. The enhanced growth of the metanephros in the same phase when the responsiveness appears could be taken to indicate that in embryonic development the metanephros plays a decisive role as it does in the adult. That the responsiveness coincides with the date when the hypophysis becomes functional would again fit in with the postulate of Halvorsen (1960) according to which the production of erythropoietin would be related to some hypophyseal factor—and so on!

It is quite obvious that these experiments raise more questions than they give answers, but they had to be done before more crucial types of experiments could be planned.

A great part of the experiments have been carried out in collaboration with Drs. Kyllästinen and Kuorinka and Mrs. Haapiainen whose help I gratefully acknowledge. My thanks are also due to the Sigrid Jusélius Foundation for a grant given to me quite in the beginning of this series of studies many years ago.

The work on which this paper is based was supported by the Finnish Medical Research Council.

REFERENCES

Astola, E., Jalavisto, E., Mustala, O. and Solanterä, L. (1959). *Acta physiol. scand.* 47, 210–217.

Case, J. F. (1952). *Ann. N.Y. Acad. Sci.* 55, 147–158.

Cruz, S. R. and Romanoff, A. L. (1944). *Physiol. Zool.* 17, 184–187.

Halvorsen, S. (1960). *Scand. J. clin. Lab. Invest.* 13, 564–575.

Jacobson, L. O., Goldwasser, E., Fried, W. and Plzak, L. F. (1957). *Nature* 179, 633–634.

Jalavisto, E., Kuorinka, I. and Kyllästinen, M. (1965). *Acta physiol. scand.* 63, 476–486.

Jalavisto, E., Haapiainen, T., Kyllästinen, M. and Kuorinka, I. (1965). *Ann. Acad. Sci. fenn. A.V.* (to be published).

Leaders, F. E., Dixon, R. L., Obsborne, J. W. and Long, J. P. (1962). *Proc. Soc. exp. Biol. Med.* 110, 436–440.

Naets, J. P. (1958). *Nature* 181, 1134–1135.

Osnes, S. (1958). *Br. med. N.J.* No. 5109, 1387–1388.

Romanoff, A. L. (1960). "The avian embryo." Macmillan, New York.

DISCUSSION SUMMARY

Chairman: C. G. Craddock

Professor Lajtha summarized for us the degree of precision which can be obtained with the quantitation of erythropoiesis. This is of considerable relevance because the red cell is the most readily sampled tissue in humans as well as many animals. The extent to which the turnover times and production rates of various proliferative and nonproliferative compartments can be assessed in the erythron is remarkable indeed. The good agreement between various measured parameters, using a variety of techniques, and the predicted values as derived from various mathematical approaches is indeed encouraging that the models of erythropoiesis he has evolved are valid.

The firmness of this quantitation of erythropoiesis makes it understandable that erythropoietin remains the single, surest demonstration of a tissue specific trophic hormone which acts by increasing the rate of cell production. The major action of this hormone appears to be at a very early level of cell development and Dr. Lajtha has reviewed the evidence that this effect is on a stem cell population which is self-sustaining. This stem cell population is generally conceived of as being pluripotential, at least insofar as the other hemic cells produced in the marrow are concerned. Erythropoietin is believed to influence the development of these cells into the differentiated erythroid line. Other influences of erythropoietin on the differentiated cell, in terms of increasing the rate of hemoglobin formation, etc., may be important but are secondary to this major action. The question was raised, and discussed at some length, as to whether the pluripotential feature of the stem cell model is a necessary part of the concept of erythropoietin in action.It is clear that at the moment the experimental findings can be explained best by this view. Nevertheless, it is equally clear that many normally regenerating cell systems must maintain themselves from a differentiated cell level. Accordingly, the view that the marrow contains cells which have the potential to develop into one of several directions, if established, may be a rather unique situation.

Doctors Rytömaa and Kiviniemi presented additional evidence of multiple factors influencing marrow granulocytopoiesis. Using a very elegant system of short term culture of rat marrow they have been able to separate effects which are stimulatory from others which are inhibitory, insofar as proliferative activity by granulocytic precursors are concerned. These effects are ascribed to a balance of factors which stimulate granulocytopoiesis and factors which inhibit cell proliferation. This balance changes in the serum of rats stimulated by the removal of granulocytes from the peritoneal cavity. The clearest effect of serum from such rats subjected to leukophoresis is the very rapid increase in DNA synthesis and granulocyte production in the myelocyte component of the marrow population. Since the myelocytes and perhaps metamyelocytes are the mitotable cells which are farthest along the differentiation pathway, it is clear that this stimulatory effect is quite different from that of erythropoietin on the pluripotential stem cell population. Doctors Rytömaa and Kiviniemi have, therefore, designated this action on myelopoiesis as representing an antichalone effect. This would presuppose that normally there is a constant inhibitory feedback on myelopoiesis, a chalone-like action, similar to that described for the epidermis. There has long existed good, though circumstantial, evidence for such a feedback mechanism involving the mature granulocyte. This concept was supported by the earlier studies of leukophoresis in dogs as carried out by Craddock and his associates. The evidence of Doctors Rytömaa and Kiviniemi, acquired by an entirely different approach, would add further support to this view. These workers suggest that the immediate effect of granulocyte (chalone) removal is to shift the balance in favor of an antichalone which promotes cell replication at the myelocyte level. If the techniques employed to promote the appearance of this antichalone can be improved and the substance obtained in a more purified state, this approach may well permit new insights into the control of granulocytopoiesis.

Professor Jalavisto, in beautifully straightforward observations, points out additional complexities in the control of erythropoiesis. There is a point in the development of the chick embryo before which the erythron is unresponsive to hypoxia and after which the erythron is responsive. Response to hypoxia is well established to be erythropoietin mediated. A possible interpretation is that erythropoiesis is subject to different control mechanisms at different points in its evolution. This type of evidence emphasizes the variabilities which may be at play in measuring the response to the same stimulus in the same tissue but at different stages in its development. There is a natural tendency to oversimplify and to generalize about such matters. This type of observation should serve to make us cautious.

Section IV

Lymphoid Cells

Observations on the Relationship of Thymus to Hematopoiesis and Lymphopoiesis

CHARLES G. CRADDOCK, ALAN WINKELSTEIN,* YOSHIHIKO MATSUYUKI,† and JOHN S. LAWRENCE

with the technical assistance of Mary Baker, Robert Yelenosky, Shirley Gordon and Marlene Hartmann.

Department of Medicine, University of California at Los Angeles, and Hematology Research Laboratory, Veterans Administration Center, Los Angeles, U.S.A.

I. Introduction

In spite of the contributions of many outstanding scholars the precise relationship of the mammalian thymus to hematopoiesis and lymphopoiesis remains mysterious in many of its aspects. Time does not permit recounting of all the fact and theory which has accumulated over the years. Instead, selected observations about this peculiar tissue will be noted with the objective of bringing together a body of evidence which, it is believed, assigns a specific and unique role to the thymus. The role which will be dealt with relates to the thymus' participation in the immune response. Other roles, less well defined, are by no means precluded by the present discussion. Many observations will be omitted, not because of any judgement by the authors as to their validity or significance, but because of lack of relevance to the particular thymic function under discussion. The most provocative suggestions that the thymus may influence the development of incoming pluripotential hematopoietic stem cells and the evidence that the epithelial components of the thymus supply a humoral secretion which promotes lymphopoiesis (Miller 1965) are exciting possibilities for future clarification.

II. Cellular Proliferation in the Thymus

It has been known since the investigations of Kindred (1940), Andreasen (Andreasen and Christensen 1949), Ottesen (Andreasen and Ottesen 1945),

This investigation was supported by Public Health Service Research Grant No. CA 07169-02 from the National Cancer Institute.

* Trainee in Hematology, Department of Medicine, on USPHS Training Grant.

† Postgraduate Research Hematologist, Department of Medicine, from Kyushu University, Japan.

and many others that the mitotic activity of the thymic cortex is 7 to 10 times that of the peripheral lymph node tissue. Furthermore, measurements of DNA renewal by several isotopic methods confirm the rapid rate of new cell formation in the thymus. This evidence has led most to accept the view expressed by Kindred that the thymus is the primary source of lymphocytes in the immature animal (including the human). This interpretation of the contribution of the thymus to the total lymphatic tissue of the body must be viewed in the perspective of more recently established facts about the kinetics of peripheral lymphocytes.

It is now well accepted, from the classical studies of Gowans and his associates, that the small lymphocyte, formed in lymph nodes, is a long lived cell with the remarkable qualities of being able to survive repeated recirculations through blood and lymph and able to migrate into and through most of the body's tissues, returning to lymphoid follicles (Gowans and McGregor 1964). Furthermore, the studies with phytohemagglutinin and various antigens provide clear evidence that the small lymphocyte can be stimulated to re-enter the proliferative phase and assume the characteristics of an immature, large lymphoblast which will undergo division and give rise to cells which manufacture immunoglobulins (MacKinney *et al.* 1962, Hirschhorn *et al.* 1963, Van Furth 1964). The work of several investigators confirms the view that this transformation, so readily observable in tissue culture, occurs *in vivo* as well (Gowans *et al.* 1962, Porter and Copper 1962, Craddock *et al.* 1964). These findings, together with the established longevity of peripheral blood lymphocytes in various species by widely differing techniques (Craddock *et al.* 1964, Hamilton 1957, Little *et al.* 1962, Norman *et al.* 1965) leave little doubt that the peripheral blood lymphocytes are: (a) self sustaining, (b) long lived, (c) capable of re-entering the proliferative cycle after very prolonged intervals (years in the human), (d) capable of antibody synthesis.

The question arises, therefore, why are such large numbers of additional cells needed from the immature thymus? This question is unanswerable in any kinetic terms if it is assumed that the thymocyte is destined to become a peripheral, long lived, self replicating cell. If this is the actual fate of the thymocyte then the vast majority of cells produced in the thymus must die, leaving only a rare cell to contribute to the growth of the peripheral lymphatic system. As will be discussed subsequently it is our view that the thymocyte is a short-lived cell which fills a different role from the long lived peripheral lymphocyte. In spite of their morphologic similarities and their mutual participation in the immune response we will develop evidence that thymocytes and peripheral lymphocytes are two separate and discrete cell populations.

III. Proliferative Response to Antigen

It is now firmly established from studies in animals reared in a "germ free" environment that antigenic exposure is a prerequisite for proliferation of peripheral lymphocytes and development of a mature immunologically effective, lymphatic structure. The thymus does not require this type of challenge for its production of new cells. From studies in germ free animals and in animals undergoing antigenic stimulation it is clear that cell division in the normal thymus is independent of exogenous antigen. Metcalf has shown convincingly that the thymus is independent of peripheral feedback inhibitory factors or peripheral stimulatory factors (Metcalf 1964). This tissue appears to have its own inherent growth stimulatory apparatus and much work has been performed to suggest that the internal milieu of the thymus is influenced by a trophic secretion of the fixed epithelial cells of the thymus (Miller 1965).

Are thymic cells (thymocytes or lymphoid cells in the thymus cortex) capable of undergoing the type of metamorphosis as a result of phytohemagglutinin (PHA) or antigenic stimulation which has been demonstrated for peripheral lymphocytes? There is much confusion about this simple but vitally important question. In Fig. 1 are data from another paper on this subject (Winkelstein and Craddock, submitted for publication) comparing the proliferative response of normal human thymus and normal human lymph node suspensions in short term culture with and without PHA. An important parameter of cellular proliferation in this system is DNA synthesis, as measured by uptake of tritiated thymidine (H^3T). Note that the normal thymus starts off with a high proliferative rate, a finding previously made in intact rats and in suspensions of rat thymocytes (Craddock *et al.* 1964). The lymph node suspension, on the other hand, starts off with a very small percentage of cells in DNA synthesis. After 3 days in culture with PHA the lymph node cells show the characteristic explosive DNA synthesis and cell division so well described for peripheral blood lymphocytes. The thymus suspension, however, has shown a gradual loss in proliferative potential and activity. It is readily apparent that the thymus cell suspension has shown very slight if any response to PHA stimulation whereas the lymph node suspension has exhibited remarkable stimulation. If one only looked at the thymus culture after 3 days without having the observation as to the initial proliferative activity in the system one could easily be misled to the conclusion that PHA had indeed stimulated thymocyte proliferation. This, we believe, accounts for much of the confusion about the proliferative response of thymocytes to PHA and antigen.

There is abundant evidence that thymocytes differ from lymph node and spleen cells in their ability to mount a graft *versus* host reaction in tolerant recipients (Billingham and Silvers 1964), to transfer immunity adoptively, and to produce manifestations of delayed hypersensitivity (Thorbecke and Cohen 1964) and to synthesize immunoglobulins *in vitro* (Van Furth 1964). The only experiments in which specific antibody production by cells in the thymus have been convincingly demonstrated have followed the direct injection of antigen into or around the thymus (Marshall and White 1961)

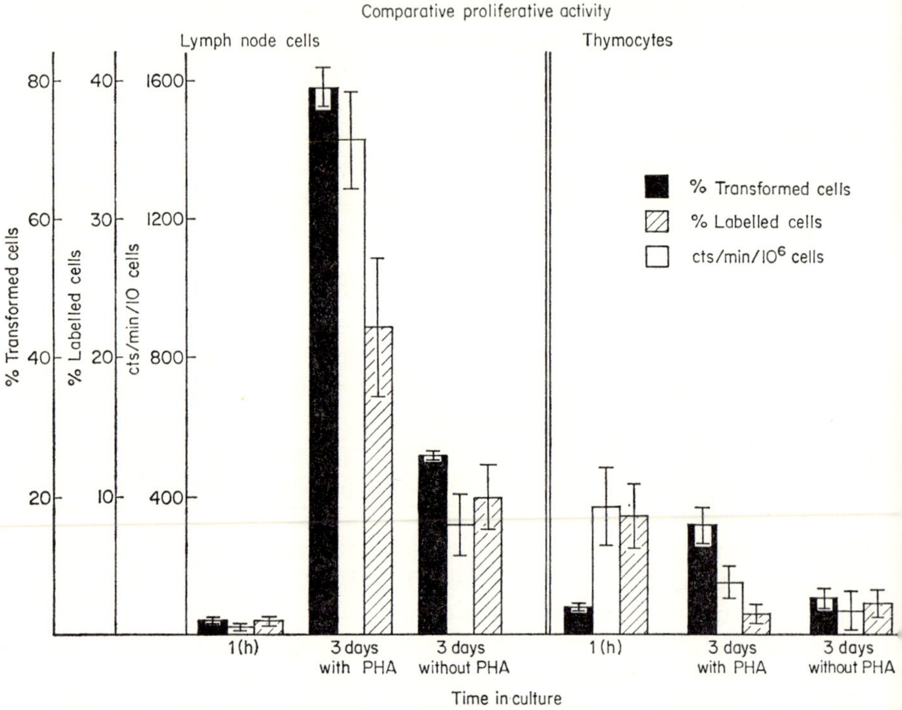

FIG. 1. The proliferative response of normal lymph node cells and normal thymocytes in culture with and without phytohemagglutinin. Fresh thymus tissue and inguinal lymph node was collected at the time of thoractomy from eight hematologically normal subjects undergoing heart surgery. The percentage of cells which were large blastoid cells and the uptake of tritiated thymidine added one hour prior to termination of the culture, determined by autoradiography and radiochemical methods, were used as the three parameters of cellular proliferation, as described elsewhere (Winkelstein and Craddock, submitted for publication). The closed line at the top of each bar is one standard deviation.

or the administration of large amounts of antigen as endotoxin, producing severe stress and depletion of thymic cellularity (Landy *et al.* 1965). There is considerable evidence that small lymphocytes formed in the periphery, believed to be immunologically competent and responsive cells, do not enter the normal thymus substance (Craddock *et al.* 1964). However, disruption of the thymus architecture might permit the influx of these cells, there to establish the proliferative response characteristic of these cells in their interaction with antigen. Under these circumstances the antibody forming cells may have derived from peripheral lymphocytes localizing in the damaged, antigen-loaded thymus rather than from thymocytes. It remains to be proved, therefore, that thymocytes have the potential for proliferative response to antigen and the production of specific antibody.

An important factor in the designation of cells produced in the thymic cortex as lymphocytes is their morphology. It is impossible, even with electron microscopic detail, to pin point any feature of the cortical thymocyte which distinguishes it from the small lymphocyte. This similarity loses its force, however, when one considers the fact that the small lymphocyte is a very "undistinguished" cell from any morphological aspect. There is a notable absence of cytoplasmic organelles which so readily identify the varieties of differentiated blood and tissue cells. This is not so surprising in view of the current conception of the small lymphocyte as a cell in prolonged, inert interphase. The thymocyte, while in the thymus, is also an inert, undifferentiated cell. Therefore, one is comparing the morphology of a null and a void. It seems unreasonable to conclude on such morphological grounds that these two cells are identical in so far as their functional destiny is concerned. Yet, the prevailing view of the similarity of these tissues is largely based on this arbitrary evidence.

Considerable data can be cited that the two tissues are different. The kinetic evidence for the thymic population being short lived has been given. The evidence for a difference in immunological capacity has been reviewed briefly. Other differences relate to cell and nuclear size and metabolism. The DNA/cell ratio of the thymus is approximately twice that of peripheral lymphocytes (Davidson 1960). The enzymes concerned with DNA synthesis are in a more active state in the normal thymus than in the spleen (Potter and Nygaard 1963). The micro-environment of the thymus, as related to pools of intrathymic thymine nucleotide precursors of DNA, is quite different from that of the spleen (Potter and Nygaard 1963). There is apparent feeding of thymine nucleotides contained in small thymocytes to large, proliferating thymocytes. These metabolic differences lead to a characteristically low uptake of exogenous labeled thymidine by each cell which is in DNA synthesis (Craddock *et al.* 1964). This pattern of labeling permits some degree of delineation of these cells from small lymphocytes formed in the mantle and marginal zones of lymph follicles by division of progenitor cells which show a high concentration of administered H^3 thymidine. Significantly, the lymphoid cells of the bone marrow show the same pattern of H^3T labeling as thymic lymphoid cells (Craddock 1965).

Tissue specific thymic antigens have been described in a variety of mice (Reif and Allen 1964), a finding which suggests thymus characteristics not shared by other lymphoid tissue. An additional difference is depicted in

TABLE I. The effect of testosterone proprionate on the lymphoid tissue of normal male rats. Note that there is a reduction in thymus and bone marrow "lymphocyte" cellularity but no reduction in the weight of other lymphatic tissue nor in numbers of circulating lymphocytes.

	Marrow Lymphoid cells (per $mm^3 \times 10^6$)	Thymus weight (mg)	Spleen weight (mg)	Lymph Node weight (mg)[a]	Blood Lymphocytes (per $mm^3 \times 10^3$)
NORMAL (Average of 11 ± S.E.)	90 ± 7·4	630 ± 30	610 ± 30	270 ± 20	7·1 ± 0·45
TESTOSTERONE[b] (Average of 5 ± S.E.)	38 ± 0·10	110 ± 20	520 ± 30	350 ± 20	6·4 ± 1·36

[a] Lymph nodes pooled for this measurement include mediastinal, axillary, mesenteric and periaortic nodes.

[b] 10 mg/100 g body weight I.M. 3 times/week for 6–10 doses.

Table I. Testosterone in large amounts produces a decrease in thymus weight and bone marrow lymphocyte count in normal male rats. No fall in lymph node or spleen weight or peripheral blood lymphocyte number was observed.

IV. Dispersion of Thymic Tissue

A feature of thymus physiology which has been a source of long standing bewilderment is the fact that thymectomy in the young adult animal or human has very little deleterious effect, in any dramatic sense, upon subsequent growth, development or immune capability. There is some late fall in blood lymphocyte count and in volume of various lymph nodes, but a negligible effect upon the immunological potential (Bierring 1960). By contrast, neonatal thymectomy in a variety of species has a dramatic and clear cut effect upon the development of the lymphatic tissue and the immune response. The classical experiments of Miller (1965) as well as the clinical associations of Good and his colleagues (Good *et al.* 1965) leave little doubt about the integral role of the thymus in this complex process. Why should the effects of neonatal thymectomy be so well defined but the effects of thymectomy in the postnatal period be so ambiguous? The current interpretation of this paradox largely centers around the view that the thymus fulfills a role at or near the time of birth which is critical. Thereafter, the lymphatic system is self sustaining and the thymus is expendable. This view is troublesome to one who believes in the wisdom of nature; it is difficult to accept that the tremendous proliferative activity that takes place in the thymus of growing animals is so much "wheel spinning" production of cells which do not serve an important function. An alternative explanation, noteworthy only because of its naivete, is the possibility that thymus function becomes disseminated at or near the time of birth. Thymectomy, removing a major source of this tissue, can be compensated for by hypertrophy of similar tissue elsewhere in the body. This simple view is supported by the observation that bone marrow "lymphocytes", a body of cells of considerable size, as pointed out by Yoffey (1960), have kinetics quite similar, if not identical, to cortical thymocytes. This conclusion about bone marrow lymphocyte kinetics was reached independently by Everett and his group, studying bone marrow "lymphocytes" of guinea pigs (Osmond and Everett 1964) and by our group studying bone marrow "lymphocytes" in rats (Craddock 1965). Osmond and Everett further demonstrated by labeling techniques combined with occlusion of the circulation to a portion of marrow, that the marrow "lymphocyte" population is autonomous and not a reflection of the influx of cells produced elsewhere. Furthermore, we have shown that the recovery of rat marrow "lymphocytes" from acute cortisol depletion occurs independently of thymic function (Fig. 2) (Craddock and Winkelstein 1965). These findings suggest that in the postnatal, growing animal, thymus function and proliferative activity may be shared with the bone marrow. Similar autonomous tissue, originating in the thymus, may be present in the spleen and lymph nodes. The experimental findings of Fichtelius long ago showed differences between thymocytes and lymph node lymphocytes insofar as their migratory routes are concerned (Fichtelius 1959). These suggestions that

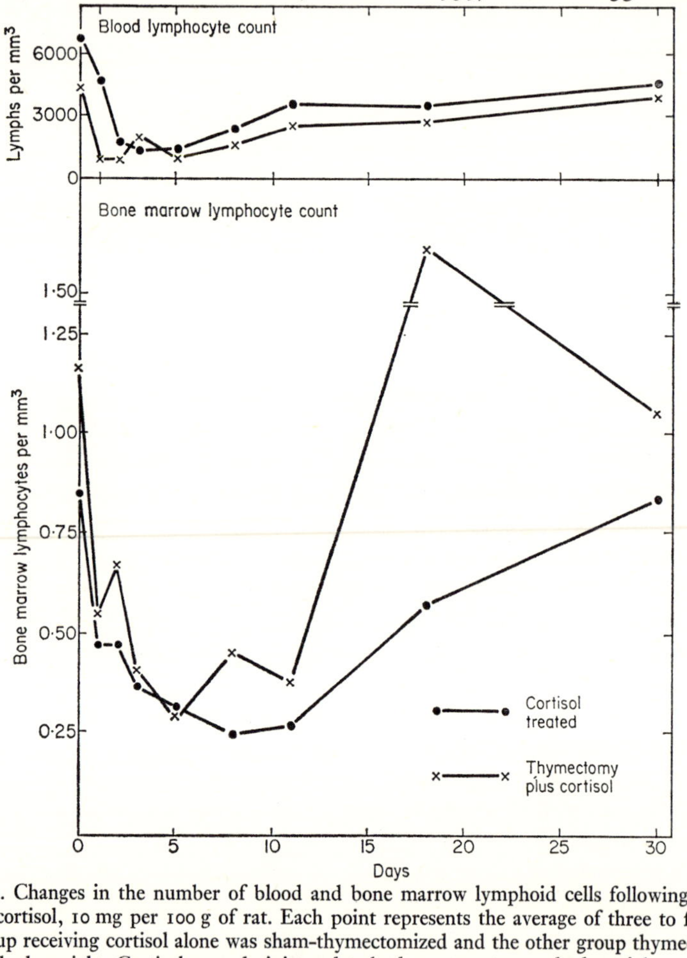

FIG. 2. Changes in the number of blood and bone marrow lymphoid cells following a single dose of cortisol, 10 mg per 100 g of rat. Each point represents the average of three to four rats. The group receiving cortisol alone was sham-thymectomized and the other group thymectomized at 100 g body weight. Cortisol was administered to both groups at 150 g body weight.

thymus tissue is widely disseminated after birth are commensurate with the interpretations of Beard in 1903 and with more recent observations such as those of Bierring (1960) who showed that *partial* thymectomy was not followed by compensatory hyperplasia and hypertrophy of the remaining tissue in the thymus. The difficulty in accepting the view that thymic tissue is distinctive and discrete from peripheral lymphatic tissue is readily understandable since both tissues may be intimately mixed in the peripheral lymphatic system of the adult animal and since both tissues are directly concerned with the immune response to antigen. Both tissues also display sensitivity to damage by X-ray and cortisone.

V. Role of Short-lived "Thymocytes" in the Response to Antigen

We believe that these two tissues play critical roles in the immune response but at different points in the process. Thymus-type cells (including cells in bone marrow, spleen and lymph nodes which have a short life span) are believed important in an early phase of response to certain primary antigens, i.e., in the processing of antigen. Lymph follicle cells of the long lived, recirculating, antigenically responsive variety are crucial for the formation of specific antibody via the pathway of cellular transformation to large proliferating lymphoblasts and their subsequent division and differentiation into cells which are distinctive by virtue of their active synthesis of immunoglobulins, as described by others.

The evidence in support of this concept of thymocyte-antigen-lymphocyte interaction will be cited. In 1948, Craddock and Lawrence reported a confirmatory observation made in 1915 by Hektoen. Rabbits were inoculated with two antigens: sheep red blood cells and killed typhoid bacilli. The level of serum antibody to these antigens was followed. Rabbits which received 250 r whole body X-ray, enough to depress blood lymphocyte counts but not enough to seriously affect the animals' nutrition or well being, were given antigens before and after X-ray. Those animals which received the antigen 1 day before X-ray developed a normal antibody titer; those animals which received the same X-ray dose 1 day before antigen injection failed to develop a significant rise in serum antibody titer. This observation has been the subject of concerted study by Taliaferro and others over the subsequent years (Taliaferro and Taliaferro 1951). The results of these studies may be summarized as follows: the radiosensitivity of the immune response to an antigen may be divided into three phases. The initial phase, occurring within hours after antigen injection, is highly sensitive to whole body X-ray. The second phase, associated with the rapid rise in antibody titer is moderately radiosensitive. The last phase of sustained antibody titer is highly radioresistant. Interestingly, and significantly, the sensitivity of the immune response to adrenal corticosteroid administration is quite similar. The crucial observation in all of these experiments is the exquisite sensitivity of the *earliest*, brief phase of the response following antigen injection.

In 1956, Berglund and Fagraeus showed that if rats were given suspensions of spleen cells or thymus cells at the time of or shortly after the administration of antigen (in this case sheep red blood cells), the crippling effect of prior adrenal corticosteroid administration was reversed. Their studies complemented the earlier demonstrations by Jaroslow and Taliaferro (1956) that the highly radiosensitive initial phase of response to antigen could be protected by such diverse substances as spleen, lymph node and HeLa cell suspensions and yeast extracts. Fagraeus and Berglund (1961) showed that the protection against corticosteroid ablation of the initial phase of the immune response could be brought about by heterologous spleen cells. These observations have been shunted into obscurity by the more recent observations and theories concerning the interrelationships between thymus, the peripheral lymphatic system and the immune response. However, if the data are reconsidered in light of the evidence for cells of thymic origin and cells of lymph node origin being two distinct cell populations, then they take on new and meaningful significance.

Experiments were undertaken in rats using sheep red blood cells (SRBC) as the antigen and cortisol (hydrocortisone) as the agent to depress the immune response. Assay of the immune response was carried out by serial measurements of the serum hemolysin titer (Fagraeus and Berglund 1961) and by the highly sensitive and quantitative assay of specific antibody forming units as developed by Jerne (Jerne and Nordin 1963). In confirmation of the work of others it was found that the immune response could be progressively depressed by increasing dosages of cortisol given prior to the antigen, as shown in Fig. 3. Other studies showed that hemolysin production occurred largely in the spleen after I.V. injection of the antigen. Splenectomy markedly suppressed the antibody titer, as shown by others (Jaroslow and Taliaferro 1956). Repeated injections of antigen by various routes (I.P., S.C., I.M., and I.V.) failed to evoke the appearance of hemolytic units in either thymus or bone marrow of normal or splenectomized recipients.

Administration of cortisol in doses of 5 mg/100 g body weight to normal rats on each day for 4 days prior to antigen injection, including the day of antigen injection (8 hours elapsed between the last dose of cortisol and antigen administration) effectively blocked hemolysin response. The ability of various cells to restore the immune response was then studied. The results are summarized in Fig. 4.

Suspensions of normal lymph node cells, containing a mixed population of reticulo-endothelial cells, exhibited striking repair of the deficit. Cells from the lymph nodes of rats which had been given 10 mg/100 g rat of cortisol 2 days before sacrifice were less active. The lymph nodes from five cortisonized donor rats were pooled to be given to a single cortisonized recipient in order to provide a number of cells comparable to the number of normal lymph node cells. Cells from the bone marrow and the thymus of normal rats were incapable of restoring the immune response in cortisonized recipients. However, in preliminary experiments it appears that, when bone marrow cells or thymus cells are administered together with cells from cortisonized lymph nodes, neither being capable of restoring the deficit when given separately, there is restoration of hemolysin production (Fig. 4).

Examination of the lymph nodes of rats 2 days after 10 mg/100 g rat of cortisol shows a striking depletion of cellularity and minimal proliferative activity. Organized lymph follicles and germinal centers in particular are absent. Nevertheless some lymphocytes remain and it was assumed that the majority of cells remaining in lymph nodes at this time after cortisol were

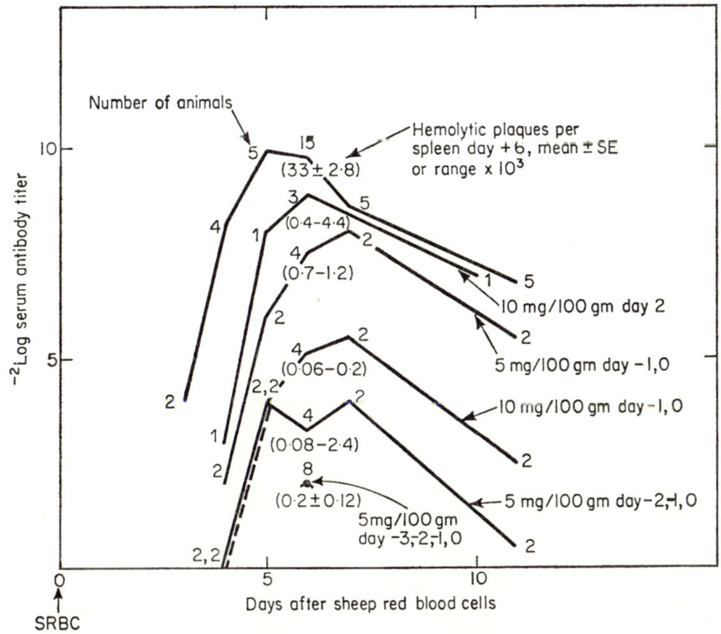

FIG. 3. The effect of cortisol administered in varying dosages and regimens on the response to sheep red blood cells (SRBC) given 8 hours after the last injection of cortisol. The figures above each point represent the number of animals in which serum hemolysin titer was determined at that point. The values in parenthesis are the mean number of hemolytic plaques per spleen in three to six animals sacrificed at day +6 after SRBC following each dose schedule of cortisol.

peripheral lymphocytes with immunological capacity; presumably any cells of thymic or marrow origin would have been eliminated by cortisol because of their exquisite sensitivity to this agent (Fig. 2).

These results are interpreted as follows: normal lymph nodes contain all the cells required to mount an immune response to SRBC manifested by hemolysin production by plasma cells in the spleens of cortisonized recipients. Partial ablation of the lymph nodes by cortisol leads to destruction of many lymphoid cells. The lymphocytes remaining in such nodes are capable of developing into antibody-forming plasma cells if they are given in combination with bone marrow or thymus cells which, by themselves, are incapable of promoting antibody formation. These findings suggest that lymphoid cells like those in normal marrow and thymus must be

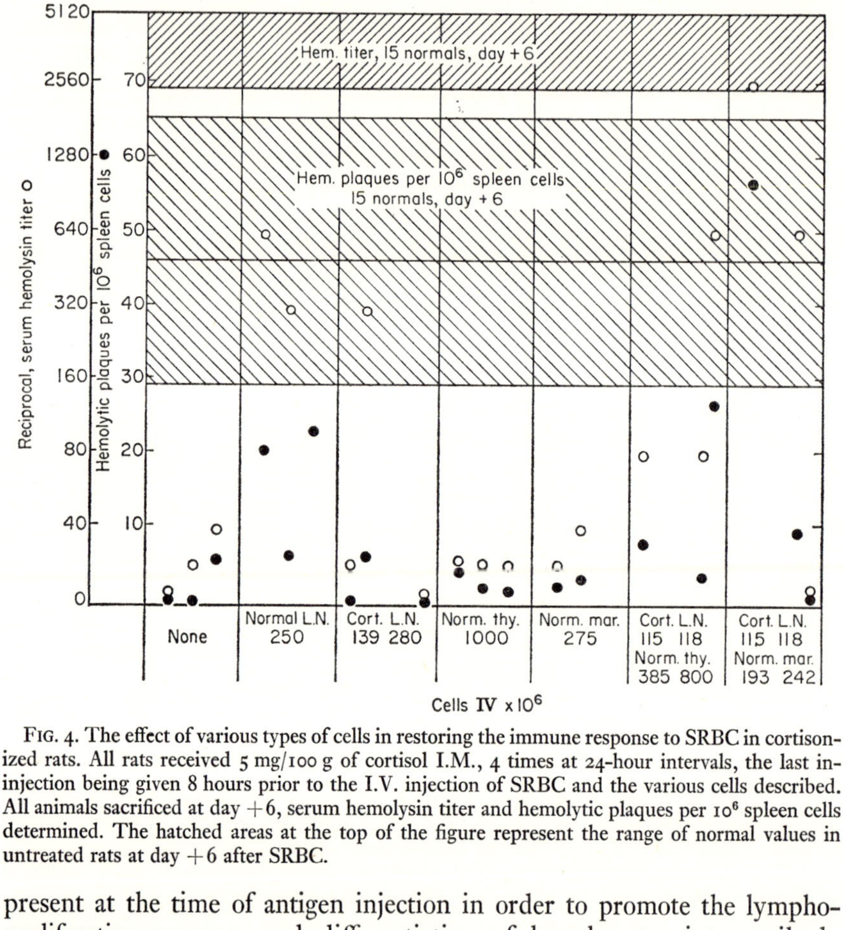

FIG. 4. The effect of various types of cells in restoring the immune response to SRBC in cortisonized rats. All rats received 5 mg/100 g of cortisol I.M., 4 times at 24-hour intervals, the last in-injection being given 8 hours prior to the I.V. injection of SRBC and the various cells described. All animals sacrificed at day +6, serum hemolysin titer and hemolytic plaques per 10^6 spleen cells determined. The hatched areas at the top of the figure represent the range of normal values in untreated rats at day +6 after SRBC.

present at the time of antigen injection in order to promote the lymphoproliferative response and differentiation of lymphocytes into antibody forming cells.

VI. Development of "Thymocytes" into Macrophages

Several observations support this interpretation. Fishman and his associates have shown the necessity for an interaction between macrophages (obtained from the rat's peritoneal exudative response to beef heart infusion-peptone broth) and antigen before antibody production by lymph node

TABLE II. The percentile distribution of lymphoid and macrophage cells in skin windows and peritoneal exudate at specified time intervals after insertion of subcutaneous cover slip or injection of beef-heart infusion, proteose mixture intraperitoneally. All animals received H^3 thymidine, 1 μC/g body weight 21 hours prior to the insertion of the window or induction of exudate and the percentile labeling and mean grain count of labeled cells is given.[a]

	Hours after implantation—skin window							
	(6)		(18)		(24)		(48)	
	Small lymphoid cells	Macrophages	Small lymphoid cells	Macrophages	Small lymphoid cells	Macrophages	Small lymphoid cells	Macrophages
% total cells	5·5–20·4 (12·5)	5·4–15·5 (10·0)	28·5–35·8 (32·2)	54·5–59·2 (56·8)	3·0–18·0 (10·5)	69–95 (82)	42·0	27·5
% labeled cells	34·6–42·5 (38·6)	64·0–72·5 (68·3)	52·0–59·8 (55·9)	67·6–74·8 (71·2)	42·0–47·0 (44·5)	57·6–81·2 (69·4)	61·0	56·0
Mean grain count, labeled cells	5·6–6·6 (6·1)	8·0 (8·0)	4·0–8·3 (6·2)	5·1–11·4 (8·4)	3·4–7·2 (5·3)	3·4–13·6 (8·5)	3·5	4·8
	Hours after intraperitoneal injection—peritoneal exudate							
Total cells recovered	29 × 10⁷	—	—	—	26 × 10⁷	—	14 × 10⁷	—
% total cells	1·8	6·0	—	—	3·4	34·4	2·8	70·4
% labeled cells	34·6	62·6	—	—	42·4	68·2	32·8	55·9
Mean grain count, labeled cells	4·7	5·8	—	—	5·2	6·2	4·2	4·7

[a] H^3T, 1 μc/g body weight 21 hours before implanting skin window or injection of beef-heart broth infusion intraperitoneally.

lymphocytes will occur (Fishman 1961, Fishman *et al.* 1963). The macrophages employed in such experiments appear to evolve from small lymphoid cells which have a rapid turnover like those seen in the normal thymus and bone marrow. Volkman and Gowans in recent studies of the macrophage response on skin windows implanted under the skin of rats have shown that these macrophages appear after small lymphoid cells. These cells, according to these workers (Volkman and Gowans 1965a, Volkman and Gowans 1965b), emanated from the bone marrow. Both morphological types (lymphoid cell and macrophage) showed a rapid turnover rate without local, *in situ* division. Table II shows the results of similar studies of the labeling kinetics of small lymphoid cells and macrophages in skin window preparations and in peritoneal exudate of normal rats given tritiated thymidine (H^3T). It will be noted that the high percentile labeling of small lymphocytes and the low mean grain count per cell is almost identical to the pattern in small lymphoid cells of the bone marrow and thymus previously published (Craddock 1965). The macrophages showed the same distribution of the DNA label as the small lymphoid cells. The striking difference between this labeling pattern of lymphoid cells and that in peripheral lymph nodes provides strong support for the view that the small lymphoid cells appearing on skin windows and peritoneal exudate have a short life span and rapid replacement rate like those from the bone marrow or thymus. The data support the view expressed by others that these lymphoid cells evolve into macrophages in these areas of inflammatory reaction. Conclusive evidence that these macrophages do not evolve from myeloid granulocytic precursors in the marrow is provided by studies in guinea pigs rendered agranulocytic by means of a specific anti-neutrophilic rabbit serum. Such animals show a normal marcophage response on skin windows (Lawrence *et al.*, submitted for publication).

These observations provide evidence for the following view of the immune response to a primary injection of SRBC in rats, as shown diagrammatically in Fig. 5. Phase I involves the interaction between macrophages, which evolve from cells which resemble lymphocytes and are formed in the thymus and bone marrow, and the sheep red cell antigen which is localized in the spleen. This phase is of brief duration, occurs predominantly in the spleen, involves phagocytosis and processing of antigen, death of macrophages and transmission of antigenic information to lymphocytes. The mechanism of this transfer has been the subject of much study and involves a low molecular weight RNA, as shown by Fishman *et al.* (1963). This phase is highly sensitive to damage by ionizing radiation or cortisol because of the documented sensitivity of the macrophage precursor, the thymus and marrow lymphoid cell, to these agents. This phase is an obligatory first step preceding the proliferative response by lymphocytes, phase II. This phase II proliferative response involves the metamorphosis of small lymphocytes into large, replicating cells, a step in the immune response which has been studied so extensively by many workers using both *in vivo* and *in vitro* techniques. Phase II is also moderately and variably injured by radiation and cortisol because of the sensitivity of the small lymphocytes to these agents. This does much to explain the difficulty which various workers have encountered in separating these two cellular events. The differentiation of some of these proliferating lymphocytes into plasmacytoid cells, actively synthesizing antibody is phase III. This last phase is resistant to injury by

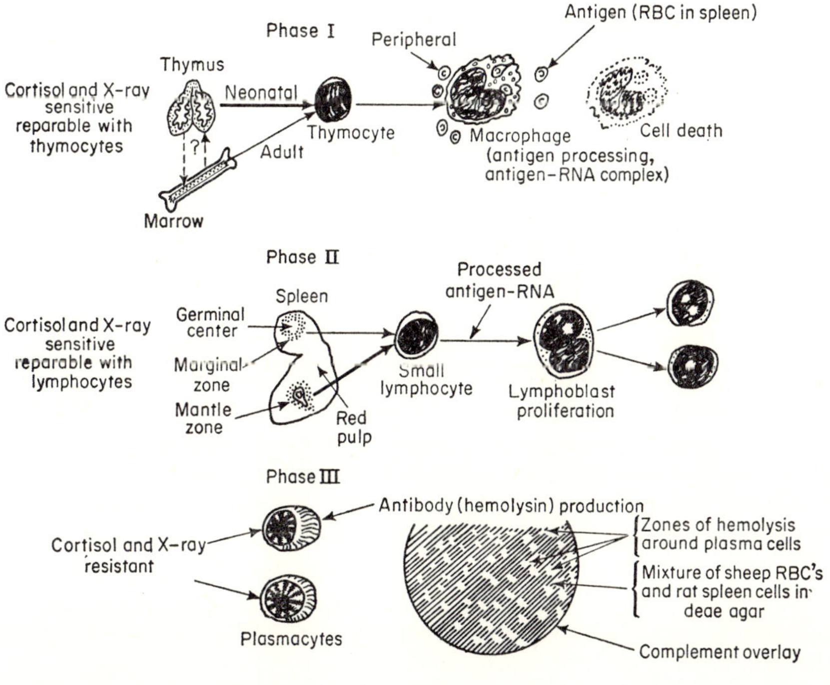

FIG. 5. Diagrammatic representation of the three phases of the immune response to sheep red blood cells (SRBC) and the postulated origin and role of thymus-type lymphoid cells and peripheral lymphocytes.

ionizing radiation or cortisol. This phase is the step at which all parts of the immune response must be assessed. It will be blocked by interference with either phase I or phase II.

Ablation of phase I can be countered by the administration of bone marrow or thymus cells, but the reparative effect cannot be demonstrated in the absence of phase II. Phase II can be restored by peripheral lymphocytes alone, but the restoration cannot be demonstrated in the absence of phase I. It is necessary to have both phase I and phase II cells present in order to respond to SRBC by antibody formation as expressed in phase III. It is our belief that this function of thymus-marrow lymphoid cells as macrophages which must be present to initiate the immune response to certain primary antigens will prove to be the major function of these controversial cells. Studies are underway to clarify whether the neonatal thymus, so essential for the development of a mature lymphatic system, is the sole reservoir of phase I cells at birth. If this proves to be the case then a straightforward explanation can be provided for the crippling effects of neonatal thymectomy. Later in life, as the bone marrow population develops as an autonomous and dominant source of phase I cells, the thymus assumes a secondary role and thymectomy does very little to suppress the immune response.

VII. Conclusions

Many facets of this concept require elaboration and modification. Nevertheless, we can state the following conclusions with respect to the thymus-marrow lymphoid cell population:

1. The thymocyte and bone marrow lymphoid (BML) cell in the young adult rat differs from the peripheral small lymphocyte in production rate, life span, physiology and function. Cells in both thymus and marrow show a rapid, autonomous replacement rate which is not influenced by antigenic stimuli. Human thymocytes, in contrast to human peripheral lymphocytes, do not undergo a proliferative response in culture with phytohemagglutinin. Peripheral lymphopoiesis is dependent upon antigenic exposure and the cell products of such lymphopoiesis are long lived. Thymus-marrow lymphoid cells and peripheral lymphocytes are sensitive to injury by ionizing radiation and cortisol. Thymus-marrow lymphoid cells are suppressed by testosterone, peripheral lymphocytes are not. 2. The predominant role of the peripheral small lymphocyte in the immune response to a primary antigenic exposure is to undergo a proliferative response and give rise to antibody synthesizing cells. We propose that the predominant role of the thymus-marrow lymphoid cells in the immune response is to encounter antigen in the periphery, evolve into macrophage (phagocytic) cells, engulf antigen, process antigen in some way and transfer this antigenic information to peripheral lymphocytes. The thymus-marrow lymphoid cell is incapable of forming antibody or of evolving into an antibody forming cell. The thymus-marrow lymphoid cell is expended in its function. This function may be the essential first step which precedes the proliferative response by peripheral lymphocytes and their development into antibody forming cells.

REFERENCES

Andreasen, E. and Christensen, S. (1949). *Anat. Rec.* **103**, 401.
Andreasen, E. and Ottesen, J. (1945). *Acta physiol. scand.* **10**, 258.
Beard, J. (1903). *Zool. J.* **17**, 403.
Berglund, K. and Fagraeus, A. (1956). *Nature* **177**, 233.
Bierring, F. (1960). *In* "Ciba Symposium on Haemopoiesis" (G. E. W. Wolstenholme and M. O'Connor eds.), pp. 185–198. Churchill, London.
Billingham, R. E. and Silvers, W. K. (1964). *In* "The Thymus" (V. Defendi and D. Metcalf eds.), p. 41. Wistar Institute Press, Philadelphia.
Craddock, C. G. (1965). *Acta haemat.* **33**, 19.
Craddock, C. G. and Lawrence, J. S. (1948). *J. Immun.* **60**, 241.
Craddock, C. G., Nakai, G. S., Fukuta, H. and Vanslager, L. M. (1964). *J. exp. Med.* **120**, 389.
Craddock, C. G. and Winkelstein, A. (1965). *J. clin. Invest.* **44**, 1038.
Davidson, J. N. (1960). "The Biochemistry of the Nucleic Acids." Methuen, London.
Fagraeus, A. and Berglund, K. (1961). *J. Immun.* **87**, 49.
Fichtelius, K.-E. (1959). *In* "The Kinetics of Cellular Proliferation" (F. Stohlman, Jr., ed.), pp. 166–172. Grune and Stratton, New York.
Fishman, M. (1961). *J. exper. Med.* **114**, 837.
Fishman, M., Hammerstrom, R. A. and Bond, V. P. (1963). *Nature* **198**, 549.
Good, R. A., Peterson, R. D. A., Martinez, C., Sutherland, D. E. R., Kellum, M. J. and Finstad, J. (1965). *Ann. N.Y. Acad. Sc.* **124**, 73.
Gowans, J. L. and McGregor, D. D. (1964). *In* "The Immunologically Competent Cell" (G. E. W. Wolstenholme and J. Knight eds.), p. 20. Churchill, London.
Gowans, J. L., McGregor, D. D., Cowan, D. M. and Ford, C. E. (1962). *Nature* **196**, 651.
Hamilton, L. D. (1957). *In* "Homeostatic Mechanisms". Brookhaven Symp. Biol., No. **10**, 53.
Hektoen, L. (1915). *J. infect. Dis.* **17**, 415.
Hirschhorn, K., Bach, F., Kolodny, R. L., Firschein, I. L. and Hashem, N. (1963). *Science* **142**, 1185.
Jaroslow, B. N. and Taliaferro, W. H. (1956). *J. infec. Dis.* **98**, 75.
Jerne, N. K. and Nordin, A. A. (1963). *Science* **140**, 405.
Kindred, J. E. (1940). *Am. J. Anat.* **67**, 99.
Landy, M., Sanderson, R. P., Berstein, M. T. and Lemer, E. M. (1965). *Science* **147**, 1591.
Lawrence, J. S., Campbell, T. N., Copeland, H. L., Craddock, C. G. and Greipel, M. A. (submitted for publication). Response of guinea pigs to anti-neutrophilic serum.
Little, J. R., Brecher, G., Bradley, T. R. and Rose, S. (1962). *Blood* **19**, 236.
MacKinney, A. A., Stohlman, F., and Brecher, G. (1962). *Blood* **19**, 349.
Marshall, A. H. E. and White, R. G. (1961). *Br. J. exp. Path.* **42**, 379.
Metcalf, D. (1964). *In* "The Thymus" (V. Defendi and D. Metcalf eds.), p. 53. Wistar Institute Press, Philadelphia.
Miller, J. F. A. P. (1965). *Br. med. Bull.* **21**, 111.
Norman, A., Sasaki, M. S., Ottoman, R. E. and Fingerhut, A. G. (1965). *Science* **147**, 745.
Osmond, D. G. and Everett, N. B. (1964). *Blood* **23**, 1.
Porter, K. A. and Cooper, E. H. (1962). *Lancet* **2**, 317.
Potter, R. L. and Nygaard, O. F. (1963). *J. biol. Chem.* **238**, 2150.
Reif, A. E. and Allen, J. M. V. (1964). *J. exp. Med.* **120**, 413.
Taliaferro, W. H. and Taliaferro, L. G. (1951). *J. Immunol.* **66**, 181.

Thorbecke, G. J. and Cohen, M. W. (1964). *In* "The Thymus" (V. Defendi and D. Metcalf eds.), p. 41. Wistar Institute Press, Philadelphia.

Van Furth, R. (1964). "The Formation of Immunoglobulins by Human Tissues in vitro" (Thesis). University of Leiden.

Volkman, A. and Gowans, J. L. (1965a). *Br. J. exp. Path.* 46, 50.

Volkman, A. and Gowans, J. L. (1965b). *Br. J. exp. Path.* 46, 62.

Winkelstein, A. and Craddock, C. G. (submitted for publication). Comparative response of normal human thymus and lymph node cells to phytohemagglutinin in culture.

Yoffey, J. M. (1960). "Quantitative Cellular Hematology." Charles C. Thomas, Springfield, Illinois.

Thymus and Cellular Growth

KARL ERIK FICHTELIUS and JUHANI LINNA

Department of Histology, University of Uppsala, Uppsala, Sweden

I. The Existence of a Thymus Hormone

After Brown-Séquard had elaborated his concept of the endocrine function of the ductless glands it seemed quite reasonable to classify the thymus with the endocrine glands, and investigate its physiology experimentally by employment of the methods that had proved highly profitable in experimental studies on the other ductless glands, such as, removal, transplantation, extraction and so on. Of the different hormones suggested in mammals one promoting growth and another leading to sexual precocity were the most popular. But all these experiments repeated again and again yielded conflicting results (Tesseraux 1953), and gradually the view about the function of the gland changed to the concept that the thymus is a lymphoid organ, a view that is justified by its structure. But even if thymus today is looked upon as a lymphoid organ the thought that thymus produces some hormone has been proved to be right.

The great break-through on thymus was made in 1961 by Archer and Pierce, students of Good in Minneapolis and by Miller in London. Thymectomy of newborn animals makes the animals immunologically incompetent later in life. Their lymphoid organs will not develop and they die early in wasting disease (Miller 1961).

The results of earlier experiments including tracing of labelled transfused thymus lymphocytes, and the thymectomy experiments point in the direction that thymus lymphocytes seed in the other lymphoid organs very early in the development of an animal and increase in number there and give rise to the immunologically competent cells of the body. Embryological and comparative-anatomical investigations lend support to this view. Once this early seeding of thymus cells has taken place the effect of thymectomy is very small.

In 1961 Miller also showed that isologous or homologous thymus grafts could restore the immunological competence in neonatally thymectomized animals. He also showed that the lymphoid organs grew to normal size as a consequence of the thymus implantation and that all mitotic cells in the other lymphoid organs were of the host-animal type after a certain time. Even the curative thymus graft became repopulated by lymphoid cells of the host type. This indicates that the thymus graft was not curing the thymectomized animals only by offering them lymphoid cells, but perhaps mainly by influencing the cells of the host in such a way that they could behave normally when stimulated with antigens. That this really is the case was proved by the fact that thymus grafts within diffusion chambers could restore immunological capacity in neonatally thymectomized animals (Osoba and Miller 1963, Levey *et al.* 1963). The diffusion chambers were made of semipermeable millipore membranes, which do not permit the passage of cells but only of smaller particles. The effect or the effects of these closed grafts could only be mediated via some humoral factor, which was produced by the thymus tissue.

Similar diffusion chamber experiments have been unsuccessful in some hands, but the humoral theory has got a strong support from the following experiment made by Osoba (1965). Neonatally thymectomized female mice were made pregnant. The fathers were chosen so that the cells of the fetuses were not tolerated by the mothers, and yet could the fetuses restore the immunological competence of the mothers. A humoral factor seems to be responsible for the curing of the mothers.

What this humoral factor is from a chemical point of view is not known. Recent experiments in this field have been as little successful as the old experiments on the illusive thymus hormone. Metcalf (1956) described a lymphocytosis stimulating factor in the thymus without defining it chemically. Szent-Györgyi (1965) has isolated a growth stimulating substance from thymus, but also from other organs and urine. He states, that this substance, which he calls promine, is a very small molecule bound *in vivo* to a hydrophilic colloid.

In looking for a thymus hormone and the mechanism behind it we may also turn to bursa of Fabricius in birds. The bursa of Fabricius is formed from the hind gut, as thymus is formed from the front gut, and in many respects the bursa is similar to the thymus, even if they give rise to two different cell lines within the lymphoid system with different functions (Cooper *et al.* 1965).

Utilizing the technique of disc electrophoresis, evidence has been presented by Woods (1965) which suggests that a basic protein with the mobility of a histone exists in higher concentrations in the spleens of bursectomized chickens as compared to unbursectomized controls. In two separate experiments the effects of multiple injections of chicken spleen basic proteins on the immune response of chickens have been studied. This basic protein solution contained the component which increased in concentration as a result of neonatal bursectomy. In the first experiment, injections of the basic protein solution resulted in an impairment of the ability of the animals to give a primary response to bovine serum albumin. In the second experiment, an increase in the survival time of homografts was associated with multiple injections of this basic protein solution. In both experiments the growth rates of the treated animals were slightly retarded.

Several investigators have studied the effect of basic proteins on metabolic processes and presented evidence which indicates that certain basic proteins are potent inhibitors of nuclear biosynthetic activity. It has thus been demonstrated that histones can inhibit the DNA-dependent synthesis of RNA *in vitro* (Huang and Bonner 1962) as well as the synthesis of DNA (Billen and Hnilica 1963). The observation by Woods in bursectomized chickens is at least something to grasp in this marsh of thymus and bursa hormones.

II. Mitotic Activity in the Thymus

The mitotic activity of the thymus cells can be assessed by direct mitotic counts or by measuring the turnover of DNA (Kindred 1942, Andreasen and Ottesen 1945 and many others). Both methods show that the thymus lymphocytes belong to the most rapidly dividing cells of the body. In young mice the mitotic index is about 10%, and at puberty there is a decrease to about 5% (Metcalf 1964). This decrease seems to be the basis for the age involution. At all ages the thymus lymphocytes are dividing 5 to 10 times more rapidly than lymphocytes elsewhere. The whole lymphocyte population in a mouse thymus is replaced in 3 days. The corresponding figure for children is 4 to 6 days (Metcalf 1964).

The mechanism behind the regulation of mitotic activity in the thymus is of course of great importance and may be of special interest to this audience. Some interesting experiments have recently been made by Metcalf and Wakonig-Vaartaja (1964). They have studied thymus grafts in inbred mice of different ages. The thymus graft has the same histological picture as has the normal thymus, and in its cortex the lymphocytes have the same high mitotic activity. With the aid of marker chromosomes it has been possible to show that the epithelial reticulum cells survive in such a graft, but the lymphocytes die, and 3 weeks after transplantation all lymphocytes are of the host type. The same is valid if the host was thymectomized before grafting. It thus seems as if lymphocytes of non-thymic origin can settle in thymus and acquire the high mitotic activity in this environment. Thus this high mitotic activity is not built in into the thymus lymphocytes, but is depending on the special environment.

The analysis of the growth of the thymus grafts is very elucidating (Metcalf *et al.* 1961). The younger the donated thymus was, the larger became the graft, independent of the age of the host. Thymus from newborn mice grew fast in old mice, whose own thymus was decreasing in weight during the same period. When the thymus graft reached the age at which involution normally should start, it underwent involution. It thus seems as if age involution were not due to environmental changes in the ageing host or to the age of the lymphocytes, but to time-related changes in the non-lymphoid elements of the thymus, the epithelial reticulum cells.

When more than one thymus graft was put into an animal, every graft grew independently of the other grafts and independently of the original thymus of the host (Metcalf 1963). This means that at least under these circumstances the animal cannot regulate the total amount of thymus tissue. This situation differs from that existing with multiple spleen grafts, in which the body is able to regulate effectively the total mass of spleen graft tissue (Metcalf 1964).

III. The Fate of Thymus Lymphocytes

Even if it seems to be proved that thymus produces a humoral factor, which can pass a cell-impermeable millipore filter and the chorionic membrane, and probably has something to do with the epithelial reticulum cells, the problem of the fate of all the lymphocytes formed in thymus remains unsolved. What is this intense lymphocyte production good for? The first questions are: How many of the lymphocytes formed within thymus die there, and how many leave thymus? These questions are still open. Some authors state that hardly any lymphocytes leave the thymus, others think that the most important function of thymus is to produce lymphocytes which in a steady flow leave thymus and do some good in the other lymphoid organs or other parts of the body as immunocompetent cells, or as trephocytes.

The last word has not been spoken on this subject, and it is difficult to state which opinion has most advocates. It may be mentioned that Ernström *et al.* (1965) have made a very direct approach, namely by counting the cells in arterial and venous thymus blood they found more lymphocytes in the venous blood than in the arterial and the lymphocytes of the venous blood contained fewer mitochondria than the lymphocytes in the arterial blood. The importance of this difference has to be elucidated by measuring the blood flow.

It could be shown that radioactively labelled thymus lymphocytes, which were injected from one animal to another, collected in the liver and the spleen and lymph nodes (Fichtelius 1953, Diderholm and Fichtelius 1959). These transfusion experiments have been followed up with other experiments in which thymus lymphocytes were labelled *in vitro* and injected back to the same animal in which they were formed (Diderholm 1961).

These transfusion experiments have certain drawbacks. The use of H^3 thymidine has made more physiological labelling experiments possible. The uptake of H^3 thymidine seems to start immediately it comes in contact with DNA-synthesizing cells. It can therefore be assumed that a certain degree of local labelling would follow local injection of H^3 thymidine.

Nossal and Gorrie (1964) reported on a method of local labelling with H^3 thymidine in the thymus. Using autoradiography they found that lymphocytes emigrate from the thymus to the mesenteric nodes in young guinea pigs. Similar results were obtained by Murray and Woods (1964), but they also found evidence for a migration of thymus lymphocytes to the spleen. Both these experiments were made with autoradiography and only heavily labelled cells were considered as migrants. Linna and Stillström (1965) have worked out a method to trace thymus DNA which is not disturbed by inevitable leakage of label into general circulation. They simply compare the distribution of label between different organs after intrathymic and intraperitoneal labelling and can conclude that a quantitatively significant transport of thymus DNA takes place to the spleen in young guinea pigs.

IV. Lymphocytes as Trephocytes

There is an old theory that lymphocytes are of importance to cell growth, that the lymphocytes are trephocytes (Carrel 1924). This theory combined with the fact that lymphocyte production in thymus is intense makes another theory: That thymus functions by exporting lymphocytes or breakdown products of lymphocytes, which are used in the formation of new cells.

Is there any real evidence that the lymphocytes as such have anything to do with the growth and division of other cells? The literature available on the subject up to 1958 is reviewed by Kelsall and Crabb (1958). That review shows that the trephocyte function for lymphocytes has not yet been fully substantiated. It may be added that the presence of healthy lymphocytes within the cytoplasm of other cells, mostly epithelial, has been recorded so often that the phenomenon must be accepted and its significance considered. Trowell (1958) has reviewed the literature in this field to 1958. In 1949 Trowell reported that lymphocytes enter fibroblasts in mixed cultures of rat fibroblasts and thymus cells. Darcy (1952) studied thyroid and submandibular gland homografts in rabbits and noted that when lymphocytes invaded the grafts some of them penetrated the epithelial cells of the graft. Humble *et al.* (1956) cultured mixtures of human malignant cells and lymphocytes and demonstrated that many lymphocytes entered the malignant cells.

V. Re-utilization of DNA from Dying Cells

Is there any evidence that something from a broken down lymphocyte can be used at the formation of new cells?

Fichtelius and Diderholm (1961) made the following experiment. A fraction of the lymphocyte population in mice was labelled with four daily injections of H^3 thymidine. Three days later, when free thymidine was non-detectable, a wound was made on the mice, and after 5 additional days, when there was a marked accumulation of lymphocytes in the wound, the animals were sacrificed and the wound and a symmetrically located area of the animals' skin was examined autoradiographically. It was shown that there were more labelled "round cells" in the wound than in control skin, and that the additional label of the "round cells" was brought to the wound from outside—probably via nucleated blood cells. In addition, it was shown that the labelling of fibroblasts and epidermal cells was higher in the wound than in control skin. The high labelling of epidermal cells in the wound was at that time interpreted as a re-utilization on some level of blood leukocyte DNA. Being lymphocyte-men we were of course thinking of the lymphocyte as the DNA-vector involved in this experiment.

Bryant (1963), at that time working in our laboratory, made the following experiment: Lymphoid cells from mice injected 54 hours and 30 hours earlier with H^3 thymidine were washed and transfused into isogenic recipients at 29 to 30 hours after partial hepatectomy. The recipients were killed 28 to 30 hours later, and liver, intestine, a Peyer's patch, spleen, and the transfused cells were examined in autoradiograms exposed for 6 months. The labelled DNA contained in the transfused cells appeared in 72 to 78% of hepatocyte nuclei, in 30 to 35% of liver reticulo-endothelial cell nuclei, and in 90 to 95% of intestinal crypt cell nuclei. The results suggest that re-utilized DNA is derived from lymphoid cells.

Everything looked fine for the lymphocyte men—the trephocyte function of the lymphocytes had been demonstrated. Independently Rieke (1962) demonstrated that lymphocyte DNA was re-utilized *in vivo* by sarcoma cells growing in the peritoneal cavity, but he also demonstrated that what happened to lymphocyte DNA also happened to sarcoma DNA. And later Fichtelius and Groth (1963) demonstrated that DNA thymidine from skin, intestinal wall, ventricular wall, liver and kidney could be re-utilized by macrophages. And Bryant (1964) demonstrated *in vivo* re-utilization of the DNA thymidine of necrotized liver cells in testis and intestine.

There we are: Broken down DNA of most dying cells can be re-utilized at the DNA synthesis of most other cells. It seems as if the transfer of DNA thymidine were mediated largely at the acid-soluble level, but a limited transfer of polynucleotides cannot at present be excluded. Hard and Kullgren (1966) in Minneapolis have recently presented evidence for a transfer of DNA at the chromosomal level. C3H mice were given 900 r total irradiation. One billion spleen cells from (C3H × T6) F1 donors were given within 24 hours. After one month CCl_4 and additional 200 millions of F1 cells were given. This combined treatment was repeated twice with a 2-week interval. Finally the animal were given one more dose of CCl_4, colchicinized and killed. Chromosome preparations were made by the trypsin, hypotonic, air-dry method.

On routine paraffin sections it can be seen that at least 90% of the cells in mitosis 54 to 60 hours after CCl_4 treatment are hepatic parenchymal cells. On the chromosome preparations the majority of the mitotic figures have the T6-marker chromosome present. No marker was found in controls: normal C3H irradiated given (A × C3H) F1 cells. The only way we can explain these results just now is that nuclear material from trephocytic cells has been transferred to rapidly regenerating cells.

VI. Export of DNA from Thymus

At a glance it would seem that thymus lymphocytes have no particular position as deliverers of precursors at the DNA synthesis of other cells. But nevertheless it may be something extraordinary with thymus in this respect. The only way to find out is to study the export of DNA from thymus from a quantitative and qualitative point of view.

A. *Export to lymphoid organs*

Since we know, that thymus DNA is probably exported to the so called peripheral lymphoid organs, the spleen and the lymph nodes, it is natural to start this discussion with the following question: Can DNA or DNA-fragments of thymus lymphocytes enhance antibody formation? This problem has been studied by Berglund *et al.* (1963) and Berglund (1965). They found that a homogenate of homologous thymus cells (or spleen cells) increased the haemolysin titres in rats immunized intraperitoneally with sheep erythrocytes if the lymphoid cells were given a few hours before or simultaneously with the antigen, but had no effect when injected 2 or 6 hours later. Differential centrifugation of disintegrated thymus cells showed that most of the enhancing activity seemed to be present in the nuclei. Incubation of thymus lymphocytes with DNA-ase deprived them of the stimulating effect.

The possibility that the enhancing factor is DNA has been investigated by Berglund *et al.* (1963) and Berglund (1965) in two ways: (1) A purified DNA from rat spleens was injected together with the antigen without effect. (2) Thymocyte nuclei labelled with H^3 thymidine were injected with sheep erythrocytes intraperitoneally into homologous mice. Hemolysin producing cells in the mediastinal node on day 5 was identified with the Jerne technique. No labelled nuclei were found in the hemolysin producing cells.

Merrit and Johnsson (1964) have reported enhancement of the primary antibody response in mice by extracts of an isologous liver and spleen cell mixture. This effect was produced not only by the DNA-rich fraction, but also by the RNA-rich. The enhancing effect was eliminated by treatment of the extracts with DNA-ase or RNA-ase respectively.

The question of the enhancing effect of thymus DNA on antibody formation needs to be further studied with special attention to the fact that the thymus dependent lymphoid cells are not the antibody forming plasma cells but the pyroninophilic cells engaged in delayed type of hypersensitivity, homograft reaction and graft versus host reactions (Cooper *et al.* 1965).

B. *Export to other sites*

The hypothesis that thymus functions by exporting lymphocytes or break-down products of lymphocytes, which are used in the formation of new cells at other sites than peripheral lymphoid organs was tested by Fichtelius and Olerud (1960).

Young guinea pigs were administered with ^{32}P. Half of the animals were subjected to subtotal thymectomy immediately before the injection of the isotope and the other half sham-operated. Twenty days after the operation two incisions were made on the backs of the animals. The animals were killed 5 days later when there was a marked accumulation of lymphocytes in the wound with surrounding skin. Equally large pieces of symmetrical skin were excised from the animals and the specific activity (spec. act.) of DNAP was determined. The relative activity (rel. act. = the number of impulses per mg DNAP from the skin or wound per cent of the number of "impulses" injected per g body weight) was also determined.

The relative activity of DNAP of the normal skin in thymectomized animals was lower than in sham-operated controls. This fact can be interpreted in the following way: Thymus lymphocytes are normally transported to the skin. By thymectomy the strongly labelled cells which should have been formed during the first days of the experiment are withheld from the skin. In sham-operated controls these heavily labelled cells are formed and within 25 days transported to the skin thus giving a higher relative activity of skin-DNAP.

The amount of DNA in the wound was about 2·5 times more than the DNA in an equally large piece of non-wounded skin. It was shown that the specific activity of additional DNAP in the wound was influenced by thymectomy in the same way as the specific activity of DNAP in the normal skin. This fact speaks in favour of the participation of thymus lymphocytes in wound healing. The highly labelled cells which should have been formed in the thymus during the first days of the experiment were probably withheld from the wound by thymectomy. In sham-operated controls these heavily labelled thymus-cells were probably formed and transported to the wound.

The incorporation of ^{32}P into DNA of different organs may be lowered by thymectomy (Gyllensten and Ringertz 1954). Such a mechanism could explain the results of the referred experiment in which the relative activity was lower in thymectomized animals, both in the skin and in the wound. Therefore the following experiment was done.

Young guinea pigs were thymectomized or sham-operated. 20 days after the operation all animals were labelled with ^{32}P and at the same time, 20 days after the operation, two incisions were made on the backs of the animals. The animals were killed after further 5 days and the relative activity of the DNAP of the wound and symmetric normal skin was determined.

In this case the relative activity of the wound was higher in thymectomized than in sham-operated animals. There is probably a dilution of the labelled DNA of the wound with "old" unlabelled lymphocytes or other nucleated cells from the blood. This dilution of labelled DNA of the wound is decreased in thymectomized animals. The results thus speak in favour of the hypothesis, that "old" unlabelled thymus lymphocytes have accumulated in the wound of sham-operated animals.

A lowered incorporation of ^{32}P into DNA in different organs by thymectomy could explain the results of the first experiment, in which the relative activity of DNAP was lower in thymectomized animals, both in the skin and in the wound. But a decrease of the incorporation of ^{32}P into DNA of the wound as a consequence of thymectomy cannot, of course, explain the results of the second experiment, in which the relative activity was higher in the wound of thymectomized animals.

Even if it is shown, however, that a reduction of the incorporation of ^{32}P cannot explain the results, the conclusion that thymus lymphocytes are transported to the wound is not safe. The results obtained can also be explained in another way: It may be that as a consequence of thymectomy, the lymphocytes formed in the lymph nodes are drawn off to other organs and reach the wound to a lesser extent than in sham-operated controls. Both interpretations involve, however, that fewer cells are mobilized and transported to the wound in thymectomized animals.

These thymectomy studies are supported by transfusion experiments. In several experiments ^{32}P-labelled lymphocytes were transfused from one animal to another and then traced within the new host (Fichtelius *et al.* 1960). In many of these experiments the epidermis or the skin of the recipient animals was also examined and it could be shown that transfused thymus lymphocytes or DNA of thymus lymphocytes accumulated in the skin, particularly epidermis.

A more direct evidence for a transport of DNA or break-down products of DNA from the thymus to a healing wound has recently been presented by Linna (1965), using intrathymic labelling with ^{3}H thymidine and tracing of the label with a quantitative chemical technique.

It has been shown by Landy *et al.* (1965) that the administration of *salmonella enteritidis* endotoxin to rabbits dramatically decreases the lymphocyte content of the thymus. As there is no histological evidence of thymic cell destruction (Landy *et al.* 1965) this was looked upon as a fine model to study the possible cell migration from thymus.

Young New Zealand albino rabbits were given ^{3}H thymidine in the thymus by subcapsular infiltration. Controls were sham-operated and given the same amount of ^{3}H thymidine intravenously. All the animals were given 5 gamma *salmonella enteritidis* endotoxin (Lab. of Immunology, NIAID, NIH, Bethesda, Md., USA) intravenously. After 48 hours, the animals were killed. DNA from different parts of the organism, among them the operation wound, was isolated and measured, and the radioactivity of DNA was measured in a liquid scintillation counter.

The distribution of label in different organs was studied in both types of labelling, and from a comparison it could be concluded that label had been brought from the thymus to the wound, probably with intact lymphocytes. These experiments are now repeated without endotoxin.

We think that the first part of our hypothesis very probably is right, i.e. that thymus DNA or break-down products of thymus DNA can be exported to areas with increased cell production. The second part of the hypothesis, that the material brought from the thymus is used at the formation of new cells, is still uncertain. Taking into account the result from the autoradiographic analysis of the wound and normal epidermis after ^{3}H thymidine labelling, which showed that epidermal cells in the wound can re-utilize the label after the first wave of labelling brought there, it seems very reasonable to assume that the broken down DNA from thymus lymphocytes can be used at the synthesis of new DNA in epidermal cells.

REFERENCES

Andreasen, E. and Ottesen, J. (1945). *Acta physiol. scand.* **10**, 258.

Archer, O. and Pierce, J. C. (1961). *Fed. Proc.* **20**, 26.

Berglund, K. (1965). *In* "Molecular and Cellular Basis of Antibody Formation" (J. Šterzl et al., eds.), Proceedings of a Symposium held in Prague 1964, p. 405. Publishing House of the Czechoslovak Academy of Sciences, Prague.

Berglund, K., Borum, K., Fagraeus, A. and Sköld, O. (1964). Transactions of the XIV Scandinavian Congress on Pathology and Microbiology, p. 271. Oslo.

Billen, D., Hnilica, L. S. (1963). *J. cell Biol.* **19**, 7A.

Bryant, B. J. (1963). *J. cell Biol.* **18**, 515.
Bryant, B. J. (1964). *Expl Cell Res.* **32**, 209.
Carrel, A. J. (1924). *J. Am. med. Ass.* **82**, 255.
Cooper, M. D., Peterson, R. D. A., and Good, R. A. (1965). *Nature* **205**, 143.
Darcy, D. A. (1952). *Phil. Trans. R. Soc. B.* **236**, 463.
Diderholm, H. and Fichtelius, K. E. (1959). *Acta paediat.* 48, suppl. **117**, 13.
Diderholm, H. (1961). *Acta path. microbiol. scand.* suppl. 146, vol. 51.
Ernström, U., Gyllensten, L. and Larsson, B. (1965). *Nature* **207**, 540.
Fichtelius, K. E. (1953). *Acta Anat.* Suppl. 19—1 ad Vol. 19.
Fichtelius, K. E., Danielsson, D. and Hallander, H. (1960). *Acta Dermatovener.* **40**, 1.
Fichtelius, K. E. and Diderholm, H. (1961). *Acta path. microbiol. scand.* **52**, 11.
Fichtelius, K. E. and Groth, O. (1963). *Nature* **200**, 587.
Fichtelius, K. E. and Olerud, S. (1960). *Acta path. microbiol. scand.* **50**, 297.
Gyllensten, L. and Ringertz, N. R. (1954). *Acta path. microbiol. scand.* **35**, 309.
Hard, R. and Kullgren, B. (1966). *Science* **152**, 349.
Huang, R. C. and Bonner, J. (1962). *Proc. nat. Acad. Sci. U.S.A.* **48**, 1216.
Humble, J. G., Jayne, W. H. W. and Pulvertaft, R. V. J. (1956). *Br. J. Haemat.* **2**, 283.
Kelsall, M. A. and Crabb, E. D. (1958). *Ann. N.Y. Acad. Sci.* **72**, 393.
Kindred, J. E. (1942). *Am. J. Anat.* **71**, 207.
Landy, M., Sanderson, R. P., Bernstein, M. T. and Lerner, II, E. M. (1965). *Science* **147**, 1591.
Levey, R. H., Trainin, N., Law, L. W., Black, P. H. and Rowe, W. P. (1963). *Science* **142**, 483.
Linna, J. (1965). To be published.
Linna, J. and Stillström, J. (1967). *Acta path. microbiol. scand.* (in press).
Merrit, K. and Johnsson, A. G. (1964). *Fed. Proc.* **23**, 139.
Metcalf, D. (1956). *Br. J. Cancer* **10**, 442.
Metcalf, D. (1963). *Aust. J. exp. Biol. med. Sci.* **41**, 437.
Metcalf, D. (1964). *Transplantation* **2**, 387.
Metcalf, D. (1964). *In* "The Thymus in Immunobiology" (R. A. Good and A. E. Gabrielsen eds.), p. 150. Hoeber Medical Division, New York.
Metcalf, D., Sparrow, N., Nakamura, K. and Ishidate, M. (1961). *Aust. J. exp. Biol. med. Sci.* **39**, 441.
Metcalf, D. and Wakonig-Vaartaja, R. (1964). *Proc. Soc. exp. Biol. Med.* **115**, 731.
Miller, J. F. A. P. (1961). *Lancet* **11**, 748.
Murray, R. G. and Woods, P. A. (1964). *Anat. Rec.* **150**, 113.
Nossal, G. J. V., and Gorrie, J. (1964). *In* "The Thymus in Immunobiology" (R. A. Good and A. E. Gabrielsen, eds.), p. 288. Hoeber Medical Division, New York.
Osoba, D. (1965). *Science* **147**, 298.
Osoba, D. and Miller, J. F. A. P. (1963). *Nature* **199**, 653.
Rieke, W. O. (1962). *J. cell Biol.* **13**, 205.
Szent-Györgyi, A. (1965). *Science* **149**, 34.
Tesseraux, H. (1953). "Physiologie und Pathologie des Thymus." J. A. Barth, Leipzig.
Trowell, O. A. (1949). *J. Physiol. Lond.* **110**, 5P.
Trowell, O. A. (1958). *Int. Rev. Cytol.* **7**, 235.
Woods, R. (1967). To be published.

Control of the Regeneration of Haematopoietic Tissues in Irradiated Mice

J. F. DUPLAN

Laboratoire Pasteur de l'Institut du Radium. Paris, France

I. Introduction

THE purpose of this paper is not to study the spontaneous regeneration or haematopoietic cells in an irradiated organism but the multiplication and differentiation of normal haematopoietic cells grafted in an irradiated host. For those who are not familiar with the radiobiological aspects of this problem I shall briefly summarize the outstanding points of this technique.

There is, at present a general agreement that a dose of X-radiation of 850 (LD 100) sterilizes more than 99% of the haematopoietic cells in the mice and in other mammals (McCulloch and Till 1962, Puck 1964). This aplasia can be controlled by injecting isologous haematopoietic cells which repopulate the haematopoietic tissues of the irradiated recipient. The probability of rejection of the isologous graft is very small, if the dose of radiation has been high enough. It has been demonstrated that the active type of cell is the stem cell and that each of them can eventually produce a clone of differentiated cells (McCulloch and Till 1964) belonging mainly to one series: erythrocytic, granulocytic, megakaryocytic or lymphocytic (Fig. 1). Moreover with few exceptions the myeloid cells are found only in the myeloid organs and the lymphoid cells in the lymphoid ones. This obviously means that if there is more than one type of stem cells (lymphoid or myeloid) each is selected and (or) induced to differentiate according to its micro-environment and that if there is only one type of stem cell the micro-environment allows the induction of only one type of differentiation either myeloid or lymphoid. Finally it must be noticed that as a consequence of the haematopoietic aplasia the new pool of stem cells and their progeny are subjected to a strong and long lasting stimulus in which are likely to participate many if not all of the factors that control homeostasis at all levels.

The present paper is an attempt to study some of the factors that regulate the regeneration of lymphatic and granulocytic cells in irradiated mice restored by foetal liver cells, bone marrow cells or spleen cells.

In all the experiments that we shall describe the mice are male and female of the inbred strain XVII and F1 hybrids between XVII and C57BL; both strains are maintained in our Institute. Irradiations are performed in lucite boxes with a 225 kV X-ray machine (11 mA, 0·3 mm Cu and 2 mm Al filters, 90 r.p.m.).

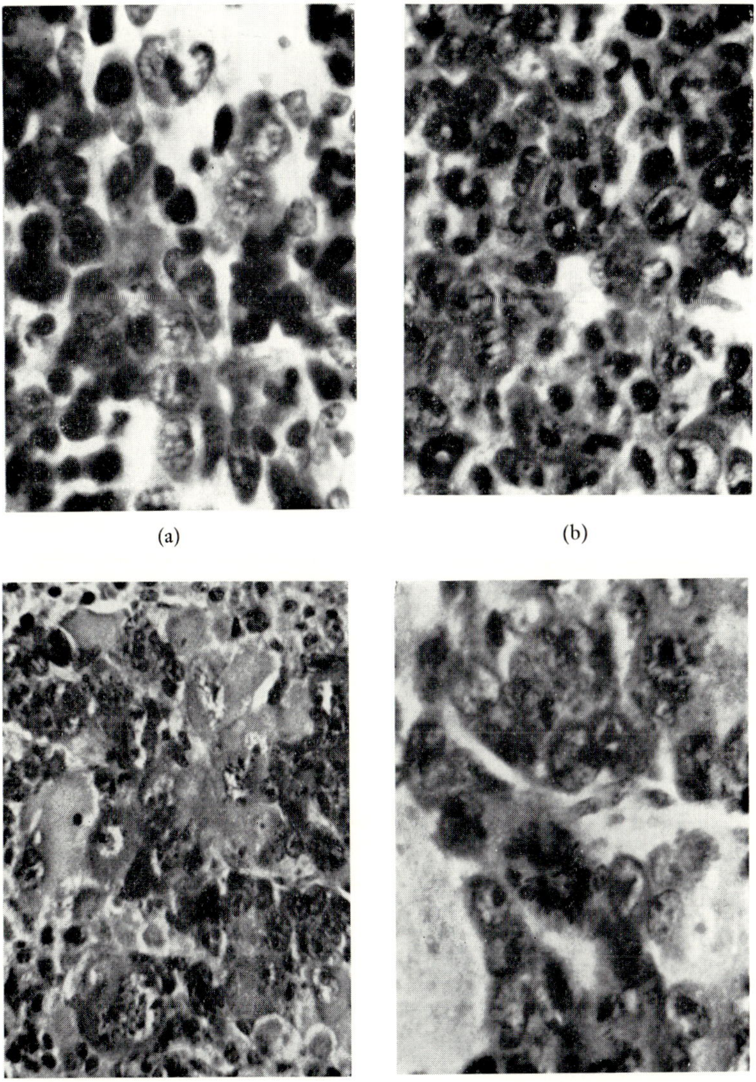

FIG. 1. Different kinds of clones in a spleen: (a) Erythrocytic cells; (b) Granulocytic cells; (c) Megakaryocytic cells; (d) Mitosis in a tetraploid megakaryocytic cells.

Nucleated haematopoietic cells are counted in a Burker haemocytometer after dilution in 0·6% acetic acid.

The number of living cells estimated by the Schrek method is 80% for the suspensions prepared with bone marrow cells, 50% for foetal liver cells from 14 to 16 day-old embryos and 65% for the spleen cells. All suspensions are prepared in Tyrode.

Blood smears are stained by May Grünwald-Giemsa and histological sections by haematoxylin and eosin.

II. Control of Lymphoid Regeneration

Thymectomy performed in newborn mice results in lasting lymphoid aplasia and lymphopenia and produces a wasting disease in some animals (Miller *et al.* 1962). When thymectomy is performed in young adult mice which later on are irradiated with a lethal dose of X-radiation and then treated by an injection of isologous bone marrow (Miller *et al.* 1963) or of foetal liver cells (Duplan 1963), the lymphoid regeneration is impaired and some cases of runt disease can develop.

Two possible ways are open to study this lymphoid regeneration: the 1st uses the haematological and histological techniques, and evaluates the recovery of the lymphatic system according to morphological criteria; the other tries to measure the immunological reactivity of the irradiated mice and to bring some light on the functional differentiation of the lymphoid cells. Among the different possibilities that immunology offers, we have chosen the graft versus host (G.V.H.) reaction as described by Simonsen (1962).

A. *Histological and haematological studies*

The changes found in adult thymectomized irradiated mice are usually of much less consequence than the ones which develop in neonatally thymectomized animals. The reason for this discrepancy is that: (a) when the dose of X-rays is sublethal the surviving lymphoid cells repopulate at least partly the lymphoid organs; (b) when a lethal dose has been delivered and followed by an injection of bone marrow or splenic cells, the lymphatic cells contained in the inoculum can restore to some degree the lymphoid organs. But when restoration is obtained by the injection of foetal liver cells where no differentiated lymphoid cells are present the lymphatic aplasia takes a very typical aspect that can be easily studied in the spleen.

1. Changes in histological structure of the spleen

During 30 to 60 days after irradiation only weak differences can be found between thymectomized irradiated mice and non-thymectomized irradiated controls as far as the structure of the splenic white pulp is concerned. Large germinative centres are found in the animals of both groups. But, while in the non-thymectomized mice the number of small lymphocytes increases

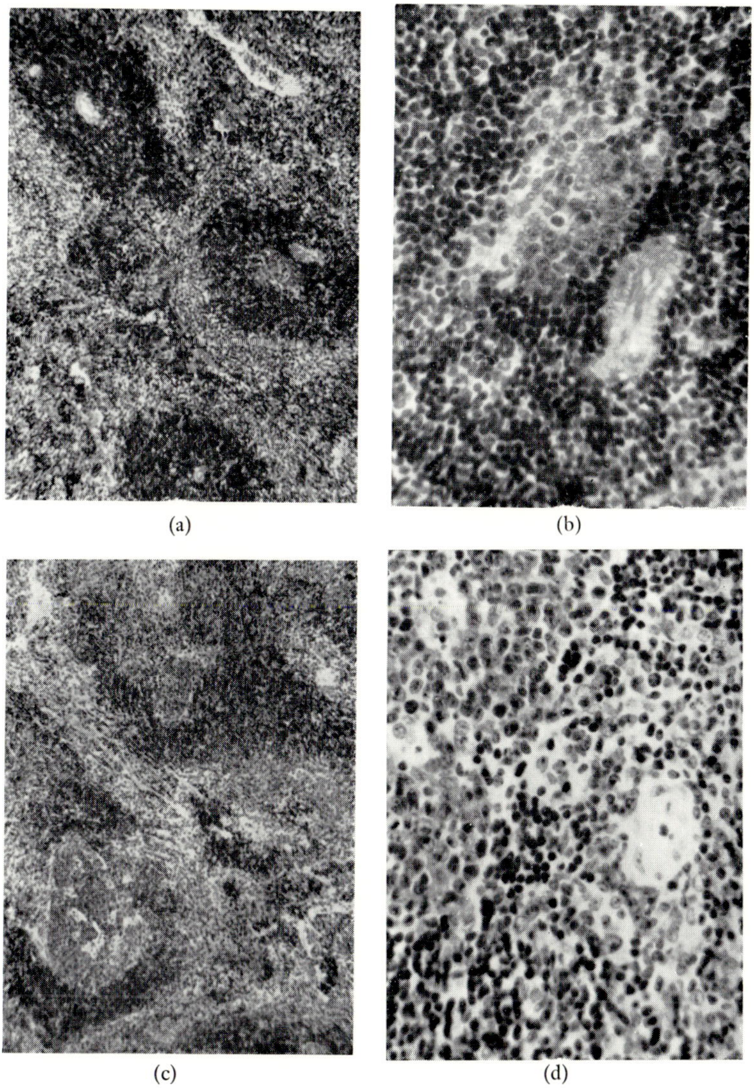

FIG. 2. Malpighian bodies (M.B.) in spleens of control (a,c) or thymectomized (b,d) irradiated mice 90 days after irradiation. (a) Large M.B. with small germinative centre; (b) small germinative centre and densely packed small lymphocytes around the artery; (c) large germinative centres increasing the size of the M.B. but very few lymphocytes; (d) scattered small lymphocytes almost completely absent in the periarteriolar area.

rapidly at the periphery of the germinative centre, no such recovery appears in the thymectomized animal. Ninety days after irradiation characteristic differences between these two groups become evident. Malpighian bodies of the controls display a typical normal structure while those of thymectomized mice have a large germinative centre with few mitotic figures; around it a thin ring of small lymphocytes exists, but the most distinctive feature is that no lymphatic cells are found in the peri-arteriolar zone (Fig. 2).

The same aspect has been described by Defendi *et al.* (1964) in rats and by Waksman *et al.* (1962) in rabbits thymectomized at birth. This is certainly linked to a lack in lymphocytes, but if these cells were randomly distributed around the germinative centre there would be no reason for such a systematic depletion of the peri-arteriolar area. The significance of this finding is not clear but it could well be related to a disturbance in cell differentiation and (or) function.

2. *Studies on circulating lymphocytes*

The reappearance of lymphocytes in the peripheral blood has been studied in two groups of mice. The first one was thymectomized, irradiated and injected with foetal liver cells from 14 to 16 day-old embryos, the second received the same dose of X-radiation (850 r) and the same inoculum of foetal cells ($1 \cdot 8 \times 10^7$) but had been sham-operated. Each point in the curves of Fig. 3, represents the mean value for 10 mice. The leucocyte level returns to normal in both groups after 35 to 55 days, while the lymphocytes reach the control range 55 days after irradiation only in the sham-operated group. Thymectomized mice never return to the control value. It follows that thymectomized irradiated mice have an absolute increase in the number of polymorphonuclear leucocytes.

Differential counts were performed to evaluate the number of small and large lymphocytes in these two groups.

At 55 days when the leucocyte count has returned to normal, it was found that: (a) it was mostly large lymphocytes which were lacking in the thymectomized mice ($3 \cdot 6 \times 10^3$ instead of 10^4 in the control group). (b) The number of small lymphocytes was normal if not increased in the thymectomized group. (c) Small lymphocytes were almost completely absent in the thymectomized mice displaying symptoms of runt disease (these were not included in Fig. 3).

These results are somewhat at variance with those of other authors who describe a strong decrease in small lymphocytes (Miller *et al.* 1963). This, in our experiments, is true only for the first 4 weeks following irradiation.

From these experiments it appears that the depletion of germinative lymphatic centres related to thymectomy performed before irradiation, is reflected in the blood. In the absence of the thymus a chronic state of lymphopenia develops, but the leucocytosis is maintained at its normal level by an increase in polymorphonuclear cells. The homeostatic mechanism that regulates leucocytosis does not seem to depend on the presence of the thymus and it seems more concerned with the total number of circulating white

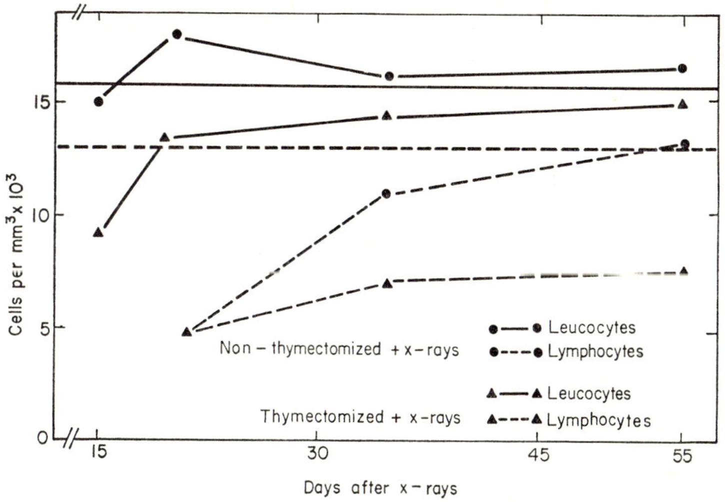

FIG. 3. Absolute number of leucocytes (—) and lymphocytes (- - -) in irradiated thymectomized (▲) or normal (●) mice restored by foetal liver cells. Straight lines represent the mean values of control mice.

blood cells. It would be interesting to consider the effect of this new equilibrium leading to chronic granulocytosis on the function of the bone marrow; it is not unlikely that under such circumstances granulopoiesis could be strongly stimulated. One must remember that in mice thymectomy can in some cases favour the development of myeloid leukemias (Gross 1960).

B. *Recovery of immunological reactivity*

The recovery of immunological reactivity was studied in the two groups of mice previously described and in four other groups that had been treated with splenic cells or bone marrow cells after irradiation. In each case one group of mice was thymectomized and the other was not. The mice were sacrificed 35, 45, 60, 80 and 110 days after irradiation; a suspension of splenic cells was prepared, and $1{\cdot}5 \times 10^6$ nucleated cells were injected i.p. into 3 (XVII ×C57BL)F1 newborns of each litter. Only 6 newborns (3 injected, 3 controls) were kept in each litter to test the G.V.H. reaction. These mice were killed 11 days after injection and their spleens were weighed.

The ratio of the weight of the spleens belonging to the injected hybrids to the weight of the spleens of the non-injected ones, measures the immunological reactivity of the cells contained in the suspension previously injected (Simonsen 1962).

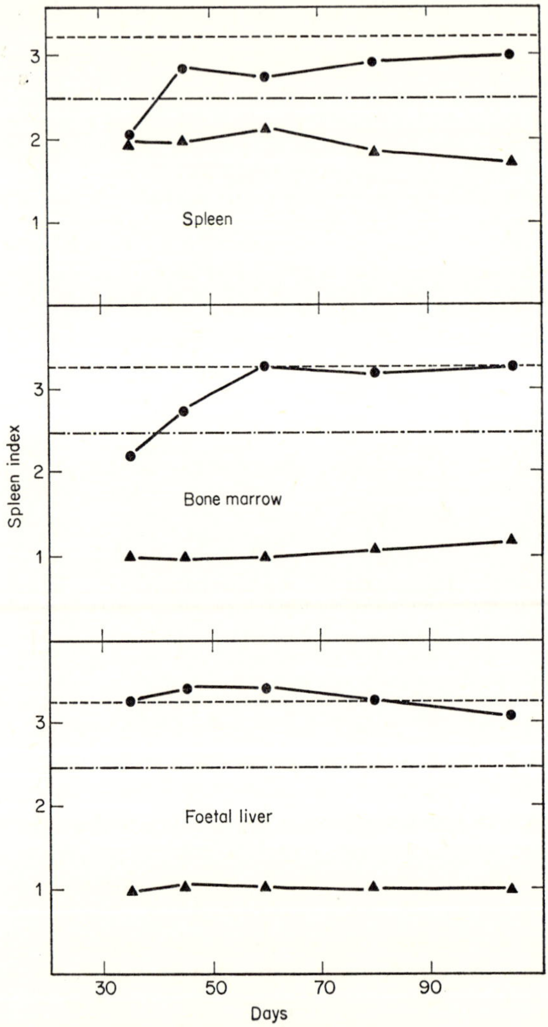

FIG. 4. G.V.H. reaction (spleen index) in thymectomized (▲ — ▲) or normal (● -- ●) irradiated mice restored by splenic cells, bone marrow cells or foetal liver cells. Straight lines: normal values for non-thymectomized (– – –) or thymectomized (– . – .) controls.

1. Thymectomized irradiated mice

The results are given in Fig. 4: this brings evidence that the thymectomized mice restored with suspensions of foetal liver cells or bone marrow cells, which contain none or very few immunologically competent cells (I.C.C.), do not recover their immunological reactivity (Zaalberg *et al.* 1963, Duplan 1965). On the contrary thymectomized mice injected with spleen cells recover a weak but stable immunological capacity. Thence it results that lymphatic cells located in the spleens of the donors are able to repopulate the recipients, and to express an immunological responsiveness even

TABLE I. Effect of the addition of thymic cells to the foetal liver cells on the G.V.H. reaction induced by spleen cells from thymectomized irradiated mice restored by foetal liver cells.

(Test performed 50 to 60 days after X-irradiation)

Treatment	Spleen index
foetal liver $1{\cdot}5$ to $3{\cdot}0 \times 10^7$	0·9 to 1·1
foetal liver $1{\cdot}5$ to $2{\cdot}0 \times 10^7$ + thymic cells $2{\cdot}0 \times 10^7$	1·2 to 1·7
foetal liver $1{\cdot}5 \times 10^7$ + thymic graft	1·2 to 1·6
foetal liver $1{\cdot}5 \times 10^7$ + irradiated (1000 r) thymic graft	1 to 1·7

when the thymus is absent. Consequently, it would mean that splenic lymphatic cells grown in a normal animal and transplanted into a thymectomized host, keep and perhaps transmit to their daughter cells the ability to respond to an antigenic stimulus, which they have previously acquired. This population of cells is likely to be less sensitive to thymic influence than to antigenic stimulation. Some cells from thymus and lymphnodes have the same characteristics. As can be seen in Tables I and II, the cells from the thymus or from the lymphnodes added to the foetal liver cell suspensions produce a weak but stable immunological responsiveness in the thymectomized irradiated hosts.

It must be noticed (Table I) that s.c. thymic grafts, whether they are irradiated prior to transplantation or not, are not more efficient than the suspension of thymic cells. The grafts belonging to normal thymus have a normal structure and certainly a normal function while the ones belonging to an irradiated thymus often show a nearly complete absence of cellular repopulation. As both kinds of graft seem able to induce the appearance of I.C.C. in their recipients, it seems that this weak restoration of the immune reactivity is not directly linked to the cellularity of the thymic grafts. This fact could be considered as an argument in favour of the existence of a lymphocytosis stimulating factor (Metcalf 1958, Levey *et al.* 1963) released by the thymus.

TABLE II. Effect of the addition of lymphnode cells to the foetal liver cells on the G.V.H. reaction induced by spleen cells from thymectomized irradiated mice restored by foetal liver cells.

(Test performed 50 to 60 days after X-irradiation).

Treatment	Spleen index
foetal liver $1{\cdot}5$ to $3{\cdot}0 \times 10^7$	0·9 to 1·1
foetal liver $1{\cdot}5$ to $2{\cdot}5 \times 10^7$ + lymph node 1×10^7	1·5 to 2·0

2. *Control irradiated mice*

But for our purpose it is also interesting to study what happens in the three different groups of mice that were not thymectomized. The recovery of the immunological reactivity is completed 35 days after irradiation in the group treated with foetal liver cells, while it takes 60 days to obtain the same result in the group restored with bone marrow. The animals injected with splenic cells do not come back to the control value during the 110 days of the experiment. As no I.C.C. are present in the foetal liver, a small number in the bone marrow and a large number in the spleen, the speed of recovery appears to be inversely related to the number of I.C.C. persent in the suspension injected after irradiation. Or perhaps the number of percursors to I.C.C. which would become active in the presence of the thymus is much greater in the foetal liver than in the spleen.

If the first statement is exact, the addition of lymphnode cells to the foetal liver cell suspension will increase the number of I.C.C. (Table II) and therefore the spleen of the mice injected with this mixture should elicit a similar response (G.V.H. reaction) as the spleens of mice restored with a suspension of splenic cells. The results summarized in Table III show that this is actually the fact. The addition of lymphnode cells to foetal liver cells decreases the rate of recovery of the immunological capacity of the irradiated hosts as compared with that of the irradiated recipients injected only with foetal liver cells.

What are the implications of these experiments as far as the regeneration of the lymphoid tissues is concerned? It appears from these results that foetal liver and adult bone marrow contain precursors to the I.C.C. and that these precursors are able to be transformed to I.C.C. under the influence

TABLE III. Effect of the addition of lymphnode cells to the foetal liver cells on the G.V.H. reaction induced by spleen cells from control irradiated mice restored by foetal liver cells. (Test performed 45 days after X-irradiation).

Treatment	Spleen index
foetal liver $1\cdot5$ to $3\cdot0 \times 10^7$	$3\cdot0$ to $3\cdot5$
foetal liver $1\cdot8 \times 10^7$ + lymph node $1\cdot1 \times 10^7$	$2\cdot0$ to 3
spleen cells $3\cdot0 \times 10^7$	$2\cdot4$ to $2\cdot9$

of the thymus. It is not possible to decide whether this organ acts through a humoral factor or whether there must be a contact between precursors and thymic tissue. Although the effects of the irradiated thymic graft favour to some extent the first possibility. Finally it is possible that both mechanisms are implicated in the control of lymphopoiesis. Anyway, when precursors have been transformed in I.C.C. they become sensitive to the antigenic stimulus and can express their immune capacity even in the absence of the thymus. Yet, in thymectomized irradiated mice, no precursors can enter the pool of I.C.C. as it occurs in the presence of the thymus and one of the problems is to know whether this cellular compartment is able or not to maintain itself without the support of precursors. If it were so, this would mean that some cells directly related to I.C.C. could behave like stem cells.

III. Control of Myeloid Regeneration

The discovery of the role played by the erythropoietin in the regulation of erythrocytosis or by the thymus in the development of lymphoid tissues has determined most of the recent progress in these fields. Unfortunately no such unequivocal finding has been made that could improve our knowledge of the regulation of myelopoiesis.

The following experiments are related to the study of myeloid regeneration in the spleen and in the femoral bone marrow of male mice X-irradiated with 850 r (LD 100) and restored by an i.v. injection of bone marrow cells. One group (I) of mice received an inoculum of $0{\cdot}4 \times 10^7$ and the other (II) of $1{\cdot}2 \times 10^7$ nucleated cells. In each group, 8 animals were sacrificed, 5, 6, 7,

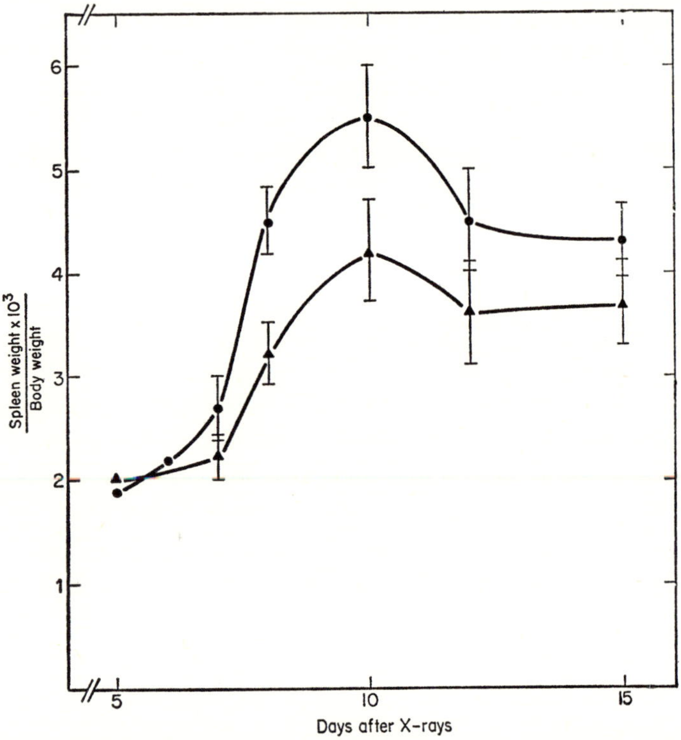

FIG. 5. Recovery of the spleen weight in irradiated mice restored by $0{\cdot}2 \times 10^7$ (● — ●) or $1{\cdot}2 \times 10^7$ (▲—▲) bone marrow cells.

8, 10, 12 and 15 days after irradiation. Blood was taken for haematological studies; the body and the spleen were weighed in order to establish the ratio spleen weight/body weight, and suspensions of femoral bone marrow in Tyrode were prepared to evaluate the cellularity of the femoral bone marrow (number of cell per femur). Histological section of bone marrow and spleen were studied after staining by haematoxylin and eosin.

A. *Regeneration of the spleen*

As we have stated at the beginning of this paper the injected cells repopulate the spleen in a very specific way. Some stem cells settle in the red pulp and give clones of erythrocytic, myelocytic or megakaryocytic cells (Fig. 1); only one kind of differentiated cell being usually present in each clone. According to Lewis and Trobaugh (1964) 42% of the clones are erythropoietic, 21% granulocytic and 21% megakaryocytic, only 16% contain mixed cellular types. From 8 to 10 days after irradiation, these clones produce small nodules beneath the surface of the spleen. When more than 8×10^4 bone marrow cells are injected the nodules become confluent and cannot be counted. At first sight, the size and weight of the spleen ought to be directly related to the number of injected cells, i.e. to the number of stem cells that give birth to clones (colony forming cells).

The curves of Fig. 5 represents the ratio of the spleen weight to the body weight as a function of the time elapsed after X-radiation. It can be seen that from the eighth to the twelfth days the weight ratios are significantly higher for the spleens belonging to the group I restored with $0{\cdot}4 \times 10^7$ bone marrow cells. Histological examinations show that the increase in weight is mainly related to an increase in cells of the granulocytic series. However, thereafter the experimental evidence is the reverse of what could have been theoretically forseen. One explanation could be that the space available in the spleen for the growth of each clone is reduced when the number of clones forming cells increases. This does not seem satisfactory because the nodules are confluent macroscopically and microscopically as soon as 8×10^4 cells are injected; thence the number of cells actually injected is 50 times over this limit in group I and 150 times in group II, i.e. group II has received only 3 times as many cells as group I. In such a situation it is unlikely that the rather small differences between group I and II could change to a large extent the very small space available for the growth of each clone. The study of the regeneration in the bone marrow will provide complementary informations.

B. *Regeneration of the bone marrow*

In Fig. 6 are plotted the mean numbers of nucleated cells counted in each femur 5, 6, 7, 8, 12 and 15 days after irradiation. From the fifth to the eighth days both curves follows a logarithmic growth. At 5 days the number of cells found in the femur of the mice of group II is about 3 to 4 times as much as the number found in the femur of the mice of Group I. This fits well with what could be theoretically expected. But it must be noticed that between 5 and 8 days the slope of the curve corresponding to group I is steeper than the one of group II. Finally, at 10 days, both groups have nearly the same cellularity. This would mean that cell divisions are more stimulated in group I that had received the smaller number of cells; indeed from a general point of view one may say that the smaller the number of cells injected the stronger the stimulus in the bone marrow.

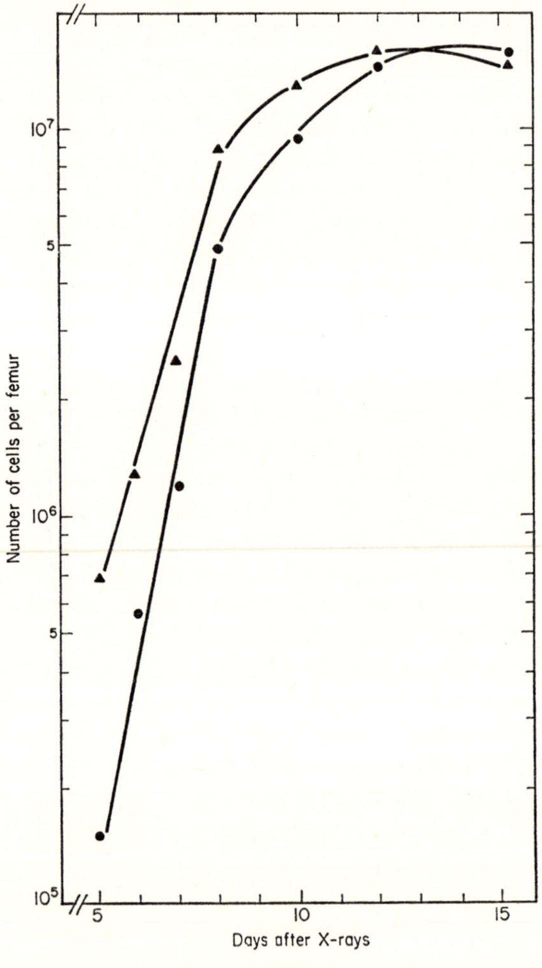

FIG. 6. Cellularity of femoral bone marrow in irradiated mice restored by $0{\cdot}4 \times 10^7$ (● — ●) or $1{\cdot}2 \times 10^7$ (▲ — ▲) bone marrow cells.

The curves of Figs. 7 and 8 demonstrate that the number of leucocytes and polymorphonuclear leucocytes in the blood is directly related to the number of cells injected. It can also be noticed by comparing these curves to those of Figs 5 and 6 that the growth of the spleen as well as the multiplication of bone marrow cells begins to stop at the eighth day when granulocytosis appears.

Thus it seems very likely that the regeneration of bone marrow is controlled by a feed back mechanism triggered not only by bone marrow cellularity itself but also to the granulocytosis in the peripheral blood. The regulation of the spleen regeneration although subjected to the

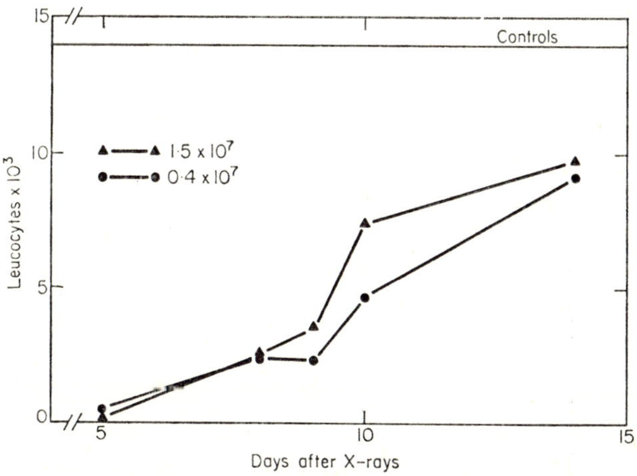

FIG. 7. Recovery of leucocytes in mice irradiated with 850 r and restored by either $0{\cdot}4 \times 10^7$ or $1{\cdot}2 \times 10^7$ bone marrow cells. Straight line: mean number of leucocytes in non-irradiated controls.

same stimulus, seems to be less accurate maybe because its clones respond only to one of the two preceding factors (Bierman (1964, Yoffey *et al.* 1964, Gordon *et al.* 1964). It must also be noticed that the myeloid cells present in the spleen can to some extent be considered as ectopic and for these reasons not accurately controlled. In our experimental conditions splenectomy performed three weeks before irradiation, does not modify the regeneration of bone marrow (see Table IV) neither does it change the survival of the irradiated animal (Playfair and Cole 1964).

Attempts to isolate in mice of group I a humoral factor responsible for the stimulation of myeloid growth in the spleen, have until now been unsuccessful. The only unquestionable effect was an inhibition of the splenic hypertrophy by the injection of plasma of normal mice.

TABLE IV. Influence of splenectomy on the regeneration of femoral marrow in isogenic radiation chimaeras.

Treatment	Femur cellularity 10 days after X-rays
850 r + $1 \cdot 5 \times 10^7$ bone marrow	$9 \cdot 0 \times 10^6$
splenectomy + 850 r + $1 \cdot 5 \times 10^7$ bone marrow	$1 \cdot 1 \times 10^6$

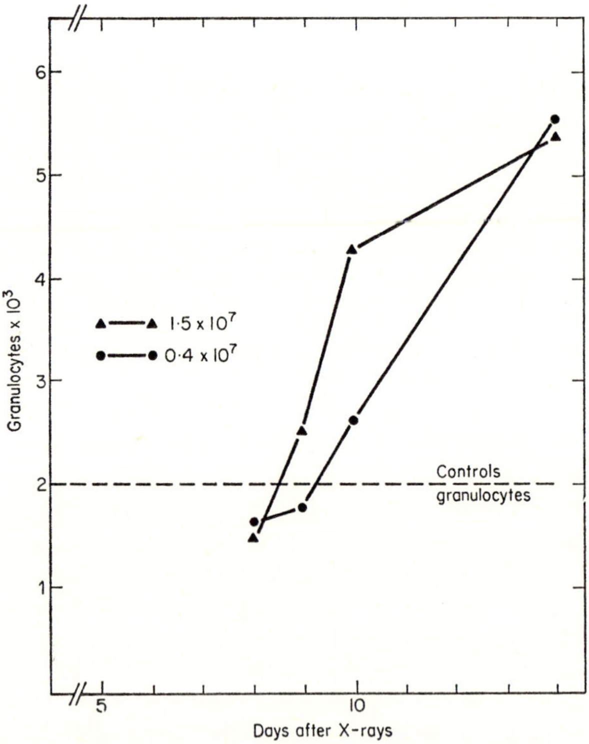

FIG. 8. Recovery of granulocytes in mice irradiated with 850 r and restored by either $0 \cdot 4 \times 10^7$ or $1 \cdot 5 \times 10^7$ bone marrow cells. Straight broken line: mean number of granulocytes in non-irradiated controls.

IV. Discussion

As the different implications of each experiment have been discussed in each part of this paper this discussion will be devoted to the problem of the stem cell in relation to the findings presented.

As we have previously seen, the possibility that two different kinds of stem cells are implicated in the lymphatic repopulation, must be taken into consideration. The main experiment that supports this hypothesis is that in thymectomized irradiated mice injected with foetal liver cells there is no recovery of the immune response contrary to the situation in nonthymectomized irradiated mice. However spleen cells and lymphnode cells bring immunological competence in both at least for a period of 4 months. As in these latter conditions, the stem cells that have not acquired a differentiation toward immunological competence do not become I.C.C., there must exist in the spleen and in the lymphnode a population which is able to maintain immunological competence. Since we know that some lymphocytes that retain a capacity to divide have a very long life span (300 days or more) it could well be that they are the ones that maintain the immunological reactivity in the thymectomized irradiated mice treated with spleen or lymphnode cells and therefore the number of I.C.C. will decrease steadily after 200 days and even disappear completely later on. From a practical point of view this fact is rather difficult to check. But our first findings show that the mice treated with a mixture of foetal liver and lymph node cells display a more rapid decrease of their immunological capacities (G.V.H. reaction) when they have been thymectomized than when they have not. This would mean that there is only one pool of precursors to I.C.C. and that the lymphatic cells which have differentiated toward immunological competence do not constitute in thymectomized chimeras a self maintaining population. The thymus would be the key to the pathway permitting the precursors cells to enter the compartment in which they will undergo differentiation toward immunological competence.

In the case of the spleen and bone marrow regeneration, one must distinguish between the number of clones belonging to each kind of cells (erythrocytic, granulocytic and megakaryocytic) and the volume of the nodule produced by the growth of each clone. In the situation which we have described, the histologic examinations of the spleen show that its hypertrophy seems correlated with an overgrowth of the granulocytic series. If this were exact it would mean that the stimulus to growth is more directed towards granulopoietic than towards erythropoietic series. This is not unlikely in view of the strong granulocytopenia that exists in these irradiated mice while no or very little anemia has yet developed. Likewise two facts can explain why the megakaryocytic clones do not increase much in volume:

(a) two endomitotic divisions take place during the cytodifferentiation of megakaryocytes leading to adult polyploid cells; then the number of these megakaryocytes, although these cells have a very large volume, is still four to sixteenfold less than the quantity of cells produced in the other series after the same number of cell cycles assuming the cell cycle has comparable length, (b) Each megakaryocyte can release a very great number of platelets without dividing.

Precursors (stem cell line) of the lymphatic I.C.C. are obviously present in the bone marrow, the foetal liver and the spleen, as are the stem cells for erythrocytic, granulocytic and megakaryocytic series. This does not necessarily mean of course that they all belong to the same stem cell population but this possibility cannot be ruled out simply on quantitative basis because many stem cells circulate in the peripheral blood and the population located in the blood forming organs constitutes only a fraction of the total of the stem cells available at any time. Then the question rises whether the stem cells constitute a unique population of multipotential elements or of various groups of cells, each one corresponding to a line of haematopoietic cells or of both of these types. In the first case a specific factor induces the multipotential stem cells to enter one of the compartments where their cytodifferentiation starts, in the second and third instances the same factors trigger only one corresponding recipient stem cell line. The finding that the stem cells existing in one clone, when transplanted in an irradiated recipient, are able to produce clones of cells belonging to the other haematopoeitic series, favours the idea either of a pool of multipotential stem cells or of 2 compartments of stem cells one of which containing multipotential cells the other committed cells. To some extent these latter hypothesis are also supported by the existence of clones containing a mixed cellular population.

This general scheme has been questioned on the following basis: when splenic clones are serially transplanted in irradiated mice the number of clone forming cells decreases rapidly and after two to three passages, clones are no longer obtained (Barnes *et al.* 1959, VanBekkum and Weyzen 1961, Duplan 1963), although the growth and the structure of the last nodules are normal. This has suggested that the colony forming cells could constitute a precursor cell population different from the stem cells. In fact it seems that the colony forming cells are actually the stem cells but that they are very fragile and selectively destroyed when the cell suspensions are prepared.

Another possibility (Bennett *et al.* 1964) which does not exclude the preceding, is that the transplanted cells cannot regenerate a normal-sized haematopoietic stem cell population because the requirements placed on them for differentiated cells are too great in relation to the available supply of transplanted stem cells.

V. Summary and Conclusions

The present experiments have been dealing with the repopulation of haematopoietic tissues in irradiated (850 r) mice injected with isogenic foetal liver cells, adult spleen cells or adult bone marrow cells.

The regeneration of lymphatic tissues has been studied in intact and in thymectomized mice both from the point of view of the cellular recovery and of the repair of the immunological competence (G.V.H. reaction). The repair of immunological competence is obtained more rapidly in non-thymectomized irradiated mice treated by foetal liver cells or adult bone marrow than in the group injected by splenic cells. This would point to a mechanism that regulates the transformation of lymphatic precursor cells into I.C.C.; the stimulation being strongest in the animals that have been restored with the suspension that contains the smallest number of I.C.C.

In thymectomized irradiated mice, injected either with foetal liver cells or adult bone marrow, no recovery of immunological competence occurs during the 110 days of the experiments. On the contrary the mice restored by spleen cells display a weak but steady immunological responsiveness. The same result can be obtained by mixing foetal liver cells with adult lymphnode cells. But the duration of this immunological responsiveness does not seem to last as long in these mice as in the non-thymectomized animals which have received the same treatment. This would mean that the I.C.C. present in lymphnodes or spleens do not constitute a self maintaining population independent of the thymus.

The regeneration of the myeloid cell population has been studied in the bone marrow and in the spleen of irradiated mice restored by bone marrow cells (4×10^6 or $1{\cdot}5 \times 10^7$). These preliminary studies indicate that the control of haematopoietic regeneration is more accurate in the bone marrow than in the spleen. Bone marrow repopulation is more stimulated in the group of animals that have received the smallest number of cells but the final cellularity reaches the same plateau at the same time in both groups. On the contrary, this stimulation produces an overgrowth of granulocytic cells in the spleen and leads to a transient hypertrophy of this organ.

REFERENCES

Barnes, D. W. H., Ford, E. E. and Loutit, J. F. (1959). *Sang* **30**, 762–765.

Bennett, M., Shearer, G. M., Upton, A. C. and Cudkowicz, G. (1964). *Nature* **204**, 351–353.

Bierman, H. R. (1964). *Ann. N.Y. Acad. Sci.* **113**, 753–765.

Defendi, V., Roosa, R. A. and Koprowski, H. (1964). *In* "The thymus in Immunobiology" (R. A. Good and A. E. Gabrielsen eds.), vol. 28, pp. 504–521. Harper & Row, New York.

Duplan, J. F. (1963). *C.R. Acad. Sci.* **256**, 3366–3368.

Duplan, J. F. (1963). *C.R. Soc. Biol.* **157**, 286–289.
Duplan, J. F. (1965). *Annls. Inst. Pasteur, Paris* **108**, 1–12.
Gordon, A. S., Handler, E. S., Siegel, C. D., Dornfest, B. S. and Lobue, J. (1964). *Ann. N.Y. Acad. Sci.* **113**, 766–789.
Gross, L. (1960). *Proc. Soc. exp. Biol. Med.* **103**, 509–514.
Levey, R. H., Trainin, N. and Law, L. W. (1963). *J. natl. Cancer Inst.* **31**, 199–217.
Lewis, J. P. and Trobaugh, F. E. (1964). Radiation Protection and Recovery Discussion Meeting, Chicago. *In* "Experimental Hematology" (Biology Division, O.R.N.L. ed.), No. 7, pp. 7.
McCulloch, E. A. and Till, J. E. (1962). *Rad. Res.* **16**, 822–832.
McCulloch, E. A. and Till, J. E. (1964). *Rad. Res.* **22**, 383–397.
Metcalf, D. (1958). *Ann. N.Y. Acad. Sci.* **73**, 113–119.
Miller, J. F. A. P., Doak, S. M. A. and Cross, A. M. (1963). *Proc. Soc. exp. Biol. Med.* **112**, 785–792.
Miller, J. F. A. P., Marshall, A. H. E. and White, R. G. (1962). *Adv. Immunol.* **2**, 111–162.
Playfair, J. H. L. and Cole, L. J. (1964). Radiation Protection and Recovery Discussion Meeting, Chicago. *In* "Experimental Hematology" (Biology Division, O.R.N.L. ed.), No. 7, pp. 8.
Puck, T. T. (1964). *In* "Cytogenetics of cells in culture" (R. J. C. Harris ed.), pp. 63–77. Academic Press, New York.
Simonsen, M. (1962). *Progress in Allergy* **6**, 349–467.
Van Bekkum, D. W. and Weyzen, W. H. (1961). *Path Biol., Paris* **9**, 888–893.
Waksman, B. H., Arnason, B. G. and Jankovic, B. D. (1962). *J. exper. Med.* **116**, 187.
Yoffey, J. M., Makin, G. S., Yates, A. K., Davis, C. J. F., Griffiths, D. A. and Waring, I. S. (1964). *Ann. N.Y. Acad. Sc.* **113**, 790–798.
Zaalberg, O. B., Weysen, W. W. H. and Vos, O. (1963). *Nature* **197**, 300–301.

Re-utilization of Thymidine-^{3}H by Mouse Kidney during Compensatory Renal Hypertrophy*

MARTTI KEKOMÄKI and TAPIO RYTÖMAA

II Department of Pathology, University of Helsinki, Finland

THE unique role played by the thymus gland and other lymphoid cells in growth and growth control still remains a mystery in many respects. Some most provocative statements and observations were presented in this symposium by Drs. Craddock and Fichtelius (Craddock *et al.* 1966, Fichtelius and Linna 1966).

We have studied the phenomenon of re-utilization of labelled thymidine by the remaining kidney during its compensatory growth after unilateral nephrectomy. In these preliminary experiments, consideration was given to the part played by the thymus gland as a possible donor organ of the re-utilized thymidine-^{3}H-label. Growing mice were injected intraperitoneally with ^{3}HTdR (specific activity 3·6 c/m mole), the amount of tritium in the three 12-hourly injections totalling 1·0 μc/g of body weight. 24 hours after the last injection the left kidney was removed from the test animals (28 mice). The mice of the control group (26 animals) were subjected to a sham operation. The accumulation of ^{3}H-label in the remaining kidney was followed for the next 120 hours (non-volatile substances; combustion technique). In addition to this, the changes in the radioactivity of the thymus gland, expressed per unit weight of the gland, were followed in both groups of mice, by taking the values of the control group as a baseline. It must be emphasized that no change in the weight of the thymus gland could be observed between the test and control animals.

The results are seen in Fig. 1. These and other information available permit the following inferences: (1) leucocyte infiltration, which occurs in many instances where the process of DNA(-base) re-utilization has been studied earlier, is apparently not in itself responsible for the findings; (2) re-utilization of label seems to begin at a time (14·7 hours after nephrectomy†), which cannot be distinguished from the beginning of increased

* Supported by grants from the Damon Runyon Memorial Fund, New York, and Sigrid Jusélius Foundation, Helsinki.

† Obtained from the "least squares" exponential regression equation, calculated by means of an electronic computer.

DNA synthesis in the compensatory renal growth (Bury *et al.* 1965); (3) the ³H-activity of the thymus gland, calculated per unit weight, is significantly ($P < 0{\cdot}02$) diminished, i.e., the DNA(-base) turnover in the thymus gland appears to be increased after nephrectomy. Further studies on this phenomenon are now under way in our laboratory.

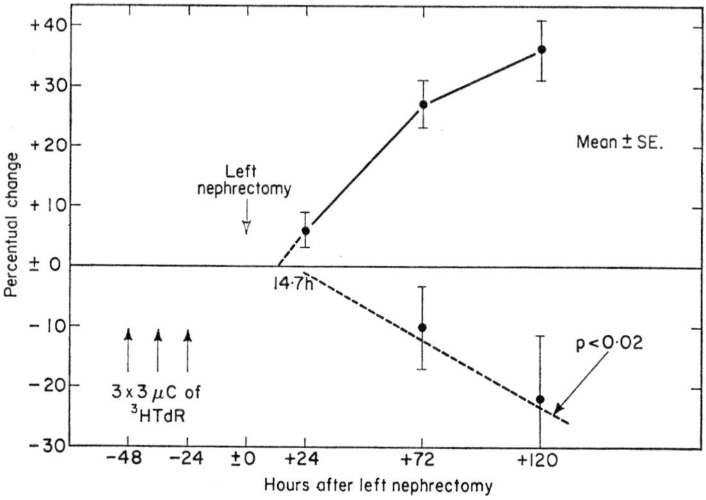

FIG. 1. The effect of unilateral nephrectomy on the labelling of the remaining kidney and the thymus gland in growing mice.

Upper delineator: Relative increase of total radioactivity in the remaining kidney (whole organ). Baseline: total activity of the removed kidney.

Lower delineator: Relative loss of the label of the thymus gland in nephrectomized litters (whole organ). Baseline: mean radioactivity of the thymus gland of the sham operated animals.

In the last few years the significance of the thymus gland in "organismal organization" has become apparent. The recognition of its importance is mainly due to the recent concepts of the development of the immune response in embryonic and neonatal life. On the other hand, the results of some experiments on the function of the thymus gland are not easy to interpret in terms of immunobiology. Thus, Fachet *et al.* (1964) have reported that significant diminution in the size of cell nuclei occurs in the regenerating liver tissue in thymectomized, adult rats. Dukor and Miller (1965), in turn, have concluded that ablation of the thymus gland, carried out either in the immunobiologically critical first 24 hours or later (up to 6 weeks), markedly decreases the mitotic index in hepatic regeneration in mice. Finally, the intensity of liver regeneration, particularly during the later stages of the reparative process, is in some way controlled by the lymphatic system (Davies *et al.* 1964). A distinct parallelism is seen between the observations cited and our results.

It may be early to speculate about the significance of intercellular shifts of DNA bases in a multicellular organism. It may also be difficult to offer any reasonable explanation for the relative hypertrophy which occurs in the thymus gland after its main function, conditioning of the immunological system, has seemingly been performed (Miller and Dukor 1964). By means of carefully planned experiments it may be possible to indicate whether the DNA(-base) re-utilization should be regarded merely as an intercellular DNA(-base) salvage route or rather as a "specific" DNA(-building block) supply during periods of increased DNA demand. As a matter of fact, the evidence obtained from *in vitro* studies (Kurita *et al.* 1959, Krauss 1961, Podgajetskaja *et al.* 1964) strongly suggests the possible existence of an intra-organismal transfer phenomenon mediated by functioning ("informative") pieces of DNA polymer. Such a phenomenon would be of fundamental value for the maintenance of genetic and tissue-immunological stability.

REFERENCES

Bury, H. P. R., Crane, W. A. J. and Dutta, L. P. (1965). *Br. J. Urol.* 37, 201–210.

Craddock, C. G., Winkelstein, A., Matsuyuki, Y. and Lawrence, J. S. (1967). This volume, pp. 153–170.

Davies, A. J. S., Leuchars, E., Doak, S. M. A. and Cross, A. M. (1964). *Nature* (Lond.) **201**, 1097–1101.

Dukor, P. and Miller, J. F. A. P. (1965). *Naturwissenschaften* 52, 189.

Fachet, J., Stark, E., Palkovits, M. and Vallent, K. (1964). *Z. Zellforsch. mikrosk.* 60, 609–614.

Fichtelius, K. E. and Linna, J. (1967). This volume, pp. 171–181.

Krauss, L. M. (1961). *Nature* (Lond.) **192**, 1055–1057.

Kurita, S., Takemura, C., Hoshino, A. and Kimura, K. (1959). *Gann, suppl.*, 49, 66–67.

Miller, J. F. A. P. and Dukor, P. (1964). "Biologie des Thymus nach dem heutigen Stande der Forschung." S. Karger, Basel and New York.

Podgajetskaja, D. J., Bresler, V. M., Surikov, I. M., Ignatova, T. N. and Olenov, J. M. (1964). *Biochim. biophys. Acta* 80, 110–115.

DISCUSSION SUMMARY

Chairman: C. G. Craddock

In the absence of our distinguished colleague, Professor Tumanishvili, Professor Teir asked me to chair this session which includes my own contribution. I present this summary with the hope that I do not deviate too widely from what Professor Tumanishvili would have considered the most significant features of the papers.

The main point which I tried to make in my presentation was that cells which, on morphological grounds are denoted lymphocytes, are comprised of two cell species with markedly differing physiology. Both of these species of lymphoid cells has a function related to the immune response. One, a long lived population comprising the majority of cells in the peripheral lymph nodes, spleen and blood, is the immunological effector cell population which elaborates specific antibody. This tissue has the characteristics of mobility and constant migration through blood and tissues. Upon appropriate stimulation, the stimulus being antigen related, each cell in this population has the potentiality of entering DNA synthesis, dividing and giving rise to antibody-producing cells. Relative to the subject of this conference is that fact that antigen in some way is involved in the stimulation of this tissue to proliferate.

The other population of lymphoid cells is short lived, is found abundantly in the thymus cortex and bone marrow and is, we suggest, a prominent component of the reactive center of the lymphoid follicle, usually denoted the germinal center. Evidence is presented that these cells have the capacity to transform into phagocytic cells after reaching the peripheral tissues. The hypothesis is presented that the macrophage population which must encounter, engulf and somehow process certain particulate antigens prior to the elaboration of specific antibody by the long-lived effector lymphocyte population may emanate from the thymus and bone marrow. Replication in the thymus-marrow lymphoid cells which give rise to peripheral macrophages is not dependent upon contact with antigen. Proliferative activity within the peripheral effector group of long lived lymphocytes, which is a necessary activity before the elaboration of specific antibody, is induced by the macrophage-antigen interaction. According to this hypothesis the short lived, thymic type lymphocyte is the crucial cell in the transmission of the proliferative signal to the peripheral lymphocyte population.

Professor Fichtelius reviewed briefly the evidence of others for a hormone elaborated by the epithelial components of the thymus which is necessary for the proliferative activity and development of the lymphatic system. Presumably this hormone would promote lymphopoiesis in a manner similar to the stimulation of erythropoiesis by erythropoietin. He suggests that a histone-like material formed in the peripheral lymphatic tissue may inhibit the action of this trophic hormone.

He points out again the evidence for reutilization of DNA or products of DNA breakdown in the thymus and between peripheral lymphoid cells and rapidly growing tissues. Most evidence favors reutilization from a nucleotide or nucleoside level. The evidence for reutilization of polymerized DNA is, as yet, fragmentary and must remain a troublesome but unlikely possibility. The likelihood that lymphoid cells may serve a less specific trephocytic function by supplying other cell constituents to rapidly growing tissues appears much more probable.

Dr. Kekomaki, in a brief report, provided additional evidence for the transfer of labeled DNA, or its constituents, from blood cells to renal tissue undergoing hypertrophy and hyperplasia after unilateral nephrectomy.

Professor Duplan presented his studies on the reconstitution of hematopoiesis and lymphopoiesis in lethally irradiated mice and the influence of thymectomy at 28 days (X-ray at 65 days) on this restoration. He also compared the effectiveness of foetal liver, spleen cells and bone marrow in providing restoration. The thymus is necessary for full restoration of this type of crippled animal. Interestingly, thymectomized animals restored with spleen cells showed very small germinal centers but normal lymphocyte cuffs in lymph follicles.

Using the Simonsen assay of spleen weight in homologous recipients as an index of immunological competence, he showed that spleen cells or lymph node cells, when used to restore irradiated animals, would lead to the development of immunologically competent lymphocytes in the spleens of such restored animals. Bone marrow and foetal liver cells did not provide such regeneration of immunologically competent cells in the restored spleen. Thymus cells were very weakly effective. These findings support other evidence that lymphoid cells of the thymus and bone marrow differ in their function from peripheral, immunologically competent lymphocytes.

In following the recolonization process in the spleen, he found discrete clones of erythropoiesis, granulocytopoiesis and megakaryocytic proliferation. He interprets these findings as supporting the earlier work of Till and McCulloch and others which indicate the presence of a single pluripotential stem cell accounting for all colonies. The differentiation observed in particular clones is presumed to be a reflection of varying peripheral demand, etc. The persistence of pluripotential stem cells in such differentiated clones can be demonstrated by subcloning. This type of evidence provides the strongest support for the existence in the adult animal of stem cells which can give rise to all varieties of myeloid cells. His evidence with respect to the ineffectiveness of the bone marrow injections to repopulate the spleen with immunologically competent lymphocytes would seem to indicate that these cells do not derive from the stem cells contained in normal bone marrow.

Section V

Liver Regeneration

Liver Tissue Growth Factors and Circadian Rhythms in Liver Regeneration

J. M. ECHAVE LLANOS*

Instituto de Patología General y Experimental. Universidad Nacional de Cuyo, Mendoza, Argentina

THE action of liver preparations on liver growth has been studied by different authors under varying conditions.

Some of them have reported their action in early stages of development of the chick (Weiss and Wang 1941, Andres 1955, Tumanishvili *et al.* 1956, Croisille 1958) and frog (Romanova 1957). Others have investigated the action of liver extracts on new-growth liver that fills holes made in the hepatic substances of frogs, rabbits and guinea-pigs (Tumanishvili 1960) or on the DNA synthesis in regenerating frog liver (Tumanishvili *et al.* 1963). Finally, most of the papers report the action of liver preparations on growth variables of normal and regenerating liver of rats and mice (McJunkin and Breuhaus 1931, Wilson and Leduc 1947, Christensen and Jacobsen 1949, Kelly and Jones 1953, Teir and Ravanti 1953, Formijne 1955, Saetren 1956, Blomqvist 1957, Paschkis *et al.* 1957, Stich and Florian 1958, Molimard 1959, Stich 1960, Yashina 1963, Fischer *et al.* 1964).

The various papers in print dealing with the action of crude liver preparations on growth variables of normal and posthepatectomy regenerating liver in mice and rats, suggest some possible effects.

Normal liver preparations would promote the increase of regenerating liver weight (Formijne 1955, Fischer *et al.* 1964). This effect would be earlier and longer lasting with regenerating liver preparations (Fischer *et al.* 1964). In other papers an inhibiting effect of normal liver preparations on the regenerating liver weight increase is reported (Molimard 1959).

The DNA content of normal (Kelly and Jones 1953) and regenerating (Fischer *et al.* 1964) liver increases after the injection of normal (Kelly and Jones 1953, Fischer *et al.* 1964) or regenerating (Fischer *et al.* 1964) liver preparations.

With respect to the control of mitotic activity, the normal young liver has a promoting effect that fades and disappears with age (Teir and Ravanti 1953, Blomqvist 1957), an inhibiting effect appearing in adult normal liver

*Present Address: Instituto de Embriologia, Biologia e Histologia, F.C. Medicas, U.N.L.P.; 60 y 120, La Plata, Argentina.

(Wilson and Leduc 1947, Christensen and Jacobsen 1949, Saetren 1956, Stich and Florian 1958). Some authors report a stimulating effect for adult normal liver (McJunkin and Breuhaus 1931, Wilson and Leduc 1947). Regenerating liver preparations have been described as growth stimulating (Blomqvist 1957) or without any effect whatever (Stich and Florian 1958).

Before ascribing such contradictory results to "biological variation" (MacDonald *et al.* 1963) it is necessary to learn more about how the results are affected by the technical and experimental factors in play. Some of these have been analyzed, others only considered and others not mentioned by the authors, despite the fact that many of them are known to affect liver growth.

In assaying liver tissue growth factors, the variables more commonly used in studying protoplasmic reproduction, have been the liver weight, the mitotic index and the DNA content of the liver. Their disadvantages have been the subject of sharp analysis by Weiss (1955). Although recognized as one of the characteristic features of the growth of the liver during regeneration (Harkness 1957), cell hyperthropy has not yet received the necessary attention (Liosner 1964) and only some papers have used it as a measure of growth (Yashina 1963, Liosner 1964).

The papers considered vary with respect to technical and experimental factors such as the type of liver weight control, the method of counting mitoses and the base lines against which they are expressed, the methods used to determine the DNA variations, and those used to prepare the liver administered, the route of injection, the dose levels used, the type of animal and its strain, inbreeding, age, sex, body weight, number of donors and receivers and the criteria taken into account for their selection. A critical survey on the bearing that these factors could have on the results obtained, has recently been prepared (Echave Llanos 1965). On the basis of this survey it can be stated that many of these factors constitute artificial sources of variation, that could, artificially, augment or diminish the scatter of the values obtained or could also augment, diminish or even produce significant differences between the values of the groups to be compared, that, really, can hardly be ascribed to the treatment under investigation.

In our laboratory we have used inbred C3H/Mza (companion issue to Mouse News letter No. 29, pp. 55-56, July 1963) female and male mice 90 days old. The homogenates were prepared by homogenizing 1 gram of a pool of livers from 4 to 6 normal or hepatectomized donors, with 10 ml saline in a Potter-Ehlvehjem (1000 rpm) (Thomas 1961) or a *Vir Tis* 45 homogenizer (40,000 rpm) during 1 or 2 minutes, using at all times an ice bath. We have not—so far—found differences between the two procedures. After that the homogenate was immediately injected, or stored for 11 to 24 hours in the freezer (—23°C), before the intraperitoneal injection to groups of 6 to 10 hepatectomized (Brues *et al.* 1936) receivers, in doses corresponding to 0·5 to 1·0 mg of wet liver per gram of body weight. All the procedures, from the extraction of the liver from donors to the injection of the homogenates to the receivers were made aseptically.

Before the experiment took place, those animals with body weights very scattered from the mean, were eliminated and those animals of the same litter, were distributed in order to equalize the groups to be compared.

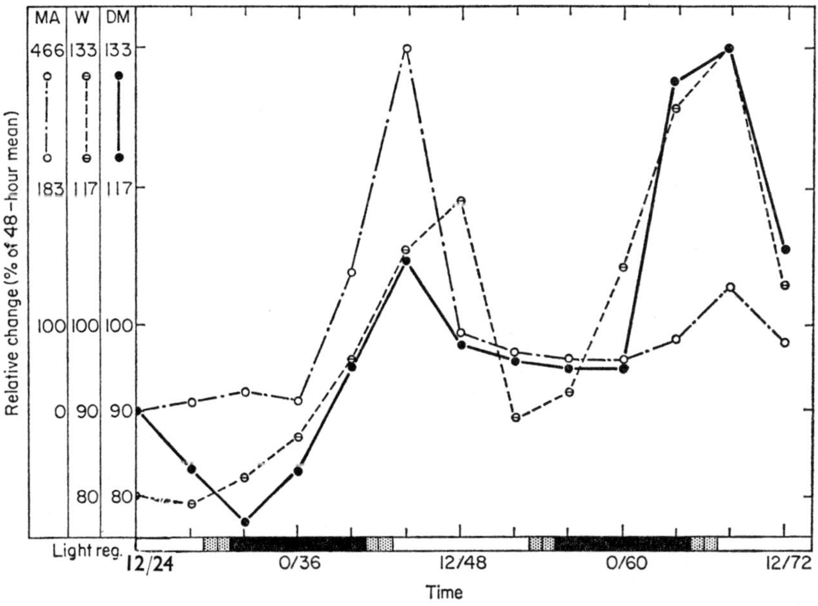

FIG. 1. Circadian variation in mitotic activity, water and dry matter content of regenerating liver. 12/24 ... etc.: Time of day/time post hepatectomy. MA: Mitotic activity: W: water; DM: dry matter (reprinted from Russo and Echave Llanos 1964).

At the time of killing we eliminated those animals with macro- or microscopical pathological processes (Echave Llanos *et al.* 1964), particularly those with necrosis of paragastric or right lobes or necrosis and inflammation of big stumps (Teir and Lahtiharju 1961, Lahtiharju and Teir 1964, Weinbren and Tarsh 1964). From those animals chosen for control, the right caudoventral lobe was extracted and fixed (Echave Llanos and Sadnik 1964) and then embedded, sectioned and stained with haematoxylin and eosin. In these sections the mitotic counts were made at a magnification of $8 \times 1{\cdot}25 \times 100$ until a number of 3000 nuclei was reached, considering the mitotic images, in their different phases, or the colchicine metaphases. The right craniodorsal lobe was used to determine the liver dry weight. After homogenization in a Potter Ehlvehjem homogenizer it was desiccated at 105°C during 12 hours. The dry weight calculated, was expressed as milligrams per 100 grams of body weight, controlled at the time of operation.

The temporal aspects of the experiments are of major importance.

From the changes produced in different variables, while liver regeneration advances, the possibility of qualitative and/or quantitative correlated changes follows, in the growth control systems of the tissue. If it is so, the posthepatectomy time at which the donors are killed and the receivers are injected and killed, could be an important factor in the variation and/or demonstration of a given effect.

In the reviewed papers the posthepatectomy time at which the donors were killed varies: 24 hours (Stich and Florian 1958) 48 hours (Blomqvist 1957, Fischer *et al.* 1964) and 120 hours (Fischer *et al.* 1964). Notwithstanding the fact that we have not found significant differences between the homogenates prepared from donors killed 43, 48, 56, 59 and 71 hours after hepatectomy (Bade 1962, Bade and Echave Llanos 1962, Echave Llanos and Bordin 1963a, 1963b) this problem must be re-investigated after we have set aright the assay procedure.

The posthepatectomy time at which the receivers are injected and killed, that is that time of regeneration at which the assay is made, also varies in the papers considered. Most of the authors dealing with the control of mitotic activity, including ourselves, have explored the first 24 to 48 hours of regeneration in which the first peak of mitotic activity takes place in rats and mice (Weinbren 1959, Bade and Echave Llanos 1963, Bucher 1963, Russo and Echave Llanos 1964).

In one of the experiments from our laboratory (Fig. 2) a relatively longer period was involved, from the time of injection, i.e. 56 hours posthepatectomy up to 108 hours after operation, sampling at 60, 72, 84, 96 and 108 hours of regeneration (Bade 1962). From this experiment it follows that short periods of action of homogenates (4-18 hours) probably show early effects of the factors they contain, but longer periods could show reactions secondary to earlier effects, which have been overlooked due to infrequent samplings. A homeostatic compensatory reaction was proposed as an explanation of this fact (Bade 1962) but other alternative explanations could be considered. This important aspect in the study of growth processes has been clearly analyzed by Weiss (1955). Where longer latencies have been used (McJunkin and Breuhaus 1931, Wilson and Leduc 1947) the stimulating effects observed are perhaps due to compensatory waves of regeneration after the changes (McJunkin and Breuhaus 1931) that the liver preparations administered could produce. As a first stage in our work we have preferred to explore short periods of the action of homogenates (up to 18 hours) sampling every 4 hours.

Most of the authors do not report the time of day at which donors and receivers have been operated and killed, or the time of day at which the receivers have been injected with the liver preparation. Some authors, however, have fixed a time of day schedule for these procedures (Christensen and Jacobsen 1949, Blomqvist 1957, Stich and Florian 1958, Stich 1960). As we shall see below a more or less arbitrary fixation of this time schedule, is not enough and could very well be the wrong one.

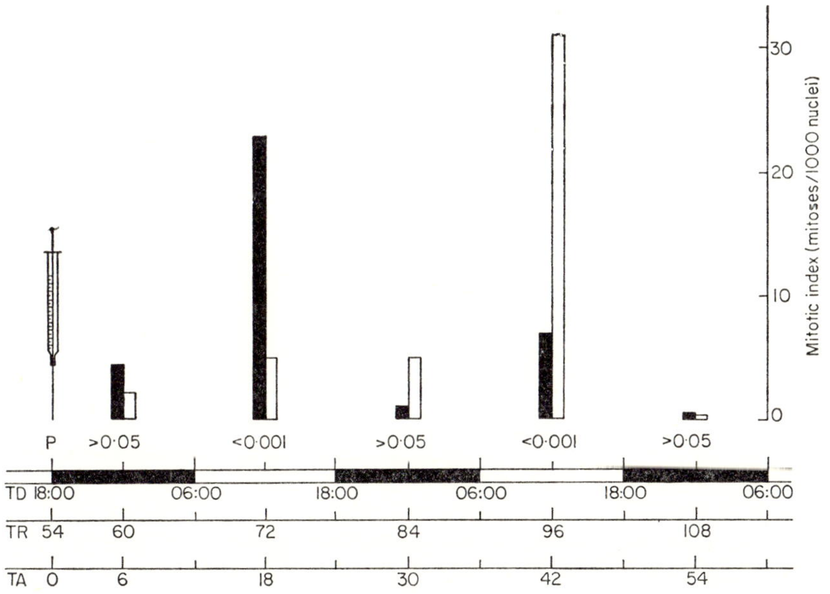

FIG. 2. Homeostatic compensation that would appear as a stimulating effect without adequate sampling. TD: time of day. ■: darkness. □: illumination. TR: time of regeneration. TA: time of action of the homogenate. (Made with data from Bade 1962).

The bearing of the time of day at which an experimental procedure is applied follows directly from the existence of circadian rhythms in hepatic variables directly related to growth, such as weight (Higgins *et al.* 1932, 1933, Hruza 1956, Barnum *et al* 1958), protein content (Higgins *et al.* 1932, Hruza 1956, Fabry and Hruza 1956, Hruza and Smetana 1958), cell size (Jackson 1959), DNA synthesis (Jardetzky *et al.* 1956, Barnum *et al.* 1957) and mitotic activity in the normal young liver (Wilson 1948, Jardetzky *et al.* 1956, Halberg 1957, Jackson 1959, Peters 1962) or in the adult liver of rats, refed after a 24-hour fasting period (Sokolova 1964).

The first report of the existence of a circadian rhythm in the mitotic activity of the regenerating liver is that of Jaffe (1954) for the rat. Several years before Wilson and Leduc (1947) stated that "the relative proportion of the various phases varied greatly in different animals, so much so that we have considered the possibility that mitosis sometimes occurs in waves rather than at a constant rate" and Christensen and Jacobsen (1949) observed that "the rats in the last series were killed at 4 p.m. and here naturally (?) fewer mitosis were found than in the series where the rats were sacrificed as usual, in the morning".

The circadian variation in mitotic activity of regenerating liver was confirmed for the rat (Peters 1962a) and in our laboratory for the mouse (Bade 1962) and then analyzed with a more frequent sampling, correlating these variations with similar variations in water and dry matter content (Bade and Echave Llanos 1963, Echave Llanos and Bade 1963, Russo and Echave Llanos 1964) (Fig. 1).

From the changes produced in these growth variables during a 24-hour period, the possibility of qualitative and/or quantitative correlated changes in the growth control systems follows.

The study of circadian rhythms requires, no matter which variable is under investigation, an adequate standardization of the animals (Halberg *et al.* 1958).

The room where the animals live 15 to 30 days before the experiment, must be isolated from the rest of the laboratory and no person must be allowed to enter it at any time other than that of control, that must be, preferably, the commencement period of activity of the animals.

It is important if the animals are caged in groups or isolated in individual cages. By this last method we have found a significant reduction in the scatter of body weight and liver weight/body weight values (Echave Llanos *et al.* 1963).

The range of the temperature in the room must not vary much, as environmental temperature can directly affect the values of mitotic activity (Peters 1962). In our experiments the temperature was $26 \pm 2°C$.

The lighting regimen is of paramount importance in the study and/or the experimental fixation of circadian variations. Light, with water and food *ad libitum*, is, particularly for mice, the primary synchronizer of their functions (Halberg 1958).

In our "rhythm room"—as we call it—the animals are given light (fluorescent, 40 W) from 6 a.m. to 6 p.m. followed by 12 hours darkness.

The importance of the lighting regimen is emphasized by the fact that a reversal (illuminating from 6 p.m. to 6 a.m.) produces, after a given number of days a complete change in the mitotic rhythm of the liver tissue (Halberg *et al.* 1958).

The quantity of food and the time of feeding the animals are of the major importance. It is of greater importance for the liver, an organ directly related with the processes that occur in the digestive tract.

In our laboratory the animals received food and water *ad libitum*. From this it could follow that the timing of feeding (time of day and length of the feeding) was irregular. Recording this timing of feeding we have found that, after some days of isolation with normal lighting (6 a.m. to 6 p.m.), the mice started feeding at 6 p.m. $\pm\frac{1}{2}$ hour and finished it at 6 a.m. $\pm\frac{1}{2}$ hour (Echave Llanos and Vilchez 1964).

They did not eat during the period of illumination and careful observations of the animals during this period indicated that the great majority of them were sleeping.

In spite of the regularity of this aspect of the food plan, different animals ate different quantities of food during the period of activity, as we diagnosed through checks made by weighing the food container every day at 6 p.m. So, when killing the animals at 8 a.m. and at noon we found that their stomachs contained very different quantities of food. This could be due to differences in the speed of emptying, but the important conclusion is that their livers are exposed to functional requirements, that are very different in quantity and duration. The results of Alov (1962) on the internal timing (Halberg *et al.* 1965) of the curves of mitotic activity and function, in different organs, suggest that these differences in functional requirements, could be an important source of variation in the results obtained.

On the basis of the experiments we have introduced recently, a fasting period of 12 hours before killing the animals. This allows us to obtain regularly empty stomachs (Echave Llanos *et al.* 1965). The 6 to 8 hour period proposed by Wilson (1958) for normal mice, is not satisfactory in our hepatectomized mice. If the animal is injected with colchicine, the fasting period must be yet longer, as this drug lengthens, significantly, the emptying period of the stomach (Echave Llanos *et al.* 1965). We do not think that this starving period could affect the mitotic values as the longer ones do (Leduc 1949).

When Jaffe (1954) operated on his rats in the morning, he observed the peaks in mitotic activity at noon and the troughs in the evening, but he reports that when operating in the evening he found that the circadian rhythm shifted by 12 hours. So, if the time of day at which the animals are operated on, is not quite the same ($\pm\frac{1}{2}$ hour for groups of 10 animals), for animals to be compared and, probably, if it is not the "right time" (Halberg 1958) for operation, it could become an important source of variation, even if the check is made at the same time of day.

The "right time" for operation could be defined as that which would produce a normal circadian rhythm in mitotic activity of regenerating liver, i.e. that of young untreated animals (Wilson 1948, Halberg 1957, Barnum *et al.* 1958). While this time is uncertain it is necessary to fix the time of operation and we have made it, by hepatectomizing our animals at noon.

Our peaks in these conditions appear at 8 a.m. (Bade and Echave Llanos 1963, Russo and Echave Llanos 1964) (Fig. 1), a time which coincides with that of Wilson (1948).

The time of day at which the donors are killed has been explored in our laboratory. In spite of a stronger inhibiting effect, of an homogenate from donors killed at 7 a.m., one of the experiments suggests (Bade 1962) we have obtained no significant differences between the action of homogenates obtained also at 11 a.m., noon, 6 p.m., 8 p.m. and 11 p.m. (Bade 1962, Bade and Echave Llanos 1962, Echave Llanos and Bordin 1963a, 1963b). We must reinvestigate this problem after we have set aright the assay procedure.

From the existence itself of the circadian rhythm in mitotic activity with variations that attain the 466% of the mean of 48 hours (Russo and Echave Llanos 1964) (Fig. 1) follows that the killing of receivers has to be made at the same time of day. Furthermore, this time of day must be the "right time"; an inhibiting effect could be easily demonstrated with groups of five animals, if control is made at 8 a.m., when the high values of mitotic activity appear; but, it is very hard to demonstrate it, if the control is made at midnight, when the values of mitotic activity are very low.

The time of day at which the receivers are injected springs from our experiments, as of paramount importance; we have found repeatedly, highly significant differences between the action of one and the same regenerating liver homogenate, when injected in the evening or in the early morning. In the first instance it produced a highly significant inhibition of the mitotic activity checked at 8 a.m. and noon of the following day. In the second one it had a significant stimulating effect on the values checked at noon. When this homogenate was injected at midnight it had no effect at all on the values controlled at 8 a.m. and noon (Fig. 3). The same results were obtained with the control of liver dry weight at noon (Bade 1962, Bade and Echave Llanos 1962, Echave Llanos and Bordin 1963a, 1963b).

These differences were not due to different periods of action of the homogenates injected, nor to different periods of the storage (—23°C) of the homogenates that were injected. Similar periods of action or storage gave different results relating to the time of injection.

It is indicative that the authors who obtained an inhibiting effect, with the injection of liver preparations (Christensen and Jacobsen 1949, Stich and Florian 1958) gave the injections on the evening of the second day and made the checks on the morning of the third one, that is to say the time schedule found for inhibition in our laboratory.

The first inference of these experiments, is that regenerating liver homogenates contain promoting as well as inhibiting factors. The "right time" for injection and check must be chosen for each one so as to demonstrate its effect. This proposition could be enlarged including normal liver homogenates. Beside their inhibiting effect when injected in the evening, with checks in the next morning (Christensen and Jacobsen 1949, Stich and Florian 1958), we have obtained a significant stimulating effect with injections of normal liver homogenate in the early morning and checks carried out at noon (Bade *et al.* 1963).

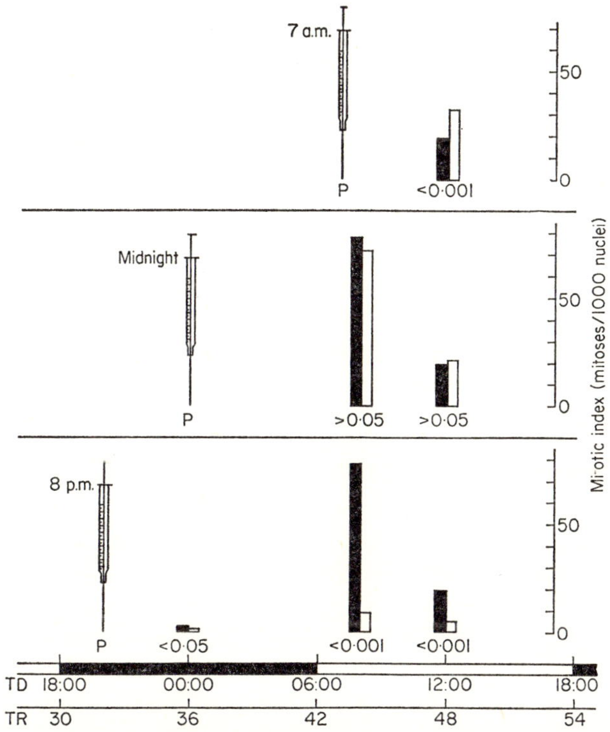

FIG. 3. Hourly variations in responsiveness to the action of one and the same regenerating liver homogenate. TD: time of day. ■: darkness. □: illumination. TR: time of regeneration. (Made with data from Echave Llanos and Bordin 1963a.)

From the above referred results it follows that, in setting aright the assay procedure for liver tissue factors, an important stage is to look for the "right times" for operation, injection and control, then being possible to fix a given time schedule for the assay.

As a second inference of these experiments we elaborated the hypothesis of a circadian variation in the responsiveness of receiving mice to the factors contained in the homogenates. The responsiveness to the inhibiting factors is high in the evening, disappearing about midnight. The responsiveness to the promoting factors is high in the morning, disappearing in the evening.

At first it was thought that this variation in responsiveness was due to a variation in the metabolic state of the liver cells (Bade 1962) but the temporal relations found afterwards (Echave Llanos and Bordin 1963a, 1963b) made us think that the circadian variation in responsiveness is due to a variation, at the level of organism, of some system through which the promoter-inhibitor system of the homogenates would act.

Bullough and Rytömaa (1965)* have suggested that the circadian variation in the output of adrenaline (Bullough 1962) is responsible for the mitotic rhythms of tissue, through the formation of an inhibiting complex, the chalone-adrenalin complex, in the period of activity, with high outputs, and its breakdown in the period of rest, with low outputs of adrenalin. Nevertheless, if this appears suitable for the skin (Bullough 1964, Bullough and Laurence 1964), some experiments would indicate that adrenalin is not so important for the liver, where corticosteroids would play a more or less similar role (Hemingway 1965), perhaps through the formation of some complex with the protein fractions that Glinos (1958, 1960) points out as having the inhibiting action on liver mitoses (Hemingway 1961). As the blood content in corticosteroids has also a well-defined circadian rhythm (Halberg *et al.* 1959), that is similar to that of adrenalin, the suggestions by Bullough and Rytömaa, should perhaps be adopted in the case of liver.

We could then suggest that, when we overload the mice with tissue factors in the evening, at the beginning of the activity period, a great quantity of some inhibitory complex would be formed and, through being stronger than the added promoting factors, it inhibits the mitotic activity of liver, which takes place normally at the beginning of the rest period. On the contrary, when the overload is made in the morning, the complex would not be formed and the promoting factors added, would be then free to show their effect.

On arriving to this point of reasoning we remember the words exchanged between Claude Bernard and Silas Weir Mitchell (Olmsted and Olmsted 1952). This one recorded: "I recall one remark of Bernard's. I said, 'I think so and so must be the case'. 'Why think,' he replied, 'when you can experiment. Exhaust experiment and then think.'" This is why we decide to stop until new experiments from ours or another laboratory could test the consequences of these or an alternative hypothesis.

The "we" used in our paper is not a literary expression of modesty. It means, literally, that the results from our laboratory, which I have reviewed, have been collected from our team. So, if perhaps I can dare to aspire to the fatherhood of some of the synthesizing ideas, it is not less true that they have risen in the innumerable talks we have had in our laboratory. Under these conditions

*This suggestion was originally made by Bullough (*Proc. Roy. Soc.* B 154, 540, 1961), and more definitely, by Bullough and Laurence (*Exp. Cell Res.* 33, 176-194, 1964).

"mine" and "yours" are rather difficult to define. For all the help and healthy criticisms for all the members of our group—in nearly ten years of working together—I am very indebted. I am specially grateful to Dr. Badrán who, reading the manuscript, made some very valuable suggestions. Many thanks are also due to our technicians Mrs. R. Cavoura (surgery), Mrs. E. V. Coria and Mrs. M. A. V. Albornoz (histology) for the valuable technical assistance given during the course of the experiments. The experiments here reported could not be carried out without the fundamental assistance of the Consejo Nacional de Investigaciones Científicas y Técnicas de la República Argentina.

REFERENCES

Alov, I. A. (1962). *Tsitologiya* 4, 297; (1963), *Fed. Proc.* **22**, 357–362.
Andres, G. (1955). *J. exp. Zool.* 130, 221–250.
Bade, E. G. (1962). Thesis for Doctor's Degree. F.C. Mèdicas. U.N. de Cuyo, Mendoza, Argentina.
Bade, E. G. and Echave Llanos, J. M. (1962). *Naturwissenschaften* 49, 351–352.
Bade, E. G. and Echave Llanos, J. M. (1963). *Naturwissenschaften* 50, 693–694.
Bade, E. G., Bordin, C. and Echave Llanos, J. M. (1963). *Sperimentale* **113**, 198–203.
Barnum, C. P., Jardetzky, Ch. D. and Halberg, F. (1957). *Tex. Rep. Biol. Med.* **15**, 134–147.
Barnum, C. P., Jardetzky, Ch. D. and Halberg, F. (1958). *Am. J. Physiol.* **195**, 301–310.
Blomqvist, K. (1957). *Acta path. microbiol. scand.* Suppl. **121**, 1–65.
Brues, A. M., Drury, D. R. and Brues, M. C. (1936). *Archs Path.* **22**, 658–673.
Bucher, N. L. R. (1963). *Int. Rev. Cytol.* **15**, 245–300.
Bullough, W. S. (1962). *Biol Rev.* **37**, 307–342.
Bullough, W. S. (1964). *In* "Cellular Control Mechanisms and Cancer" (P. Emmelot and O. Mühlbock eds). Elsevier. Amsterdam.
Bullough, W. S. and Laurence, E. B. (1964). *Expl. Cell Res.* **33**, 176–194.
Bullough, W. S. and Rytömaa, T. (1965). *Nature* **205**, 573–578.
Christensen, B. G. and Jacobsen, E. (1949). *Acta med. scand.* Suppl. **234**, 103–108.
Croisille, Y. (1958). *Archs. Anat. microsc. Morph. exp.* **47**, 360–400.
Echave Llanos, J. M. (1965). Unpublished.
Echave Llanos, J. M. and Bade, E. G. (1963). *Naturwissenschaften* **50**, 97–98.
Echave Llanos, J. M. and Bordin, C. (1963a). *Naturwissenschaften* **50**, 501.
Echave Llanos, J. M. and Bordin, C. (1963b). *Acta physiol. latino am.* **13**, 337–346.
Echave Llanos, J. M. and Sadnik, I. L. (1964). *Stain Technol.* **39**, 289–294.
Echave Llanos, J. M. and Vilchez, C. A. (1964). Unpublished results.
Echave Llanos, J. M., Bade, E. G. and Bordin, C. (1963). *Acta physiol. latino am.* **13**, 385–387.
Echave Llanos, J. M., Bordin, C. I. and Badran, A. F. (1964). *Revta. Soc. argent. Biol.* **40**, 40–46.
Echave Llanos, J. M., Strelecki, A. and Badran, A. F. (1965). Unpublished results.
Fabry, P and Hruza, Z. (1956). *Physiologia bohemoslov.* **5**, 142–148.
Fischer, A., Takacs, L. and Varga, S. (1964). *Acta hepato-splenol.* **11**, 89–93.
Formijne, P. (1955). *Acta clin. belg.* **10**, 81–87.
Glinos, A. D. (1958). *In* "The chemical basis of development" (McElroy and Glass eds) pp. 813–842. Johns Hopkins Press. Baltimore.
Glinos, A. D. (1960). *Ann. N.Y. Acad. Sci.* **90**, 592–602.
Halberg, F. (1957). *Experientia* **13**, 502.
Halberg, F. (1958). Lecture given at 1st. Internat. Symp. on submarine and space Me., New London, Conn.
Halberg, F., Barnum, C. P., Silber, R. H. and Bittner, J. J. (1958). *Proc. Soc. exp. Biol.* **97**, 897–900.

Halberg, F., Peterson, R. E. and Silber, R. H. (1959). *Endocrinology* **64**, 222–230.
Halberg, E., Galicich, J. H., Ungar, F. and French, L. A. (1965). *Proc. Soc. exp. Biol. Med.* **118**, 414–419.
Harkness, R. D. (1957). *Br. med. Bull.* **13**, 87–93.
Hemingway, J. T. (1961). *Nature* **191**, 706–707.
Hemingway, J. T. (1965). *J. Physiol.* **176**, 30 P.
Higgins, C. M., Berkson, J. and Flock, E. (1932). *Am. J. Physiol.* **102**, 673–682.
Higgins, G. M., Berkson, J. and Flock, E. (1933). *Am. J. Physiol.* **105**, 177–186.
Hruza, Z. (1956). *Čslká Fysiol.* **5**, 40–44.
Hruza, Z. and Smetana, R. (1958). *Physiol. bohemoslov.* **7**, 80–86.
Jackson, B. (1959). *Anat. Rec.* **134**, 365–377.
Jaffe, J. J. (1954). *Anat. Rec.* **120**, 935–954.
Jardetzky, Ch. P., Barnum, C. P. and Halberg, F. (1956). *Am. J. Physiol.* **187**, No 3.
Kelly, L. S. and Jones, H. B. (1953). *Am. J. Physiol.* **172**, 575–578.
Lahtiharju, A. and Teir, H. (1964). *Expl. Cell Res.* **34**, 205–207.
Leduc, E. H. (1949). *Am. J. Anat.* **84**, 397–429.
Liosner, L. D. (1964). *Folia biol., Praha* **10**, 472–479.
MacDonald, R. A., Rogers, A. E. and Pechet, G. S. (1963). *Ann. N.Y. Acad. Sci.* **111**, 70–84.
McJunkin, F. A. and Breuhaus, H. C. (1931). *Archs. Path.* **12**, 900–908.
Molimard, R. (1959). Thesis, pp. 1–120. R. Foulon, Paris.
Olmsted, J. M. D. and Olmsted, E. H. (1952). "Claude Bernard and the experimental method in medicine." Henry Schuman. New York, pp. 69.
Paschkis, K. E., Cantarow, A. and Goddard, J. W. (1957). *Fed. Proc.* **16**, 98.
Peters, R. (1962a). *Z. Naturf.* **17**, 164–168.
Peters, R. (1962b). *Z. Naturf.* **17**, 169–172.
Romanova, L. K. (1957). *Bull. exp. Biol. Med. U.S.S.R.* **43**, 99–101.
Russo, J. and Echave Llanos, J. M. (1964). *Z. Zellforsch. mikrosk. Anat.* **61**, 824–828.
Saetren, H. (1956). *Expl Cell Res.* **11**, 229–232.
Sokolova, L. V. (1964). *Bull. exp. Biol. Med.* (*U.S.S.R.*) **57**, 98–102.
Stich, H. F. (1960). *Ann. N.Y. Acad. Sci.* **90**, 603–609.
Stich, H. F. and Florian, M. L. (1958). *Can. J. Biochem. Physiol.* **36**, 855–859.
Teir, H. and Ravanti, K. (1953). *Expl Cell Res.* **5**, 500–507.
Teir, H. and Lahtiharju, A. (1961). *Expl Cell Res.* **24**, 424–428.
Thomas, A. H. (1961). Scientific Apparatus. General Catalogue, 358, 4288c.
Tumanishvili, G. D. (1960). *J. Embryol. exp. Morph.* **8**, 226–238.
Tumanishvili, G. D., Jandieri, K. M. and Svanidze, I. K. (1956). *Dokl. Acad. Nauk. SSSR, Otd. Biol.* **106**, 1107–1109.
Tumanishvili, G. D., Mandzhgaladse, V. P. and Dzhanelidse, N. (1963). *Bioquímica. U.S.S.R.* **28**, 942–950.
Weinbren, K. (1959). *Gastroenterology* **37**, 657–668.
Weinbren, K. and Tarsh, E. (1964). *Br. J. exp. Path.* **45**, 475–481.
Weiss, P. (1955). *In* "The hypophyseal growth hormone, nature and actions". (R. W. Smith, O. H. Gaebler, and C. N. H. Long eds.) pp. 3–16. The Blakiston Division, McGraw-Hill, New York.
Weiss, P. and Wang, H. (1941). *Anat. Rec.* Suppl. 2, **79**, 62.
Wilson, J. W. (1948). *Anat. Rec.* **101**, 672.
Wilson, J. W. (1958). *In* "Symposium on Liver Function", pp. 175–197. American Institute of Biological Sciences, Washington.
Wilson, J. W. and Leduc, E. H. (1947). *Anat. Rec.* **97**, 471–493.
Yashina, I. N. (1963). *Bull. exp. Biol. Med. U.S.S.R.* **55**, 101–105.

Humoral Regulation of Liver Regeneration

H. WRBA and H. RABES

German Cancer Research Center, Heidelberg
Department of Experimental Pathology
and
Department of Pathology, University of Munich, German Federal Republic

DESPITE the great number of papers on the subject the problem of humoral control of liver regeneration is still undecided (Bucher 1963, Leduc 1964). A point particularly in dispute is the organ specificity of the emerging factors. There are positive as well as negative results, depending on experimental arrangements and differing from case to case. In recent years we have developed a technique of quantitative determination of certain effects on an organ. Our method is to place explants of the organ to be examined in a medium containing radioactive matter (Wrba and Rabes 1962). The amount of labelled matter incorporated by the explant during a given period is measured. This has proved to be an objective measure of the vitality of the explants and of experimental effects. Despite the other methods of organ explantation based on morphological evaluation, the turnover of labelled matter yields significant insights even without any visible morphological change (Wrba *et al.* 1955). However, the weight of the explants is of decisive importance. Large pieces incorporate more labelled matter than smaller ones do. But there is an inverse proportion between large and small explants as regards the counts-weight ratio. Due to differing surfaces and greatly varying central necrosis, large explants show a great variation range in incorporation. In order to avoid central necrosis the ideal case would be an explant with dimensions ensuring a sufficient supply of O_2 and nutritive compounds to the central portions. This ideal size is below the limit of reliable manipulation in the case of the liver, the organ studied by us. The smallest size that can be produced with a fair degree of accuracy is equivalent to 1 to 1·5 mg wet weight. We have used cubic explants of this order (Wrba and Rabes 1962).

As a rule eighteen explants were weighed for each experiment, and incubated in a medium composed of salt solution, embryonic extract, and small amounts of labelled matter (Activity added: $^{32}PO_4$ = 0·2 μc/ml, ^{14}C compounds = 0·1 μc/ml, ^{3}H compounds = 1·0 μc/ml). By various modified methods the explants are kept in suspension mechanically (Wrba and Rabes

1962). The maximum duration for one culture is 48 hours. The usual duration is 24 hours, with only 10 hours for thymidine. After the period of cultivation the explants are washed several times in salt solutions containing the corresponding inactive compounds. Next, they are dried for counting.

TABLE I. Action of serum of partially hepatectomized rats on the incorporation of phosphate-32 by adult rat liver explants.

pulses/min/mg wet weight
n=6

Serum of normal rats A	Serum of rats 48 h after partial hepatectomy B	Difference per cent B against A
505±55	626±113	+23% $P<0{\cdot}05$

The activity incorporated is measured in an end-window counter as far as ^{32}P is concerned. Explants containing ^{14}C and ^{3}H are burnt in an oxygen atmosphere (Kalberer and Rutschmann 1961). After combustion the radioactive matter is present as $^{14}CO_2$ or as tritium-labelled water. These substances are absorbed in cooled methanol and estimated in a liquid scintillation spectrometer.

TABLE II. Action of serum of partially hepatectomized rats on the incorporation of phosphate-32 by embryonic rat liver.

p/min/mg wet weight
n=8

Serum of normal rats A	Serum of rats 48 h after partial hepatectomy B	Difference per cent B against A
2782±722	4032±529	+45% $P<0{\cdot}005$

Activity measurements in explants reveal the rate of phosphorylation with the aid of labelled phosphate, while protein synthesis is estimated by the rate of incorporation of labelled amino-acids, and labelled thymidine indicates the rate of DNA synthesis. Explants are extracted with cold trichloracetic acid, next with alcohol-ether, and finally with hot TCA, which renders separate fractions of acid-soluble cell contents, lipoids, nucleic acids, and proteins, each of which is measured individually!

In order to estimate humoral processes in liver regeneration we explant liver with controls. Blood serum of normal rats is added to the medium of the controls, while the experimental cultures receive serum from partially hepatectomized animals. With this arrangement we regularly find an increase in the intake of ^{32}P in the cultures with serum of partially hepatectomized animals.*

If adult liver is explanted and serum taken from partially hepatectomized rats 48 hours after surgery is added, increased incorporation is generally in evidence. But the increase is not highly significant (Table I). The situation is different if embryonic liver is explanted and the serum added is taken from partially hepatectomized rats 24 hours after surgery. This results in a significant increase with $P<0{\cdot}005$ (Table II). No action on kidney explants is demonstrable under the same experimental conditions (Table III).

TABLE III. Action of serum of partially hepatectomized rats on the incorporation of phosphate-32 by embryonic kidney.

p/min/mg wet weight.
n=19

Serum of normal rats A	Serum of rats 48 h after partial hepatectomy B	Difference per cent B against A
259±9	268±48	+3% $P<0{\cdot}5$

Another series of experiments served for investigation of the species-specificity, or otherwise, of the serum effect. It was found that serum from partially hepatectomized rats also stimulates the metabolism of mouse liver explants ($P<0{\cdot}025$). Conversely, serum from partially hepatectomized mice induces an increase of the same order in ^{32}P incorporation by rat liver explants. Even serum from partially hepatectomized golden hamsters induces a significant increase in the metabolism of rat liver explants as compared with controls in serum from untreated hamsters. No cross-action on kidney from other species is seen under these conditions (Table IV).

Concentration is a decisive factor in this system. The optimal concentration of serum for our purpose is 20% (Table V). Moreover, the enhancing action depends on the volume of liver resection. The stimulating action of blood serum is directly proportional to the weight of liver removed. The

* The degree of the increase varies within a very wide range. Remarkably enough the same serum, applied at the same time to tumor tissue cultures, reduces the incorporation of phosphate by tumor *in vitro* (Wrba and Rabes 1963).

TABLE IV. Incorporation of phosphate-32 by explants of embryonic organs of rat and mouse under the influence of serum of partially hepatectomized and untreated animals of other species.

p/min/mg wet weight
Mean values of n experiments with standard deviations.

Serum	Explants	Normal serum added	n	Serum 48 h after partial hepatectomy added	n	P
rat serum	mouse liver	382± 35	9	448± 50	6	<0·025
	mouse kidney	1138±109	7	1223±109	7	<0·5
mouse serum	rat liver	230± 25	9	260± 27	9	<0·025
	rat kidney	253± 34	8	276± 50	9	<0·2
golden hamster serum	rat liver	299± 46	4	448± 65	4	<0·01

TABLE V. Incorporation of phosphate-32 by rat liver explants under the influence of different concentrations of serum from partially hepatectomized rats. 18 explants per experiment. Percentual difference of the activities (p/min/mg wet weight) of the experimental culture (with serum from partially hepatectomized animals) against controls (with serum from untreated animals).

Serum concentration in nutrient medium			
33%	25%	20%	10%
−21	+23	+31	+22
−30	+ 7	+ 7	+ 4
−19	+ 8	+19	
− 6	+ 9	+36	
+ 4	+16	+23	
− 5		+19	
+ 4		+10	
− 3		+62	
− 5		+19	
− 5		+21	
		+71	
		+60	
		+15	
		+33	
		+11	
		+57	
		+47	

liver-stimulating factor presents in blood serum of partially hepatectomized animals is thermostable. No change in the stimulating effect is observed after heating the serum to 100°C for 5 min (Table VI). In order to prevent coagulation, the serum was diluted with distilled water before heating, with restitution of physiological salt concentration by addition of Hanks' solution concentrated tenfold. The humoral factor is independent of glucose concentration. Excess of glucose added to the nutrient medium causes no demonstrable change in the action of the factor.

TABLE VI. Incorporation of phosphate-32 by rat liver explants under the influence of native and heated blood serum of normal and partially hepatectomized rats (48 h after surgery). Percentual difference of activities (p/min/mg wet weight) of experimental cultures (with serum from partially hepatectomized rats) against controls (with normal rat's serum). 18 explants per experiment.

With native serum	With serum heated to 100°C for 5 min
+16%	+11%
+ 9%	+10%
+12%	+12%
+23%	+14%
+21%	+39%
+50%	+65%

Metabolism of organ explants is specific for the functional state as existing *in vivo* prior to surgery (Wrba and Rabes 1961). Incorporation of labelled precursor of DNA synthesis by embryonic liver greatly exceeds that of adult liver of the same species, obviously reflecting the emphasis on growth in the embryonic organ. If the mutilated liver itself is explanted after partial hepatectomy, its position *in vitro* is intermediate between embryonic and adult livers. Explants of regenerating liver display intense phosphorylation, and protein synthesis as well as DNA synthesis remain enhanced. By comparison with normal liver of untreated animals a measure is obtained *in vitro* for the metabolic deviation induced in the organ. Valine-^{14}C is incorporated in the protein fraction in a definitely higher degree as compared with normal adult liver (Fig. 1). As regards incorporation of thymidine-2-^{14}C by liver explants, regenerating liver shows distinct activity in nucleic acid synthesis, whereas normal adult liver incorporates practically no thymidine in the nucleic acid fraction (Fig. 2).

This technique of organ explantation lends itself to conclusions regarding the condition of the organ *in vivo*. We have applied it to another study in humoral factors. The question was: If we have animals in parabiosis, does partial hepatectomy of one mate induce a proliferation of liver cells in the other mate? There are divergent reports on the problem.

A comparison of liver explants from suitable animals results in significant differences *in vitro*. Three kinds of adult rat liver were used: 1. liver of normal parabiosis animals, 2. liver of partially hepatectomized parabionts, and 3. liver of normal mates of partially hepatectomized parabionts. Total

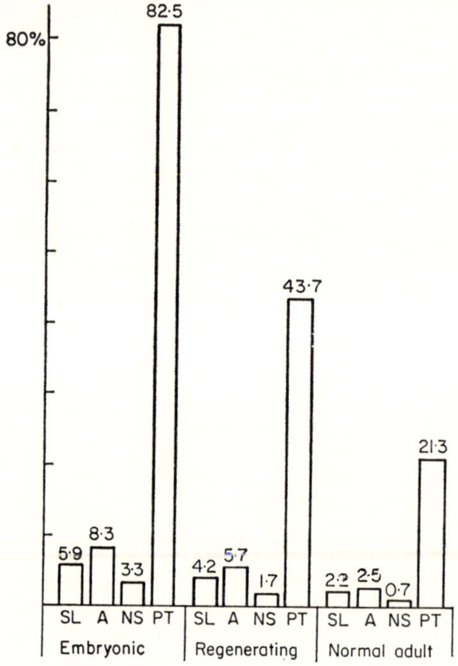

FIG. 1. Incorporation of L-valine-1-^{14}C by rat liver explants. Comparisons of embryonic, regenerating and normal adult livers. (Regenerating liver, 48 h after partial hepatectomy.)
Total incorporation by embryonic liver explants = 100 %.
Fractions: SL = soluble in 10 per cent trichloroacetic acid at 4°C
A = soluble in alcohol-ether (lipoids)
NS = soluble in 5 per cent trichloroacetic acid at 90°C (nucleic acids)
PT = insoluble residue (proteins)

incorporation of ^{32}P by liver explants from partially hepatectomized parabionts is increased by 104 per cent ($P < 0{\cdot}005$) over that by liver explants from normal parabiosis animals. Liver from the untreated mate of a partially hepatectomized animal incorporates *in vitro* more ^{32}P than normal liver from parabiosis rats does. The increase amounts to 45% ($P < 0{\cdot}01$) (Table VII).

Estimation of thymidine-^{3}H incorporation by liver explants yields even more definitive results with increased significance. The increase in thymidine

TABLE VII. Incorporation of phosphate-32 by liver explants from normal and partially hepatectomized parabiotic rats (48 h after surgery) and from mates of partially hepatectomized parabionts. Roller culture, 24 h p/min/mg wet weight. With standard deviations.

n=number of experiments with 18 explants each.

	Untreated parabiotic rats	Partially hepatectomized parabiotic rats	Untreated mates of partially hepatectomized parabionts
	n=8	n=4	n=4
p/min/mg wet weight	265±85	540±105	384±44
percentual deviation of B, C from A		+104% $P<0.005$	+45% $P<0.01$

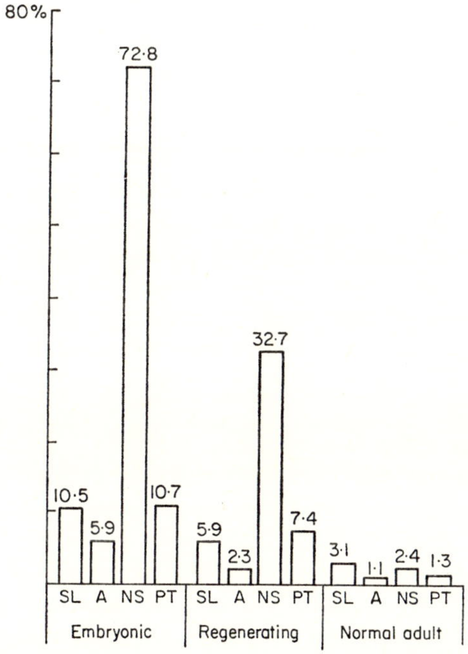

FIG. 2. Incorporation of thymidine-2-^{14}C by rat liver explants. Comparison of embryonic, regenerating and normal adult livers. (Regenerating liver, 48 h after partial hepatectomy.)
Total incorporation of embryonic liver explants = 100 %.
Symbols of fractions as in Fig. 1.

incorporation is 300% with $P<0{\cdot}001$ in liver explants from partially hepatectomized animals 48 hours after surgery as compared with liver of untreated parabiosis rats. DNA synthesis in the livers of untreated mates of partially hepatectomized parabionts is increased by 85% ($P<0{\cdot}001$) over that of liver explants of untreated parabionts (Table VIII).

TABLE VIII. Incorporation of thymidine-6-T (n) by liver explants from normal and partially hepatectomized parabiotic rats (48 h after surgery), and from mates of partially hepatectomized parabionts. Roller culture, 20 h p/min/mg wet weight with standard deviations.

n=number of experiments with 18 explants each.

	A	B	C
	Untreated parabiotic rats	Partially hepatectomized parabiotic rats	Untreated mates of partially hepatectomized parabionts
	n=12	n=6	n=6
p/min/mg wet weight	7033±1925	28667±5029	13040±1964
deviation percentual of B, C against A		+308% P<0·001	+85% P<0·001

Contradictory results of earlier studies seem to originate in the numerous individual sources of disturbance to be expected in the animals after dual surgery. Our indirect method may seem rather complex, but it has obvious advantages. It is highly sensitive, and above all the events in the explant cannot be disturbed by any processes in the rest of the body. Hence our results seem to indicate strongly that there is indeed humoral stimulation of DNA synthesis in the liver of the parabiosis mate of a partially hepatectomized animal.

The stimulating effect is temporarily maintained *in vitro* after removal of the humoral factor. The growth process, once started persists for some time.

As regards the origin of this liver-specific regenerative substance, it may be in the liver or elsewhere, but one thing is certain: the factor must be concentrated in the regenerating liver to some degree before it can become effective. No doubt it originates in some cellular synthesis and is released into the circulation. Now we have evolved a method of confrontation for explants of different organs or of animals treated differently. The explants are placed in a liquid medium, facing each other through a membrane filter. Since these membrane filters are obtainable with definitive pore sizes, it should be possible to estimate the size of the effective molecule passing the membrane. We have confronted explants of regenerating liver (taken out 20 hours after partial hepatectomy) with explants of normal juvenile rat liver (weight of animals, 100 g) in this manner which ensures that there is only humoral contact between the two kinds of explants. The upshot was an increase in nucleic acid synthesis in the normal liver explants (Table IX).

TABLE IX. Incorporation of thymidine–6–T (n) by liver explants from rats weighing 100 g after 20 h confrontation with liver explants from rats partially hepatectomized or laparotomized (controls) 20 h before. Mean values of 11 experiments with 25 liver explants per culture. Decay/min/mg dry weight of explants. Standard deviations.

Liver after confrontation with		
Liver from laparotomized rats (controls) A	Liver from rats partially hepatectomized 20 h before B	Difference of B against A
3425 257	3928 361	+15% P 0·005

Confronting kidney with partially hepatectomized liver under the same conditions, we saw no significant effect (Rabes and Wrba 1965) (Table X).

TABLE X. Incorporation of thymidine–6–T (n) by kidney explants from rats weighing 100 g after 20 h confrontation with liver explants from rats partially hepatectomized or laparotomized (controls) 20 h before. Mean values of 3 experiments with 25 kidney explants each. Decay/min/mg dry weight of explants. Standard deviations.

Kidney after confrontation with		
Liver from laparotomized rats (controls) A	Liver from rats partially hepatectomized 20 h before B	Difference of B against A
8032 ± 640	7716 ± 700	−4% $P<0{\cdot}5$

The last question is: How far is liver regeneration subject to central control? There have been indications to the effect that hypophysectomy delays the start of the mitotic wave in liver regeneration (Hemingway and Cater 1958). We performed hypophysectomy about four weeks before partial hepatectomy. In normal regeneration of rat's liver a maximum of DNA synthesis is reached about 20 hours after surgery. In hypophysectomized animals we found after 20 hours no measurable increase of the specific activity of DNA on application of thymidine-6-^{3}H (Fig. 3). This finding is confirmed morphologically by autoradiography. The onset of DNA synthesis

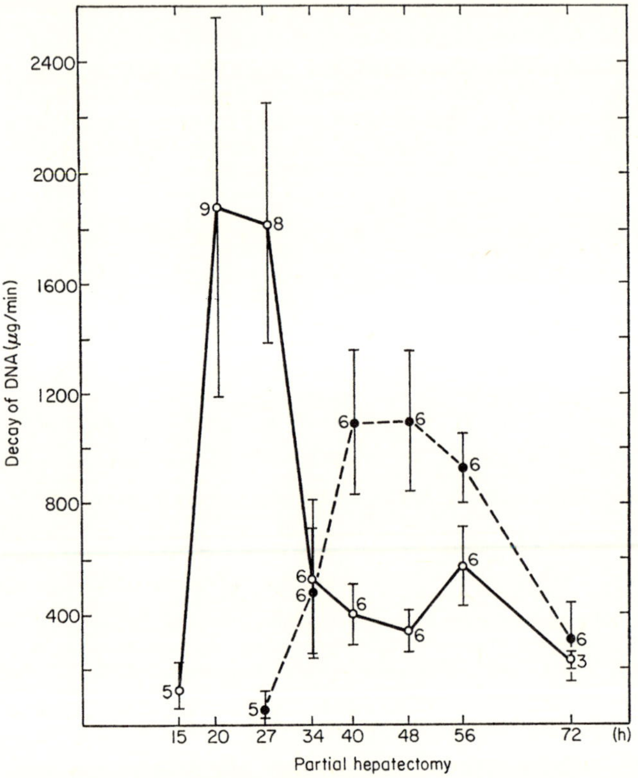

FIG. 3. DNA synthesis in regenerating liver of normal (● ——— ●) and hypophysectomized (○ - - - - - - ○) rats at different intervals after partial hepatectomy.

Abscissa: time of injection of 0.8 μC thymidine-6-T (n) (specific activity 1.9 C/mM) per body weight. Liver taken out 60 min later. DNA extracted.

Ordinate: decay/min/μg DNA. Mean values of number of animals noted at graph points. Verticals = standard deviations.

in the hypophysectomized rat is 27 to 34 hours after partial hepatectomy, while in normal liver regeneration DNA synthesis has already come to an end after this time.

Summing up, we would say that liver regeneration is controlled by a complex synergism of the endocrine system and organ-specific humoral factors which are concentrated, or formed, in the liver.

REFERENCES

Bucher, N. L. R. (1963). *Int. Rev. Cytol.* 15, 245–300.
Hemingway, J. T. and Cater D. B. (1958). *Nature* 181, 1065–1066.
Kalberer, F. and Rutschmann, J. (1961). *Helv. chimica Acta* 44, 1956-1966.
Leduc, E. H. (1964). *In* "The Liver" (Ch. Rouiller ed.). Academic Press, New York and London.
Rabes, H. and Wrba, H. (1965). *Exp. Cell Res.*, 39, 669–677.
Wrba, H. and Rabes, H. (1961). *Naturwissenschaften* 48, 646.
Wrba, H. and Rabes, H. (1962). *Exp. Cell Res.* 26, 62–69.
Wrba, H. and Rabes, H. (1963). *Cancer Res.* 23, 1116–1120.
Wrba, H., Seidler, E. and Allmann, K. (1955). *Naturwissenschaften* 42, 260.

Humoral Factors in Liver Cell Proliferation

M. VIROLAINEN

*II Department of Pathology, University of Helsinki, Helsinki, Finland**

I. Introduction

THE mechanism which regulates the compensatory growth of the liver after partial hepatectomy is unknown. Many different approaches to the problem of demonstrating the existence of a humoral control mechanism have been tried, but the results have been contradictory (Bucher 1963, Poole 1966). However, the observation that liver autograft responds to partial hepatectomy by cell proliferation offered strong evidence for the involvement of a humoral effect in the compensatory growth of the liver (Leong *et al.* 1964, Virolainen 1964a).

The aim of the present experiments was to elucidate further the mechanism that controls liver growth. A more detailed study was made of the effects of partial hepatectomy on liver autograft, with special reference to the specificity of the humoral stimulus. An investigation was also made into the cause of the inconsistent results in experiments in which parabiosis technique had been used, in the hope that this could shed light on the problem as a whole. Because changes in the plasma proteins occur after hepatectomy, and because these changes have been postulated to have a role in the regulation of liver growth (Glinos 1958), it seemed reasonable to study these changes in both parabiotic members after hepatectomy in one partner. Experiments were also performed in order to study further the stimulative effect of the dilution of plasma on liver cell division by plasmapheresis (Glinos and Gey 1952). Tentative speculations are presented on the relation of this functional demand to liver size and to post-hepatectomy liver growth.

II. Transplantation Experiments

Autotransplantation of small pieces of liver tissue to the peritoneum in mesorchium and omentum was performed on young adult Sprague-Dawley rats. A slightly modified two-stage pedicle grafting method developed by Myren and Vinje (1952) was used. This method has been described in more

* Present address: The Wistar Institute of Anatomy and Biology, Philadelphia, Pennsylvania.

detail elsewhere (Virolainen 1964a). In one series small pieces of liver tissue were also implanted subcutaneously. The size of each graft was only a few millimeters in diameter. One month after transplantation the hepatic architecture with cords of hepatocytes and sinusoids could be discerned in many

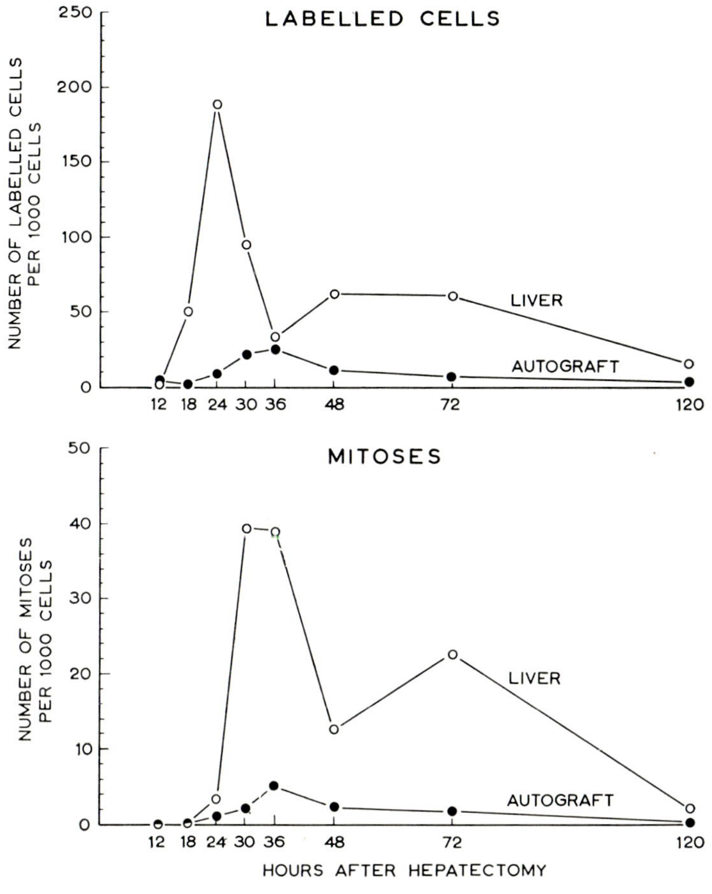

FIG. 1. Mitotic and labelling indices in liver and liver autograft at various intervals after two-thirds hepatectomy.

grafts, but bile duct proliferation and collagen formation was often pronounced. In older grafts the number of both hepatocytes and bile ducts decreased, but viable liver cells were found in all peritoneal grafts even after 250 days.

The effects of two-thirds and one-fourth hepatectomy and of sham operation on grafts about 30 days old were studied at various intervals after the operation. One hour before killing, 0·3 μc of tritiated thymidine per gram of body weight was injected intraperitoneally. The animals were always killed between 8.30 and 10.30 a.m. Autoradiograms were prepared using the stripping film technique (Kodak AR 10). Mitoses were counted in at least 5000 cells and labelled nuclei were counted in at least 2000 cells in each liver and autograft. The results of these counts after two-thirds hepatectomy are given in Fig. 1. Each point in this diagram and in subsequent diagrams represents the mean value obtained from 6 rats. Curves for the time course

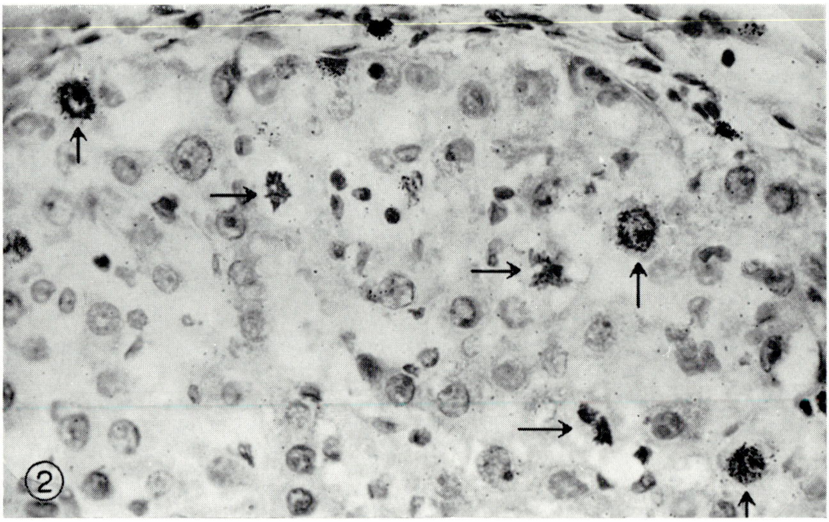

FIG. 2. Autoradiogram of peritoneal liver autograft 36 hours after two-thirds hepatectomy. Horizontal arrows indicate mitotic figures and vertical arrows indicate labelled hepatocytes. Hematoxylin stain. ×520.

of mitoses and of labelled cells in the remaining liver are in agreement with results reported by others (e.g. Grisham 1962, Edwards and Koch 1964). The peak in labelled cells occurred at 24 hours, and the peak in mitoses occurred at 30 to 36 hours. In the autografts, too, a highly significant increase in the number of mitoses and DNA synthesizing cells was found after partial hepatectomy. There seemed to be a delay in the reactions of the autografts as compared to those of the liver. The highest activity of cell proliferation in the autograft was about one-seventh that of the liver.

Mitoses and labelled cells in liver autograft 36 hours after hepatectomy are shown in Fig. 2. In the cytoplasm of the hepatocytes distinct changes were observed after hepatectomy. In the grafts of the unoperated and sham operated rats, the cytoplasm appeared loose and contained vacuoles (Fig. 3). After hepatectomy the cytoplasm of the hepatocytes in the grafts was finely granular and evenly basophilic (Fig. 4). The basophilic

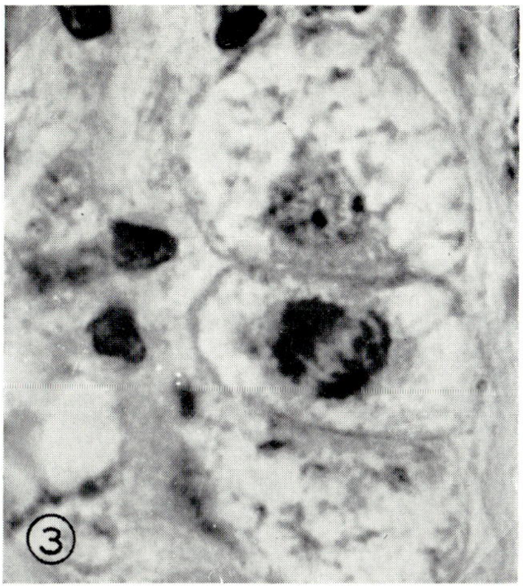

FIG. 3. Subcutaneous liver autograft 8 days after implantation. Cytoplasm is heavily vacuolated. One hepatocyte is in mitosis. Hematoxylin and eosin stain. × 1200.

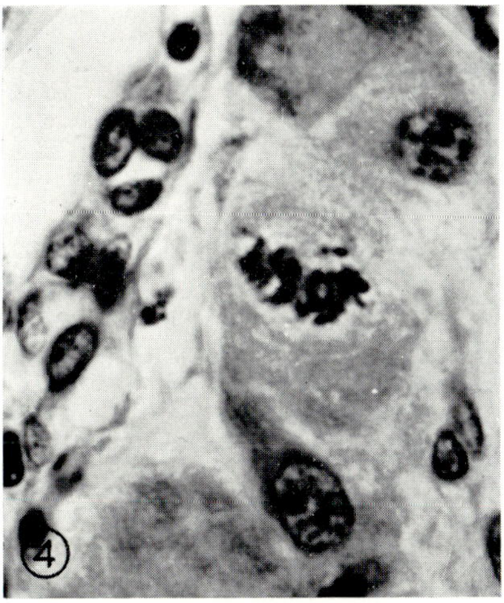

FIG. 4. Subcutaneous liver autograft from the same rat as the graft illustrated in Fig. 3 10 days after implantation and 48 hours after two-thirds hepatectomy. Cytoplasm is finely granular and evenly basophilic. One hepatocyte is in mitosis. Hematoxylin and eosin stain. × 1200.

bodies containing RNA could not be distinguished to the same extent as they could in the control grafts. The changes in the appearance of the basophilic bodies were parallel to the changes found in the hypertrophic liver (e.g. Glinos 1958, Bade 1964).

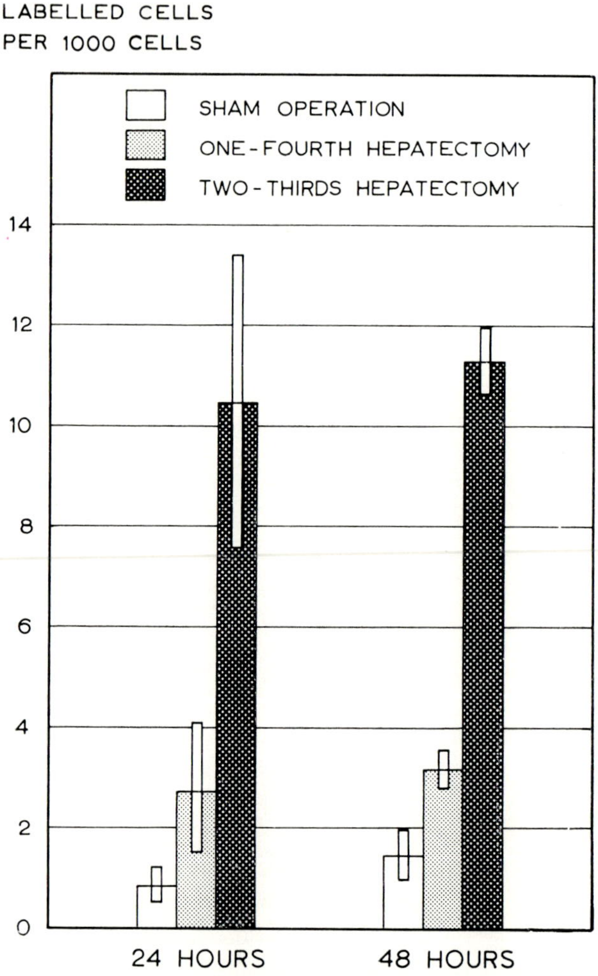

FIG. 5. Labelling indices (mean ±SE) of hepatocytes in liver autografts 24 and 48 hours after sham operation, one-fourth hepatectomy, and two-thirds hepatectomy.

When only one-fourth of the liver (the left lateral lobe) was removed instead of two-thirds, then the proliferative response in the graft was weaker (Fig 5).

In another experiment the mitotic response after hepatectomy was studied as a function of the age of the graft. At various times after grafting, 2 to 250 days, two-thirds hepatectomy was performed, and one liver graft was removed. Forty-eight hours later the animals were killed and the second graft removed. Figure 6 gives the mitotic indices in the grafts. It can be seen that the grafts responded to hepatectomy by proliferation and that the mitotic response was higher when the age of the graft was longer.

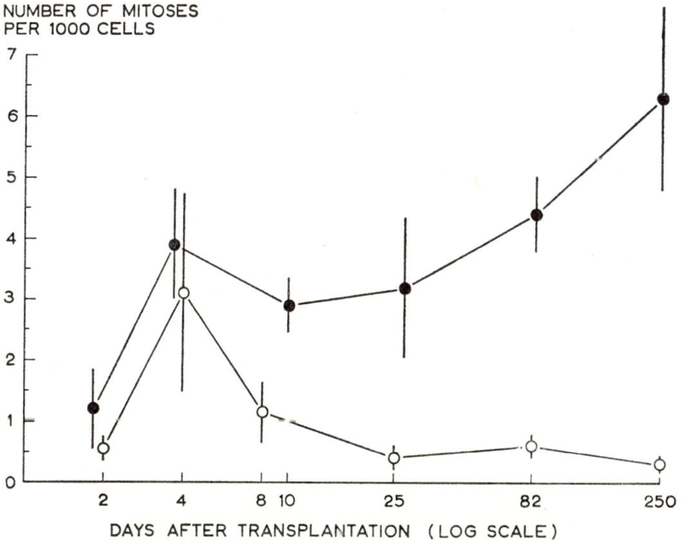

FIG. 6. Mitotic indices in mesorchium grafts at various intervals after transplantation. The lower curve (white circles) indicates the grafts removed at the time of two-thirds hepatectomy, the upper curve (black circles) indicates the grafts 48 hours after two-thirds hepatectomy. Vertical lines indicate SE.

No detectable difference in the response could be found in the grafts transplanted by a similar method to the mesorchium and omentum. The subcutaneous grafts could not always be found after longer periods of time. In some of these grafts all hepatocytes had atrophied. The mitotic response to hepatectomy was somewhat lower in subcutaneous grafts than in peritoneal grafts at all times after transplantation.

In order to study the specificity of the proliferative response in the autografts, the number of labelled biliary ductal cells was counted both in livers and grafts (Fig. 7). In the liver the labelling of ductal cells lagged behind that of the hepatocytes, in agreement with the results obtained by Grisham (1962). In the grafts no significant differences in labelling of ductal cells were found, either at different times after two-thirds hepatectomy or after the sham operation.

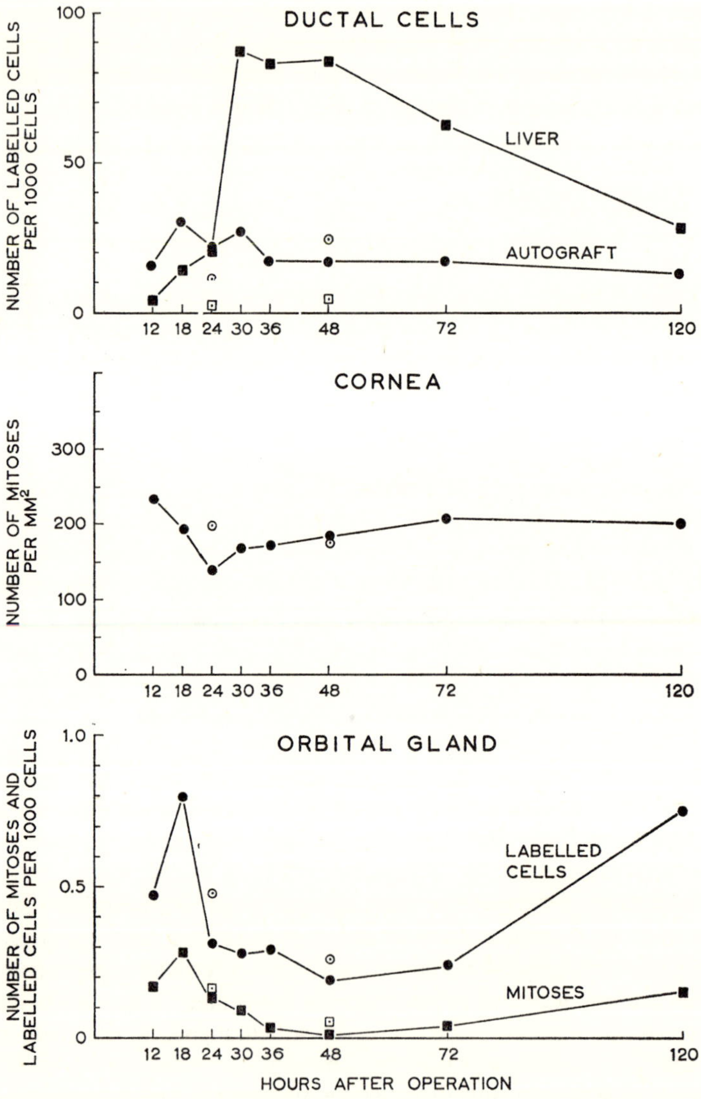

FIG. 7. Labelling indices in biliary ductal cells of liver and liver autograft, mitotic indices in corneal epithelium, and mitotic and labelling indices in outer orbital gland at various intervals after hepatectomy and after sham operation. Black circles and squares indicate the mean values after two-thirds hepatectomy, and white circles and squares indicate respective values after sham operation.

Mitoses and labelled cells per 50,000 acinar cells in the outer orbital gland, and mitoses in the corneas prepared by the method of Gay and Kaufmann (1950), were also counted. Contrary to the reports of others (Paschkis *et al.* 1959, Vinogradova 1960, Rucci *et al.* 1964), no increase was detected in the mitotic indices in the corneal epithelium (Fig. 7). In the orbital gland a decrease was observed in the indices of dividing and DNA synthesizing cells at the 2nd and 3rd day after hepatectomy, but no significant difference was found in either the mitoses or the labelled cells between hepatectomized and sham operated animals.

III. Parabiosis Experiments

The results and conclusions from the experiments in which parabiotic technique was applied to studies of regulation of the compensatory hypertrophy of the liver are conflicting. The detected stimulation has generally been either inconsistent or relatively weak (Bucher *et al.* 1951, MacDonald *et al.* 1963, Fisher *et al.* 1963), or no stimulation has been observed at all (Islami *et al.* 1959, Rogers *et al.* 1961, Asamoto 1964). In some experiments in which stimulation has been reported, the adequacy of the controls appears questionable (Christensen and Jacobsen 1949, Wenneker and Sussman 1951). Inefficacy in the exchange of blood has often been considered the cause of the lack of or the weakness of the response in these experiments. However, Alston and Thomson (1963) failed to detect stimulation even in a cross circulation experiment in which the rate of blood exchange was calculated to be about a quarter of the blood volume of each animal per minute and about 40 times as rapid as in the standard parabiotic procedure.

Experiments were planned in order to study the possibility, that even though a humoral factor existed in its circulation, the intact liver of the partner of the hepatectomized rat would for some reason be incapable of proliferative response. This lack of response could be due to local factors, such as portal pressure and the volume flow of portal blood. Autotransplanted liver tissue offered a way to study this problem, because the grafts in each parabiont would supposedly be in a similar position in regard to local factors.

Parabiosis was established between randomly bred male litter mates by the open coelomic method. Liver grafting was performed on the mesorchiums and omentum of each parabiont. The actual experiments were performed after 3 months of parabiotic union and 1 month after grafting of the liver fragments. All 11 pairs used demonstrated effective cross circulation as determined by the phenolsulfonphthalein method (Rogers *et al.* 1961). Two-thirds hepatectomy was performed on one member and a sham operation on the other. A biopsy was taken of one graft from each parabiont in connection with these operations. The pairs were killed 48 hours later. Results are summarized in Table I.

Increase in cell division could not be detected in the livers of the sham operated partners, but in the hepatectomized parabiotic rats the restoration of liver mass was significantly decreased in comparison to single rats, expressed either as liver weight indices (ratio of liver weight to body weight) or as the ratio of liver weight at killing to the weight of resected lobes.

TABLE I. Indices of mitosis and liver weight 48 hours after two-thirds hepatectomy and sham operation in single and parabiotic rats.

	Single rats		Parabiotic rats	
	Hepatectomy	Sham operation	Hepatectomy	Sham operation
Number of rats	25	16	11	11
Mitoses per 1000 cells[a]	15·7±2·4	0·01±0·01	21·7±2·3	0·02±0·01
Liver weight at killing[a] / Weight of resected lobes	0·98±0·03	.	0·79±0·02	.

[a]Mean ±SE.

However, the mean mitotic index from livers of the parabiotic rats did not differ significantly from that of the single rats. When the mitotic indices and weight indices of individual parabiotic rats were compared, it appeared that there existed a strong positive correlation ($r = +0·97$) (Fig. 8). On the other hand, no correlation was detectable in single rats. Precise explanation of this finding cannot be given from the available data, but apparently the frequency curve for mitoses after hepatectomy in parabiotic rats differs from that in single rats.

The mitotic indices in the liver grafts of parabiotic rats and comparable data for single rats from earlier experiments are shown in Fig. 9. The mitotic indices at the time of operation were of the same magnitude. The increase in mitoses in the grafts of the hepatectomized partners was similar to that of the single rats. There was no increase of the mitotic index in liver grafts of the sham operated partners in comparison to that of the single sham operated controls.

In two of the pairs there was an accidental opportunity to study the role of humoral and local factors in another way. In these pairs the tip of the triangular part of the right liver lobe was attached by fairly strong adhesions to the scar tissue in the abdominal muscles at the site of coelioanastomosis, and these livers belonged to the sham operated partners. One of these two specimens showed only one mitosis per 50,000 cells, but in the other one

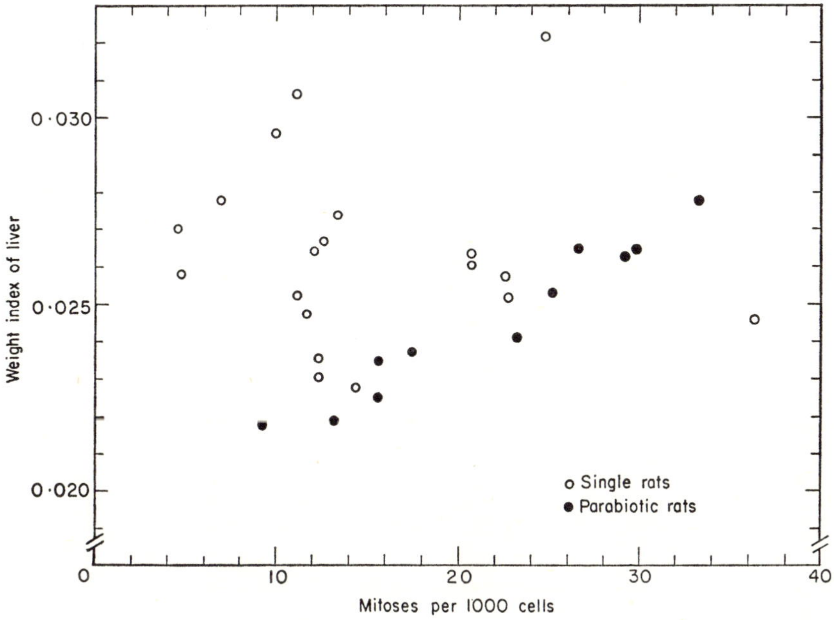

FIG. 8. Relationship between mitotic indices and weight indices of the liver in single and parabiotic rats 48 hours after two-thirds hepatectomy.

there were numerous mitoses both in the parenchymal and littoral cells. The mitotic index of the hepatocytes was 0·2%, whereas in the ordinary liver sample taken from the middle of the same lobe no mitoses were found in the counted 50,000 cells.

These solitary findings, as well as the results obtained from the mitotic counts in the grafts of parabiotic rats, are interpreted to show that local factors do not play any substantial role in the mitotic response. It seems probable that the lack of response in the partners of hepatectomized rats is solely due to an insufficient blood exchange between the partners.

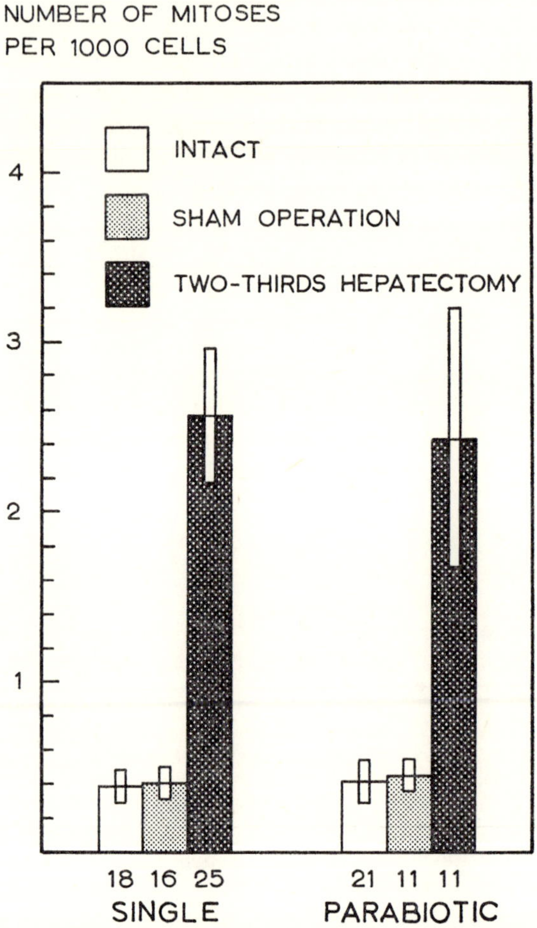

FIG. 9. Mitotic indices (mean ± SE) in liver autografts in single and parabiotic rats at the time of operation (intact), and 48 hours after hepatectomy and sham operation. Numbers under the columns indicate the number of animals.

IV. Serum Protein Determinations

Serum protein determinations were made from the same 11 parabiotic pairs used in the experiments described above, and in addition from 14 single rats, of which 4 had been hepatectomized, 5 had been sham operated, and 5 had not been operated on. Total serum protein concentration was determined by the biuret method, and quantitative analyses of serum protein fractions were performed with cellulose acetate electrophoresis method.

Albumin, α_1- and α_2-fractions separated fairly well. The β_1- and β_2-globulins did not separate distinctly and were therefore taken as a unit. Table II shows the relative amounts of the various serum protein fractions both in single and parabiotic rats 48 hours after the operation. In single hepatectomized rats the most distinct change, in comparison to both the sham operated and intact rats, was the decrease in the α-globulins. This

TABLE II. Relative amounts of serum proteins 48 hours after sham operation and two-thirds hepatectomy.

	Single rats				Parabiotic rats†		
	Intact	Hepa-tectomy	Sham operation	Difference‡	Hepa-tectomy	Sham operation	Difference‡
Number of rats	5	4	5	·	11	11	·
Albumin	42·3	35·3	34·8	+1%	40·3	40·2	0%
α_1-globulin	13·7	11·4	15·4	−26%	16·6	16·2	+2%
α_2-globulin	6·3	5·2	7·3	−29%	6·6	6·9	−4%
β-globulin	23·8	29·8	27·1	+10%	26·0	25·6	+2%
γ-globulin	14·0	18·4	15·4	+19%	10·5	11·0	−5%

† One member of each pair was hepatectomized and the other one sham operated.
‡ Difference in hepatectomized rats in comparison to sham operated rats.

finding is in agreement with the results reported by Morgan and Brackenridge (1962). There were fairly clear differences in the globulin fractions between hepatectomized and sham operated single rats, but not between the hepatectomized and sham operated parabiotic partners (Table II). Between the pairs there were differences, but in all cases the electrophoretic pattern between the members of each pair was remarkably similar. The total protein concentration was lower in the hepatectomized partners, but the difference lacked statistical significance.

These results indicate that the vascular anastomosis is sufficient to maintain the major protein fractions in equilibrium, and they therefore seem to exclude the possibility that in the post-hepatectomy hypertrophy the total protein concentration or any of the major protein fractions could be a direct regulator for liver cell division.

V. Plasmapheresis Experiments

In experiments in which the regulation of liver growth has been studied, the experimental aim has usually been to study the effect of different experimental conditions on the response after partial hepatectomy. Another approach would be to try to stimulate liver cells to proliferate without any surgical intervention. Studies in which serum or plasma from hepatectomized animals has been injected into intact animals, or in which different liver extracts or homogenates have been injected into intact recipients, have yielded equivocal results (reviewed, e.g. by Bucher 1963, Poole 1966). With the exception of the effect of exposure to certain drugs, the dilution of plasma by plasmapheresis has been the only way to get a substantial increase in cell division in the intact liver of adult rats (Glinos and Gey 1952).

The elevation of the mitotic count as well as the reported increase in the incorporation of ^{32}P into liver DNA after plasmapheresis (Naora and Naora 1957) are not reliable measures of the net growth. An increased mitotic index does not indicate any increase in the mitotic rate as such, and an elevated mitotic rate in turn could result from an increased turnover rate, e.g. due to cell degeneration and necrosis. In plasmapheresis, in which blood is withdrawn in appreciable amounts from the animal, cell death could result from anoxia. Indeed, an increased mitotic activity has been observed in the human liver in connection with ischaemic renal tubular necrosis, although histologic hepatic changes could not be distinguished (Bywaters 1946, Simson and Pepler 1960). Survis and Kennedy (1963) have reported an elevation in the mitotic index without associated increase in liver mass after various experimental procedures.

It thus seemed necessary to establish whether the reported increase in liver cell mitoses after plasmapheresis is associated with net growth of the liver mass, by studying the effect of prolonged lowering of the concentration of plasma constituents. A preliminary report on this study has been published earlier (Virolainen 1964b).

Twenty-eight young adult male rats were used in this experiment. In one group plasmaphereses were performed daily for 7 days; another group was sham treated; and still another was not treated at all. The rats on which plasmapheresis was to be performed were anesthetized with ether, a small skin incision was made, and blood corresponding to 2·5% of body weight was withdrawn from the jugular vein, alternating the side of the puncture. After separation of plasma by centrifugation, the cells were suspended in an equal volume of 6% clinical dextran solution and reinjected into the vena dorsalis penis. The procedure lasted about 20 minutes. At the time of the 4th treatment, a laparotomy was made, and a small biopsy, about 3 mm in diameter, was taken from the margin of the left liver lobe. The sham treated animals underwent the same procedures, but the blood was reinjected without substitution of plasma by dextran. One day after the 7th treatment, the animals were killed by exsanguination. The weight indices of liver, kidneys, and spleen, and the percentage of dry material in the liver were

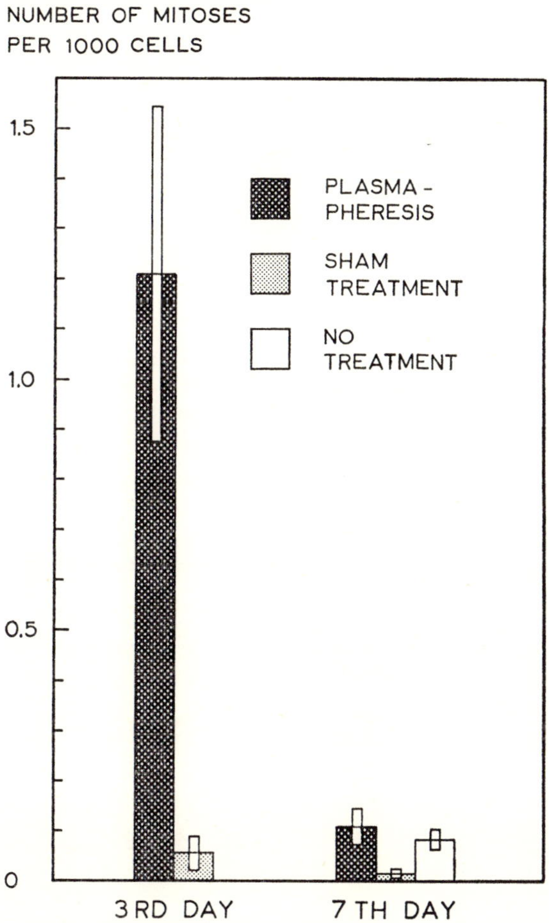

FIG. 10. Mitotic indices (mean ± SE) in liver biopsies on the third day after the beginning of daily plasmaphereses and sham treatments, and on the seventh day, when the animals were killed.

determined. Plasma protein concentration was determined in the plasmapheresis group after each treatment, and in all groups at the time of exsanguination.

The results of this experiment are summarized in Table III. Untreated rats gained weight during the experimental period. In both treated groups there was a decrease in weight during the two first days and after the

TABLE III. Results from the plasmapheresis experiment

	No. of rats	Change in body weight (%)	Plasma protein (%)	Liver weight index† (%)	Liver dry material† (%)	Estimated change in liver weight‡ (%)	Mitoses per 1000 hepatocytes†	
							3rd day	7th day
Plasmapheresis	9	−5·6	4·44	4·10±0·09	28·7±0·1	+7·4	1·21±0·34	0·11±0·04
Sham treatment	8	−6·0	5·14	3·48±0·06	29·6±0·3	−9·2	0·05±0·04	0·01±0·01
No treatment	10	+6·4	5·15	3·60±0·06	30·1±0·2	+6·4	..	0·08±0·02

† Mean ± SE.
‡ Explanation in text

laparotomy, but the changes in weight were equal in both groups. The plasma protein concentration declined in the plasmapheresis group by about 15% after the first treatment and maintained a fairly constant level thereafter.

In the plasmapheresis group there was a distinct elevation in the mitotic indices in biopsy specimens taken after 3 plasmaphereses, but at the time of killing the index did not differ significantly from the indices of the other groups (Fig. 10). The mitotic activity had also resulted in a significant increase in the liver weight index. The livers of the plasmapheresis group contained a little more water, but the increase in the weight index remains significant even if determined on the basis of dry material. If the actual change in liver weight is estimated, assuming that the liver weight index in all groups at the beginning of the experiment was 0·0361, which is the index of the untreated rats at the time of killing, it appears that in the sham treated animals the liver weight had decreased, and the change was relatively greater than the decrease in body weight. In the plasmapheresis group the liver weight had increased, and the gain in weight was even greater than in the untreated animals in spite of the reverse changes in body weights.

In the kidney weights there were no detectable differences between the groups. The spleens of both of the treated groups were larger than those of the untreated rats. This might be attributed to the operative procedure.

To rule out the possibility that the increase in liver weight would be caused by accumulation of dextran in liver, which has been reported to occur (Friberg *et al.* 1953, Turner and Maycock 1958), the amount of dextran in the livers was determined by the method of Turner and Maycock (1958). The livers of the plasmapheresis group contained on the average 0·6% dextran, which is only a fraction of the observed difference of about 17% between the liver weights of plasmapheresis and sham treatment groups.

Some of the rats of this and of another plasmapheresis series have liver autografts. No stimulation of liver cell division was detected in any of these grafts though cell division in the liver itself was increased.

VI. Summary and Conclusions

An attempt to elucidate the regulatory mechanisms in the compensatory hypertrophy of the liver was made by different experimental methods. Experiments in which the effect of partial hepatectomy on autotransplanted liver tissue was investigated gave strong support to the existence of a humoral control mechanism which is not restricted to the portal blood. The proliferative response was found to be proportional to the size of the stimulus. The effect was limited to the parenchymal liver cells. The ductal cells of the autografts did not respond to hepatectomy. This indicates that there exists a cellular specificity within the liver. No stimulation of cell division that could have been attributed to hepatectomy was found in extra-hepatic tissues.

The results of the experiments in which parabiotic rats were used did not give support to the postulated contribution of local factors in the regulation of compensatory growth. The findings in the serum proteins in parabiotic rats were interpreted to exclude the possibility that in the post-hepatectomy hypertrophy the total protein concentration or any of its major fractions could influence the proliferative response to any significant degree.

The observation that liver cell division can be induced by dilution of the plasma constituents following plasmapheresis was confirmed, and it was also found that the increase in cell proliferation results in net growth of the liver mass. However, repeated plasmaphereses seemed to cause only one mitotic wave in spite of the constantly lowered level of plasma proteins. This finding supports the conclusion made from studies in parabiotic rats that the decrease in protein concentration could not be a dominant trigger for liver cell division after hepatectomy. Under non-hepatectomy conditions, low protein concentration probably acts as a functional demand provoking cell division until a new equilibrium between liver mass and protein concentration is established. This seems to indicate the existence of a second feedback link, comparable to the link between low oxygen tension and erythropoiesis (Bullough and Rytömaa 1965). Grafted liver tissue possibly responds in a different way to hepatectomy and to plasmapheresis which could indicate a difference in the mechanism by which these stimulants of liver cell proliferation operate, but further experimentation is needed before definite conclusions on this point can be drawn.

It is thus postulated that there exists a specific humoral factor in the compensatory hypertrophy of the liver after partial hepatectomy, and also another link which regulates liver mass in response to physiological needs. The humoral factor obviously is not any of the major plasma protein fractions, and the factor is obviously very short-lived. The results do not indicate whether this factor is an inhibitor or a stimulator for cell division.

These studies have been supported by grants from Damon Runyon Memorial Fund (DRG 644), President J. K. Paasikivi's Foundation and Sigrid Juselius Foundation.

REFERENCES

Alston, W. C. and Thomson, R. Y. (1963). *Cancer Res.* 23, 901–905.
Asamoto, I. (1964). *Nagoya J. med. Sci.* 27, 119–141.
Bade, E. G. (1964). *Virchows Arch. path. Anat. Physiol.* 337, 503–514.
Bucher, N. L. R. (1963). *Int. Rev. Cytol.* 15, 245–300.
Bucher, N. L. R., Scott, J. F. and Aub, J. C. (1951). *Cancer Res.* 11, 457–465.

Bullough, W. S. and Rytömaa, T. (1965). *Nature, Lond.* **205**, 573–578.
Bywaters, E. G. L. (1946). *Clin. Sci.* **6**, 19–39.
Christensen, B. G. and Jacobsen, E. (1949). *Acta med. scand. Suppl.* **234**, 103–108.
Edwards, J. L. and Koch, A. (1964). *Lab. Invest.* **13**, 32–43.
Fisher, B., Fisher, E. R. and Saffer, E. (1963). *Cancer Res.* **23**, 914–920.
Friberg, U., Graf, W. and Åberg, B. (1953). *Acta pharm. tox.* **9**, 220–234.
Gay, H. and Kaufmann, B. P. (1950). *Stain Technol.* **25**, 209–216.
Glinos, A. D. (1958). *In* "The Chemical Basis of Development" (W. D. McElroy and B. Glass eds.), pp. 813–848. Johns Hopkins Press. Baltimore.
Glinos, A. D. and Gey, G. O. (1952). *Proc. Soc. exp. Biol. Med.* **80**, 421–425.
Grisham, J. W. (1962). *Cancer Res.* **22**, 842–849.
Islami, A. H., Pack, G. T. and Hubbard, J. C. (1959). *Surgery Gynec. Obstet.* **108**, 549–554.
Leong, G. F., Grisham, J. W., Hole, B. V. and Albright, M. L. (1964). *Cancer Res.* **24**, 1496–1501.
MacDonald, R. A., Rogers, A. E. and Pechet, G. S. (1963). *Ann. N.Y. Acad. Sci.* **111** 70–84.
Morgan, E. H. and Brackenridge, C. J. (1962). *Q. J. exp. Physiol.* **47**, 66–73.
Myren, J. and Vinje, T. (1952). *Acta path. microbiol. scand. Suppl.* **93**, 173–181.
Naora, H. and Naora, H. (1957). *Gann* **48**, 255–261.
Paschkis, K. E., Goddard, J., Cantarow, A. and Adibi, S. (1959). *Proc. Soc. exp. Biol. Med.* **101**, 184–186.
Poole, B. (1966). *Advances in Morphogenesis* **5**, 93–119.
Rogers, A. E., Shaka, J. A., Pechet, G. and MacDonald, R. A. (1961). *Am. J. Path.* **39**, 561–578.
Rucci, F. S., Satta, U. and Tota, G. (1964). *Boll. Soc. ital. Biol. sper.* **40**, 1183–1186.
Simson, I. W. and Pepler, W. J. (1960). *J. Clin. Path.* **13**, 210–213.
Survis, J. and Kennedy, R. (1963). *Fedn Proc. Fedn Am. Socs exp. Biol.* **22**, 192.
Turner, P. A. and Maycock, W. d'A. (1958). *J. Path. Bact.* **75**, 145–150.
Vinogradova, G. A. (1960). *Bull. Eksp. Biol. Med.* (*U.S.S.R.*) **50**, 105–108.
Virolainen, M. (1964a). *Expl Cell Res.* **33**, 588–591.
Virolainen, M. (1964b). *In* "Proceedings of the XIV Scandinavian Congress of Pathology and Microbiology", pp. 183–184. Universitetsforlaget. Oslo.
Wenneker, A. S. and Sussman, N. (1951). *Proc. Soc. exp. Biol. Med.* **76**, 683–686.

Studies with Autoradiography and Cytophotometry in the Rat Liver Following Partial Hepatectomy

E. GRUNDMANN

Institut für experimentelle Pathologie der Farbenfabriken Bayer A. G.
56 Wuppertal-Elberfeld, German Federal Republic

I. Introduction

ONE of the most fascinating and, at the same time, least understood phenomena in the growth of differentiated tissue in man, animals, and plants is that of tissue-specific polyploidy or polyteny. It means that the nuclei of differentiated parenchyma can be grouped according to size, DNA content, and number of chromosomes into nuclear class series in which growth takes place by doubling. Thus, for example, the vascular buds of root tips of plants regularly contain both diploid and polyploid nuclei, and in mammalian tissues we often find a very characteristic polyploidy. Most of the epithelial cells of the rat liver has twice as much DNA as the diploid cells of the same animal, and the majority of rat hepatocytes are thus tetraploid; fewer of them—as has been shown among others by Hoffman *et al.* (1955)—are diploid or octoploid. This "tissue-specific ploidy pattern" does not appear in the respective organs until post-natal functional activity begins. The liver epithelial cells of the newborn mouse are diploid, but many tetraploid and octoploid nuclei are present as early as the second week of life (Inamdar, 1958).

II. Alterations in Somatic Polyploidy within 24 hours after Partial Hepatectomy

In a first test series we studied the effect of two-thirds partial hepatectomy on tissue-specific polyploidy in the rat. The animals used were 45 Albino rats of both sexes, weighing 100 to 130 g, and fed on Latz dry feed with water *ad libitum*. The two large, anterior liver lobes, representing 65 to 70% of the parenchyma of the organ, were removed under ether anaesthesia. The animals were killed, by groups, ½, 1, 1½, 2, 5, 6, 12, 18, 24, 30, 42, 48, 54, and 72 hours later, and pieces of residual liver immediately immersed in neutral formalin for 48 hours for fixation. They were then washed for 24 hours in running water and slowly embedded in paraffin by a standard technique over a period of eight days. From the paraffin blocks sections for autoradiography and for cytophotometry were prepared. (For details of the techniques used see Grundmann and Bach (1960), and Oehlert *et al.*

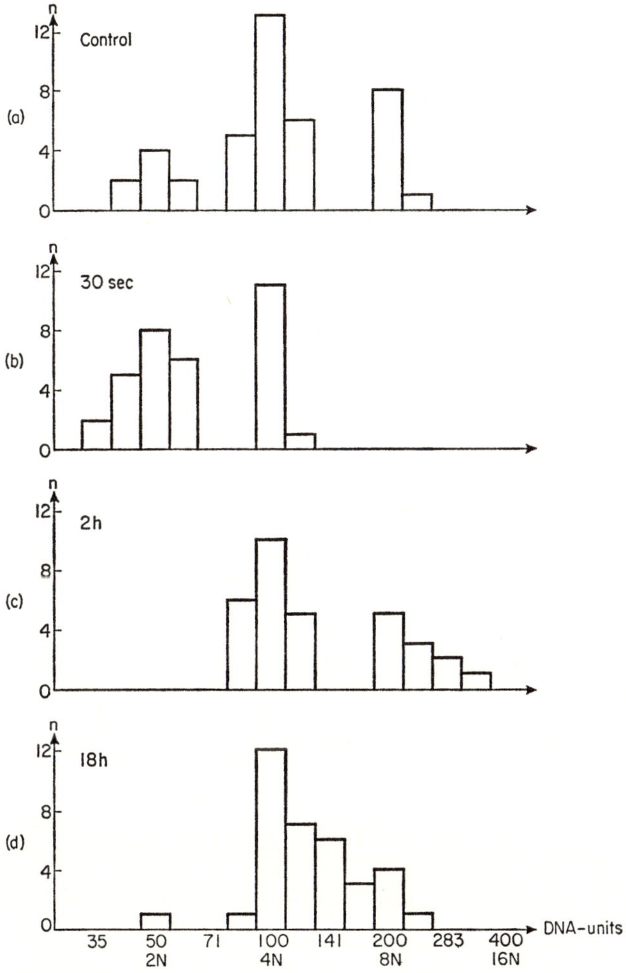

FIG 1. Distribution of the DNA values in arbitrary units in control animal, ½, 2, and 18 hours after partial hepatectomy.

(1962).) The cytophotometric studies were carried out by the two-wave-length method. Histoautoradiography was performed with Kodak stripping film; the animals had been given 250 μc of ^3H-thymidine intravenously one hour before sacrifice.

Cytophotometry (Feulgen technique) of the untreated control livers showed the typical three peaks at 50, 100, and 200 arbitrary DNA units, with the majority of the nuclei in the medium of 4N class (Fig. 1a). Half an hour after hepatectomy the ploidy distribution is quite different, the number of

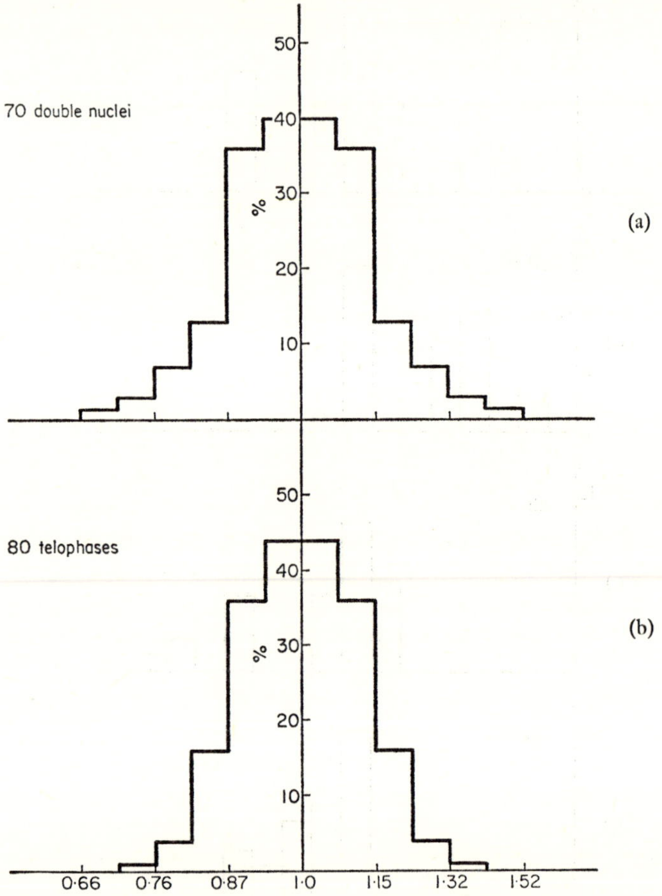

FIG. 2. (a) DNA-quotients between the two nuclei of double-nucleated hepatocytes in rat liver, (b) DNA-quotients between telophase-halves in rat liver after partial hepatectomy. (Two-wavelength cytophotometry, Feulgen-technique.)

cell nuclei of the diploid class having increased (Fig. 1b). At 2 and at 18 hours after the hepatectomy these small nuclei have largely disappeared, and tetraploid and octoploid nuclei dominate (Figs. 1c and 1d). The index of the cells labelled with tritiated thymidine—the ^{3}H-index—remains unchanged (below 0·6%) during the first 10 hours after partial hepatectomy, and it is not until after 18 hours that it rises to 1·58% (cf. Oehlert *et al.* 1962, Grisham 1962). This shift in somatic polyploidy during the first hours after partial hepatectomy, without concurrent DNA synthesis, is not significant. We therefore plan another test series.

The DNA content of each of the two nuclei of binucleate hepatocytes was measured—by cytophotometry based on the Feulgen reaction, using a technique described in detail by Grundmann *et al.* (1963)—and the results compared. The distribution of DNA quotients between the two nuclei is shown in Fig. 2a. It will be noted that in 76% of all binucleate liver cells the DNA quotients of their two nuclei lie between 1 and 1·15. This means—allowance being made for the range of error of the method—that in these 76% of cells the DNA content is the same in both nuclei. In the remaining 24% of cells the DNA quotient is over 1·15. It is of interest to note that the DNA values for 80 telophases in the rat liver after partial hepatectomy show approximately the same distribution as for the double nuclei (Fig. 2b). Thus the distribution of DNA values in the double nuclei is not essentially greater than in the mitotic telophases following partial hepatectomy.

III. Alterations in Somatic Polyploidy 24 to 72 hours after Partial Hepatectomy

The DNA content of the majority of the hepatocytes 24 hours after partial hepatectomy lies between the tetraploid and the octoploid range (Fig. 3a), without clear formation of the normal peaks for the tetraploid and the octoploid classes. The values lying between 4N and 8N indicate that many

TABLE I. ^{3}H-indices of hepatocytes after partial hepatectomy (above) and after partial hepatectomy and diethylnitrosamine (below), differentiated in the ploidy-classes. On the right the ^{3}H-indices of mesenchymal cells of the same livers.

	hepatocytes	hepatocytes of different classes diploid	tetraploid	octoploid	mesenchymal cells
hepatectomy	4·8%	0·8%	3·2%	0·8%	20·3%
hepatectomy +DENA	0·05%	0·05%	—	—	2·8%

epithelial cell nuclei are just at the stage of DNA synthesis. The ^{3}H-index at this point of time is 30% (Oehlert *et al.* 1962). After 48 hours the bulk of the hepatocytes have octoploid values of DNA (Fig. 3b), while after 72 hours diploid, tetraploid, and octoploid cell groups can once again be distinguished (Fig. 3c). At 48 hours the mitotic index has already fallen to 4%, indicating that the first wave of mitosis is almost over. There is not yet, however, a return to normal status, for over 50% of all nuclei are still in the octoploid stage or in DNA synthesis between the octoploid and the 16-ploid level. Counts of ^3H-labelled hepatocytes made 72 hours after partial

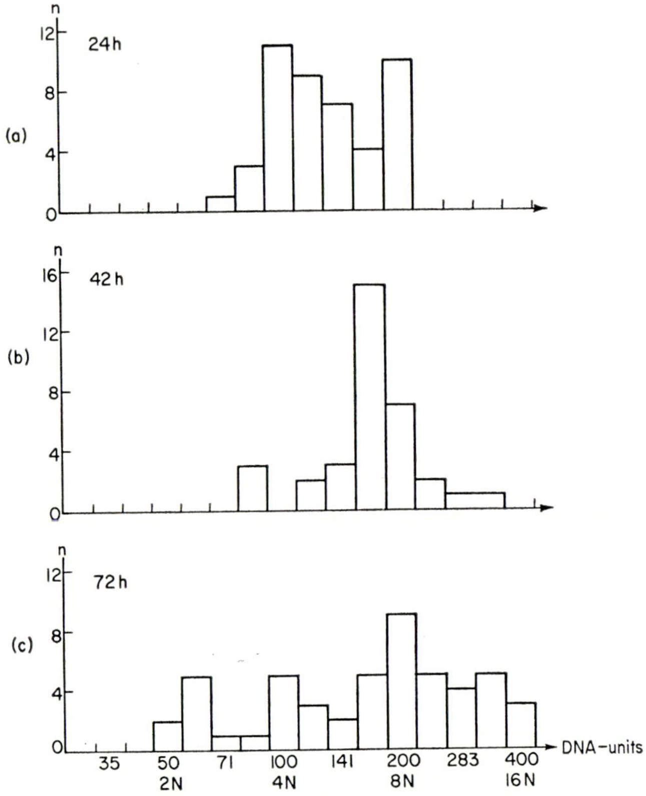

FIG. 3. Distribution of DNA values in arbitrary units, 24, 42 and 72 hours after partial hepatectomy.

hepatectomy have shown that many cells of the tetraploid group are then in DNA synthesis, but that this also applies to some cells of the diploid and octoploid groups (Table I). It should be noted that octoploid nuclei do not prefer DNA synthesis.

The liver cells 72 hours after partial hepatectomy are still carrying an increased functional burden, and this is manifested in the increased number of cells with a high ploidy. It has been deduced (chiefly by Geitler (1953)), from many observations in plants and animals, that increased functional demands on the affected tissue is an essential pre-requisite for polyploidy formation.

IV. The Effects of the Carcinogen Diethylnitrosamine on Liver Regeneration

The regenerating tissue of the rat liver now constitutes an eminently suitable test object for studying the effect of substances which exert a carcinogenic action on that organ. For this purpose we selected diethylnitrosamine. This substance, which is a variant of the hepatocarcinogen

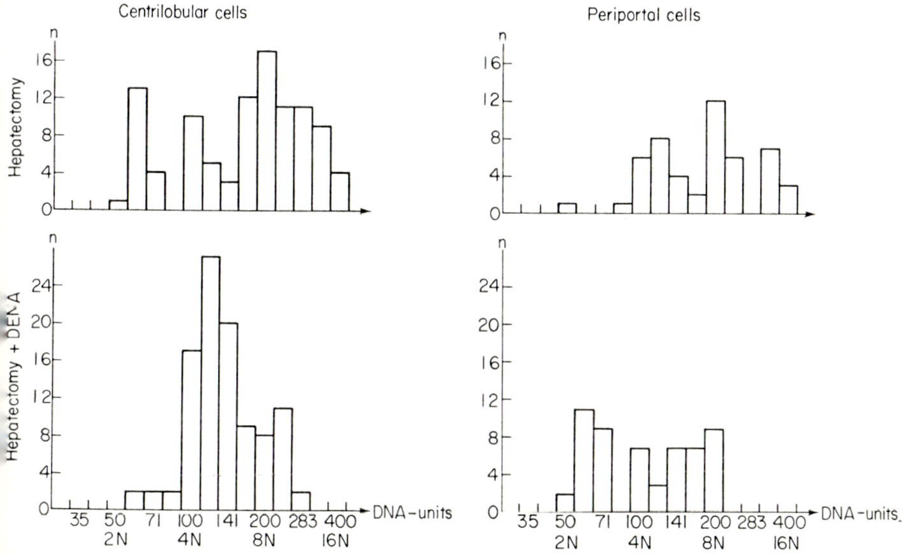

FIG. 4. Distribution of the DNA values in arbitrary units in rats 72 hours after partial hepatectomy (above) and 72 hours after partial hepatectomy and application of diethylnitrosamine (below), separated in centrilobular cells and periportal cells.

dimethylnitrosamine discovered by Magee and Barnes (1956), has been shown by Schmähl *et al.* (1960) to possess highly potent carcinogenic properties. Diethylnitrosamine was administered by stomach tube in dosage of 200 mg per kg body weight to 4 female rats of body weight 140 g 48 hours after partial hepatectomy, between 9·20 and 11·15 a.m. After a further 24 hours the animals were given 250 μc of ^{3}H-thymidine intravenously (see above) and killed one hour later. Their livers were processed and studied as described above.

In the non-hepatectomized rat liver, diethylnitrosamine produces an increase of the octo- and 16-ploid nuclei after having been fed for a few days only (5 mg per kg body weight) (Hobik and Grundmann 1962). Since the changes which lead to cancerization of liver cells in the rat are exclusively centrilobular, it is events in the centrilobular epithelial cells that are of special interest. In the test series in which diethylnitrosamine was fed after partial hepatectomy, therefore, the epithelial cells of the central and peripartial parts of the lobule respectively were studied separately.

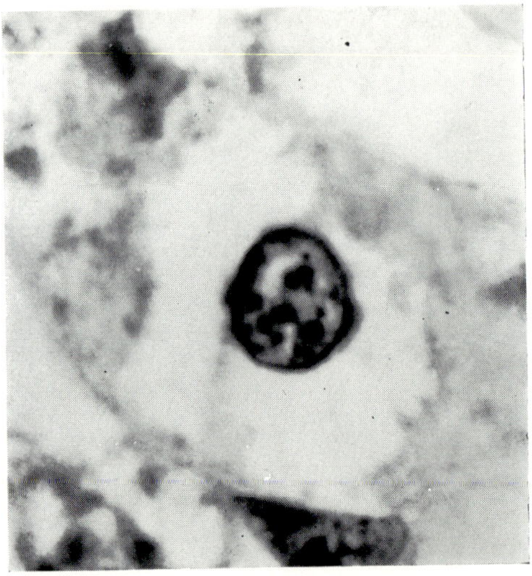

FIG. 5. Centrilobular hepatocyte 72 hours after partial hepatectomy: swollen cytoplasm with large fat droplets. Gallocyanin chromalaun, 2000×.

The distribution curves obtained with cytophotometry using the Feulgen technique are shown in Fig. 4. In both the central and peripheral parts of the lobule the high-ploidy cells are not increased in number, but diminished. In the central parts tetraploid cells predominate, and diploid and octoploid cells are seldom. In the periportal zone the group of diploid cells have increased. The supplementary strain thrown on the liver by the carcinogenic toxin seems to have thus reduced somatic polyploidy.

Morphologically, the centrilobular epithelial cells no longer contain coarse fat droplets. The cells are swollen, their cytoplasm is interspersed with fine vacuoles, and the accumulations of ribonucleic acid containing clusters in the centre of the lobules have increased (Figs. 5 and 6). At the same time many epithelial cells in the centre of the lobules have undergone necrosis and present as eosinophil agglomerations with pyknotic nuclear

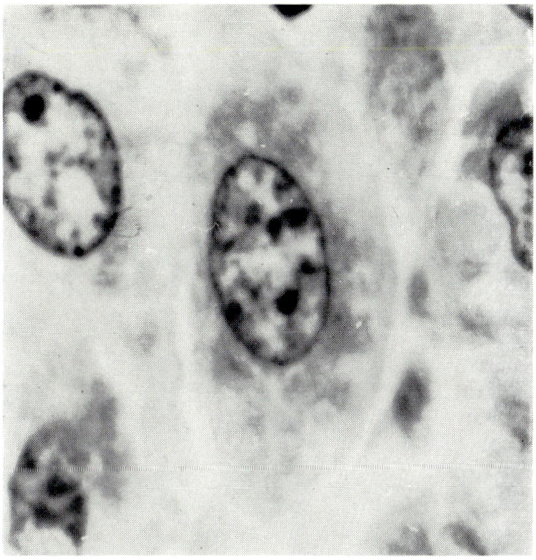

FIG. 6. Centrilobular hepatocyte 72 hours after partial hepatectomy and 24 hours after 200 mg/kg body weight diethylnitrosamine: fine vacuoles in the cytoplasm, no fat droplets. Gallocyanin chromalaun, 2000×.

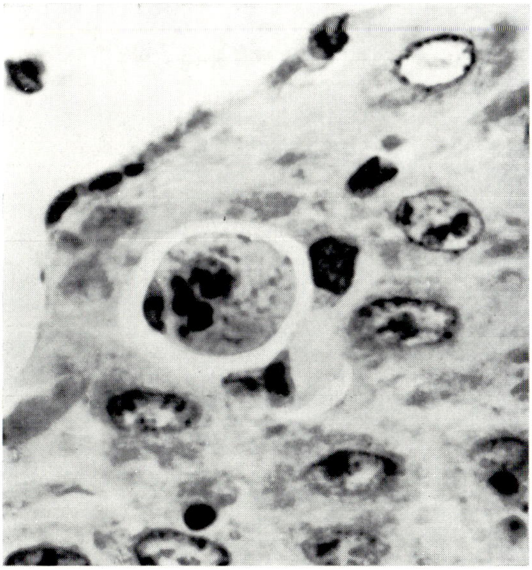

FIG. 7. Centrilobular hepatocytes 72 hours after partial hepatectomy. Above the central vein. In the centre necrotic hepatocyte with pykontic nuclear debris. Gallocyanin chromalaun, 2000×.

debris (Fig. 7). This marked increase in hepatic necrosis shows that diethylnitrosamine has led, not to improvement, but to further deterioration in liver metabolism.

Autoradiographic studies have thrown further light on reduction in somatic polyploidy by diethylnitrosamine (Table I). The ^{3}H-index has fallen from 4·8% to 0·05% and DNA synthesis has thus been largely suppressed. Very few nuclei are in the phase of DNA synthesis, and DNA measurements accordingly point to a reduction in somatic polyploidy. The number of nuclei belonging to the tetraploid or octoploid groups present at the end of the phase of DNA synthesis can be deduced approximately by comparing Fig. 4 above with Fig. 4 below.

Brief reference should be made to the work of Magee and Farber (1962) and other more recent studies suggesting that the nitrosamines may, after decomposition in the body, act as alkylating agents or cause methylation of nucleic acids in the rat liver.

V. Conclusions

The somatic tissue-specific polyploidy of the rat liver can be influenced experimentally. The number of high-ploid nuclei 24 to 72 hours after partial hepatectomy is significantly increased. Administration of the carcinogen diethylnitrosamine 48 hours after partial hepatectomy reduces somatic polyploidy, probably as a sequel to the inhibition of DNA synthesis by the carcinogen.

These results, together with the findings in other animal species and in plants, point to the existence of a specific "ploidy-equilibrium" in the liver. This "ploidy-equilibrium" between diploid, tetraploid, and higher-ploid nuclei undergoes change depending on the functional strain imposed on the tissue of the organ. These shifts in the "ploidy gradation" of the cell nuclei are a reaction to metabolic strain or at least to metabolic changes in the cytoplasm of the liver cells. We believe, however, that "ploidy-equilibrium" is also a growth process characteristic of highly differentiated parenchymatous tissues, and that there is a difference between cell increase in differentiated tissue and those which take place in undifferentiated mesenchyme.

REFERENCES

Geitler, L. (1953). *In* "Protoplasmatologia" VI C. Springer. Wien.
Grisham, J. W. (1962). *Cancer Res.* **22**, 842–849.
Grundmann, E. and Bach, G. (1960). *Beitr. path. Anat.* **123**, 144–172.
Grundmann, E., Hobik, H. P. and Schühly, A. (1963). *Dtsch. med. Wschr.* 88, 98–103.
Hobik, H. P. and Grundmann, E. (1962). *Beitr. path. Anat.* **127**, 25–48.

Hoffman, J., Himes, M. B., Lapan, S., Riszki, R. and Post, J. (1955). *Arch. Path.* **59**, 429–438.
Inamdar, N. B. (1958). *J. Morph.* **103**, 65–86.
Magee, P. N. and Barnes, J. M. (1956). *Br. J. Cancer* **10**, 114–122.
Magee, P. N. and Farber, E. (1962). *Biochem. J.* **83**, 114–123.
Oehlert, W., Hämmerling, W. and Büchner, F. (1962). *Beitr. path. Anat.* **126**, 91–112.
Schmähl, D., Preussmann, R. and Hamperl, H. (1960). *Naturwissenschaften* **47**, 89.

Biological and Ultrastructural Aspects of Early Stages in Liver Regeneration

E. G. BADE*

Instituto de Patología General y Experimental, Universidad Nacional de Cuyo, Mendoza, Argentina

THE first mitotic wave after a 60–70% hepatectomy (Harkness 1952, Weinbren 1959) or from the production of a parenchymatous wound (Podwyssozki 1886) elapses in the rat 24 to 30 hours after operation. In the mouse, the first mitotic wave occurs 42 to 50 hours after partial hepatectomy (Barnum *et al.* 1957, Trotter 1961, Bade and Echave Llanos 1963, Bade 1964). This first mitotic wave is followed, both in the rat (Jaffe 1954, Weinbren and Tarsh 1964) and in the mouse (Bade 1962, Russo and Echave Llanos 1964) by changes in the *maxima* and *minima* of mitotic activity that shape a definite circadian rhythm, similar to those reported for the normal liver (Halberg 1957, Jackson 1959, Peters 1962). Disregard of these mitotic rhythms can become a very important factor of error when analyzing the action of factors able to modify the liver's growth (Bade 1962, Peters 1962, Echave Llanos 1965). One and the same homogenate from regenerating liver can exert both an inhibitory and a stimulating effect on the mitotic activity of the regenerating liver, when injected at different times of the day (Bade 1962, Bade *et al.* 1963, Echave Llanos and Bordin 1963). This effect could only be attributed initially to differences in the sensitivity of the liver cells towards the various factors contained in the homogenates (Bade 1962). A more detailed study of the metabolic processes underlying these changes in sensitivity became necessary. As the periodical repetition of "peaks" and "troughs" in mitotic activity during all of the regenerative process was supposed to be due to the periodic repetition of the same *stimuli* that lead to the first mitotic wave (Bade 1962), the period between hepatectomy and this first mitotic wave was chosen for analysis.

The first histological change observed after hepatectomy is the early disorganization of cytoplasmic ribonucleoproteins (RNP) of the liver cells (Glinos 1958), which sub-microscopically corresponds to a disorganization of the ergastoplasm. The RNP changes progress as regeneration advances,

* Carrera del Investigador Científico, Consejo Nacional de Investigaciones Científicas y Técnicas, República Argentina.

according to a gradient from the periphery to the centre of the lobule (Glinos 1958, Oehlert *et al.* 1962), the same as DNA synthesis (Grisham 1960, 1962, Oehlert *et al.* 1962) and mitosis (Harkness 1952). Dilation of ergastoplasm follows or accompanies the disorganization, with a later diminution in the quantity of membranes (Fig. 1) (Davis 1962), and the appearance of free ribosomes, either isolated or grouped in rosettes or helical patterns (Fig. 1). Similar changes in cytoplasmic RNP have been described

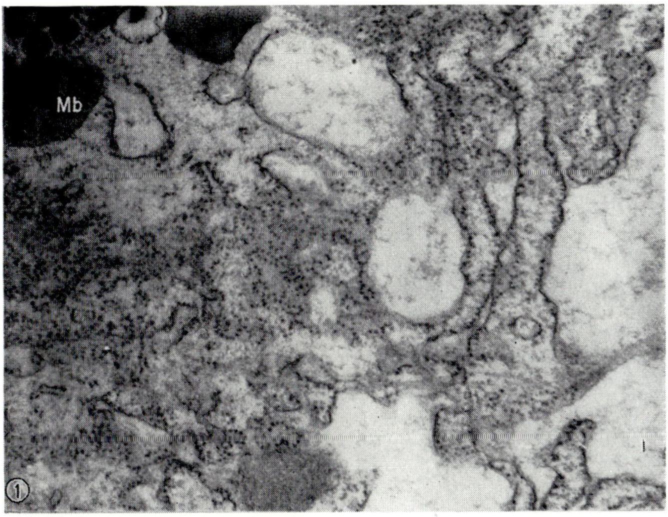

FIG. 1. Mouse liver cell 5 hours after partial hepatectomy. Dilation of ergastoplasm, disappearance of membranes and great amount of free ribosomes, in part in helical arrangement. Microbody (Mb). Palade, Vestopal. Uranyl-"staining". 41,000 ×.

by Glinos after plasmapheresis, which might be inhibited by the immediate reinjection of albumin or fibrinogen, but not by globulins (Stenram 1954a, 1954b, 1957). Nuclear and nucleolar changes and a later increase in mitotic activity have also been described after plasmapheresis. Disorganized ergastoplasm is a relatively frequent finding in the normal liver, which is emphasized by fasting (Bernhard *et al.* 1952, Fawcett 1955). Dilation and fragmentation of the ergastoplasm has been found after administrating different toxic substances (Emmelot and Benedetti 1961, Smuckler *et al.* 1962, Mukherjee *et al.* 1963, Lafontaine and Allard 1964), or during hypoxia (Bassi *et al.* 1960), after a transient ischaemia (Hübner and Bernhard 1961, Bade and Schumacher 1964) or during biliary obstruction (Steiner *et al.* 1962a). It seems to be, then, an unspecific reaction of the liver cells in the face of different situations, one of the common denominators being the functional over-loading due to the administration of toxic products or the accumulation of metabolites. The ergastoplasmic changes during posthepatectomy regeneration could be reasonably explained as being due to the intense and sustained circulatory changes (Daniel and Prichard 1951, Grindlay and Bollman 1952, Rabinovici and Wiener 1963) and to functional overloading. Liver microsomes incorporate labelled amino acids more actively than the ribosomes do (Campbell 1961, Campbell *et al.* 1964) and ergastoplasm

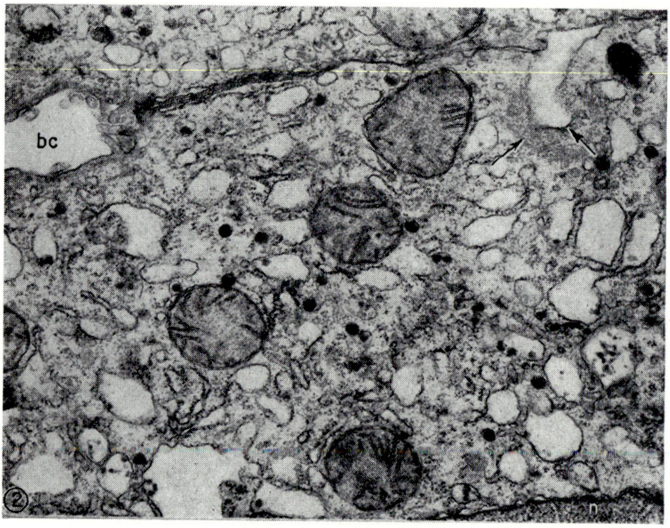

FIG. 2. Mouse liver cell 5 hours after partial hepatectomy. Dilation of ergastoplasm. Bile capillary (bc). Invagination of the cell surface (arrows), with change in density of the surrounding cytoplasm. Nucleus (n) and relatively normal mitochondria. Palade, Vestopal. Uranyl-"staining". 24,000 ×.

dilation and disappearance of membranes indicate an alteration in protein synthesis of the liver (Smuckler *et al.* 1962, Mukherjee *et al.* 1963). The decrease in plasm albumin concentration 24 hours after partial hepatectomy (Chanutin *et al.* 1938) the same as any change in other extracellular proteins, produced normally by the liver cells can be adequately explained on the basis of these findings and the loss of two thirds of the parenchyma.

Simultaneously with the change in cytoplasmic RNP, glycogen disappears gradually from the liver cells. A breakdown of glycogen is not an unspecific reaction to the operation and handling of the liver, and their possible causes have been analyzed elsewhere (Bade 1964). Both in the rat (Novikoff and Potter 1948, Aterman 1952) and in the mouse, very low levels of glycogen persevere up to 10–14 hours after hepatectomy. Eighteen hours after operation glycogen gradually reappears, reaches a maximum at 26 hours (noon) and decreases again to very low levels between 30 and 38 hours (night). The first maximum in glycogen content of the liver cells, and the subsequent decrease correspond exactly to changes in the dry weight of the liver previously observed (Echave Llanos and Bade 1963). About 46 hours after operation (noon), another maximum in glycogen content is observed. The hours of maximum and minimum glycogen content of liver correspond, almost exactly, to the rhythm described by Barnum *et al.* (1958) in untreated animals. This result seems to confirm our previous opinion that the operation at noon permits the re-assumption of a normal rhythm of liver functions.

Almost immediately after hepatectomy (Trotter 1965) a very active pinocytosis of fat particles (Bade 1965) through the liver cells commences, and large cytoplasmic vacuoles appear (Aterman 1952, 1961, Doniach and Weinbren 1952). These vacuoles are produced by invagination of the cell surface (Fig. 2), the same mechanism by which hypoxic vacuoles are formed (Oudea 1963). As time elapses, the vacuoles become increasingly PAS-positive and appear farther away from the cell surface. Aterman attributed these vacuoles to similar causes as the hypoxic ones, for which an alteration of energy metabolism (Pichotka 1942, Trowell 1946) and an increased or at least normal intrasinusoidal pressure is necessary (Trowell 1946). Both factors are undoubtedly present immediately after removal of two-thirds of the liver, and induce the liver cells to take up in this anomalous way substances that they normally incorporate through pinocytosis—or even do not incorporate them.

In the mouse livers from 2 to 10 hours after hepatectomy, oedematous mircovilli, with their loosening and subsequent appearance in the space of Disse or in the sinusoid (Fig. 3) can be found. This regressive change was initially described by Steiner (1961) in the course of hepatic lesions due to immunological aggregates, and under various other experimental situations (Steiner *et al.* 1926b), and attributed to a damage of the cell membrane. During posthepatectomy regeneration no toxic substance has yet been identified, so that the oedema is probably due to the inability of the liver cell to keep up the wholeness of that structure in an abnormal metabolic situation. A direct damage of the membrane seems therefore not always to be necessary. Oedema of liver cell microvilli has also been found by Barka *et al.* (1964) during the acute phase of copper intoxication, together with the formation of vacuoles, and alterations of the endoplasmic reticulum. These authors also found, after a phase of apparent functional adaptation to the toxic, a noteworthy increase in mitotic activity.

Between 2 and 6 hours after operation, both in the rat (Aterman 1952) and in the mouse, the cytoplasm of the littoral cells, and the space of Disse become increasingly PAS-positive, probably due to a secretory process of the littoral cells (Fig. 4) as shown by electron micrographs. According to Aterman this intensified staining reaction persists during the whole period of rapid growth, and has probably a direct participation in it.

Eight hours after the hepatectomy, disorganization of cytoplasmic RNP is complete (Glinos 1958). At this time an active incorporation of labelled amino acids into the microsomes begins, which becomes evident 16 hours after operation (Clerici *et al.* 1965), when ergastoplasma commences reorganizing (Davis 1962). Data contributed by Ferrari and Harkness (1954)

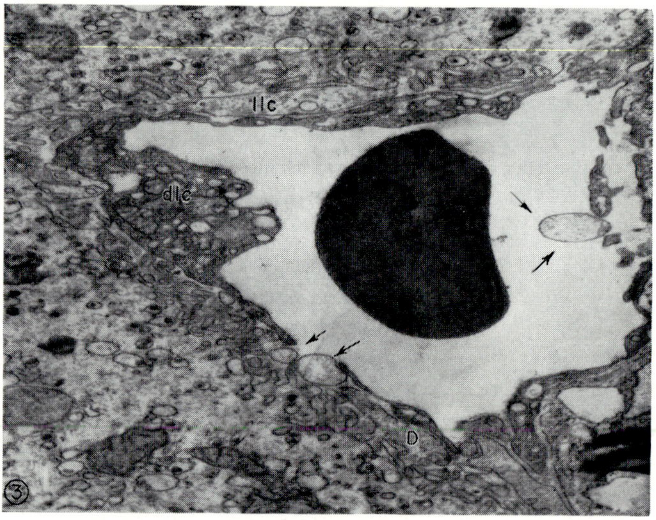

FIG. 3. Sinusoid of mouse liver, 5 hours after partial hepatectomy. Dark (dlc) and light (llc) littoral cells. Space of Disse (D) with normal and swollen microvilli (arrows) with loosening into the sinusoid. Palade, Vestopal. Uranyl-"staining". 17,000 ×.

indicate that 6 hours after hepatectomy—shortly before the beginning of active protein synthesis and the increase in cytoplasmic RNA (Glinos 1958) —there is a remarkable increase in the quantity of free amino acids in the liver. This fact becomes especially important in view of Munro and Clark's (1959) findings, that the synthesis of cytoplasmic RNA depends upon amino acids reaching the liver cells.

From 16 to 18 hours after hepatectomy in the rat (Harkness 1957, Glinos 1958) and between 24 hours and 28 hours in the mouse (Barnum *et al.* 1957, Bade and Pilgrim 1965) the incorporation of ^{32}P and thymidine-^{3}H into the nuclei of liver cells begins. The maximum of incorporation is reached in the rat 24 hours after operation (Grisham 1962) and in the mouse between 32 and 36 hours after, with a 52% of labelled nuclei (Bade and Pilgrim 1965).

This difference of time in the start of DNA synthesis in the rat and the mouse, and the slowness of the initial reaction of the mouse liver after partial hepatectomy has not yet been understood, but might be related to the differences in the biliary system, and necessarily associated physiological differences of the two animals.

The remaining third of the liver reacts then, almost immediately with microscopic and submicroscopic regressive changes to the circulatory and

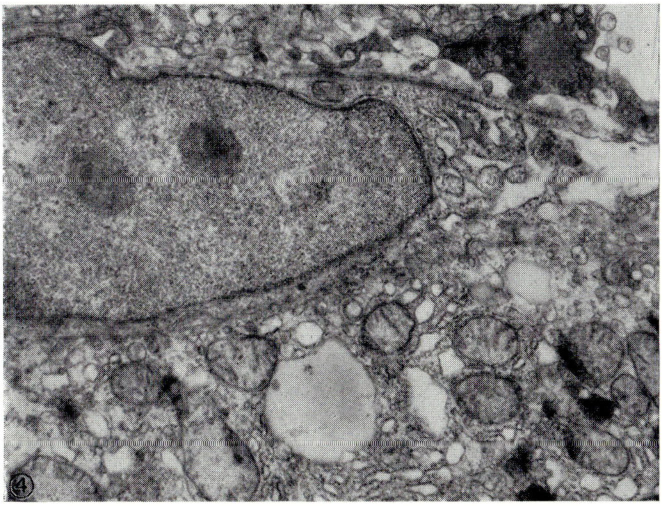

FIG. 4. Mouse liver 2 hours after partial hepatectomy. Littoral cell with two nucleoli and dilated ergastoplasm. Palade, Vestopal. Uranyl-"staining". 15,000 ×.

functional overloading brought about by partial hepatectomy. Functional alterations have been demonstrated during the early stages of liver regeneration (Royer and Bogetti 1939, Weinbren and Billing 1956), the relatively slow changes in the plasma albumins being principally due to their half-life of about three days (Maurer 1960). The regressive changes of the liver cells, which correspond mainly to changes at the cell surface and the endoplasmic reticulum, occur simultaneously with definite alterations of the littoral cells and the extracellular space.

The early "r phase" (Needham 1952) is followed in the mouse by a phase of apparent functional adaptation, with circadian changes in the glycogen content of the liver cells. This step continues with the "progressive phase" (Needham 1952), characterized by reorganization of the ergastoplasm and net increase in cytoplasmic RNA. The first "p phase" culminates with the duplication of DNA and a mitotic wave.

If work is done under "standard" conditions (Halberg *et al.* 1958) and hepatectomy is always performed at the same time, the regenerating liver is made up by a highly synchronized cell population, as evidenced by the sharp peaks in DNA synthesis, mitosis, and the various rhythmic changes. From the data obtained under these conditions, a first attempt to define, in terms of the normal cell cycle (Howard and Pelc 1953) the events of the first two days of regeneration, can be done. On this basis, a working hypothesis, which considers that the mitotic inhibition obtained by administering the homogenates from normal and regenerating liver 18 to 12 hours before the mitotic wave (Bade 1962, Bade and Echave Llanos 1962, Echave Llanos and Bordin 1963) is due to an action on the synthesis of DNA, and that the injections administered shortly before the mitotic peak, probably acting at the antephase (Bullough 1952, 1962) or G_2 of the synchronized cell population may come forward.

Duplication of DNA and probably the synthesis of some special RNA are characteristic for growth, whereas none of the changes which take place during the "r phase" can be considered as being growth specific. Only the sequence, intensity and length of each of these changes may characterize a growth reaction. The changes after a single dosis of thyroxin (Sternheimer 1936) or after repeated doses of copper sulphate (Barka *et al.* 1964) suggest that what is essential for a growth reaction to occur, is persistence of the causes leading to the "r phase" for a certain time, thus producing the described changes in a definite sequence. High intensity of one stimulus, or its long duration, could be expected to substitute those local or general reactions not initially present. The unspecific changes during the "r phase" and the fact that the degree of hypertrophy consecutive to sustained albuminuria (Lippman *et al.* 1952, Marsh and Drabkin 1958), pregnancy (Campbell and Kosterlitz 1949), lactation (Kennedy *et al.* 1958, Bade and Echave Llanos 1963), hypothalamic hyperphagia (Kennedy and Parrot 1958), and also posthepatectomy regeneration are proportional to the relative deficit of functioning parenchyma (length of albuminuria: Lippman *et al.* 1952; number of suckling rats: Kennedy *et al.* 1958; caloric intake: Kennedy and Parrot 1958; amount of liver excised: MacDonald *et al.* 1962, Bucher and Swaffield 1964), may be considered as further evidence for the conclusion (Doljanski 1960) that all these apparently different types of liver growth are only variations from a common scheme. The report by Sternheimer, that some of the changes previous to increased mitotic activity found after a single dosis of thyroxin are mainly periportally located, point in the same direction.

Diffusible tissue factors as defined by Weiss (1955) underlie these growth reactions. Specially in the case of post-hepatectomy regeneration, the stimulation of growth of responsive hepatomas (Trotter

1961) or of liver autotransplants (Virolainen 1964, Leong *et al.* 1964) are convincing proofs for this assumption. But whether this growth is due to the appearance of stimulating substances (Paschkis 1958) or to the inhibition in the production of specific inhibitors during the "r phase" (Bade 1964), is not clear from these results. Specific stimulation (Teir and Ravanti 1953, Lahtiharju and Teir 1964) and inhibition (Saetren 1956, Stich and Florian 1958) and unspecific stimulation (Paschkis 1958) by tissue factors have all been reported. Only the definitive standardization of the assay methods would really improve our possibility of resolving these questions gradually.

Much of the work reported has been realized during an Alexander von Humboldt fellowship. To Prof. H. Meessen and Prof. W. Maurer I am very indebted for encouragement and criticism during this time.

REFERENCES

Aterman, K. (1952). *Arch. Path.* 53, 197–208.
Aterman, K. (1952). *Arch. Path.* 53, 209–216.
Aterman, K. (1961). *J. Path. Bact.* 82, 367–369.
Bade, E. G. (1962). Estimulación e inhibición de la regeneración hepática por factores tisulares. Influencia del ritmo de 24 horas. M.D. Thesis. Facultad de Ciencias Médicas. Mendoza. Argentina.
Bade, E. G. and Echave Llanos, J. M. (1962). *Naturwissenschaften* 49, 351–352.
Bade, E. G. and Echave Llanos, J. M. (1963). *Acta physiol. latinoam.* 13, 193–194.
Bade, E. G., Bordín, C. and Echave Llanos, J. M. (1963). *Sperimentale* 113, 198–203.
Bade, E. G. and Echave Llanos, J. M. (1963). *Naturwissenschaften* 50, 693–694.
Bade, E. G. (1964). *Virchows Arch. path. Anat. physiol.* 337, 503–514.
Bade, E. G. and Schumacher, A. (1964). *Frankf. Z. Path.* 74, 29–45.
Bade, E. G. (1965). *Virchows Arch. path. Anat. physiol.* 338, 237–244.
Bade, E. G. and Pilgrim, C. (1965). III. Sesiones Científicas de Biología, Rosario. Argentina. Abstr. 120.
Barka, T., Scheuer, P. J. Schaffner, F. and Popper, H. (1964). *Arch. Path.* 78, 331–349.
Barnum, C. P., Jardetzky, C. D. and Halberg, F. (1957). *Texas Rep. Biol. Med.* 15, 134–147.
Barnum, C. P., Jardetzky, C. D. and Halberg, F. (1958). *Am. J. Physiol.* 195, 301–310.
Bassi, M., Bernelli-Zazzera, A. and Cassi, E. (1960). *J. Path. Bact.* 79, 179–183.
Bernhard, W., Hagenau, F., Gautier, A. and Oberling, C. (1952). *Z. Zellforsch. Mikrosk. Anat.* 37, 281–300.
Bucher, N. L. R. and Swaffield, M. N. (1964). *Cancer Res.* 24, 1611–1625.
Bullough, W. S. (1952). *Biol. Rev.* 27, 133–168.
Bullough, W. S. (1962). *Biol. Rev.* 37, 307–342.
Bullough, W. S. (1964). *In* "Cellular control mechanisms and cancer" (P. Emmelot and O. Mühlbock eds.). Elsevier, Amsterdam.
Campbell, P. N. (1961). *In* "Protein biosynthesis" (R. J. C. Harris, ed.). Academic Press, London.
Campbell, P. N., Cooper, C. and Hicks, M. (1964). *Biochem. J.* 92, 225–234.
Campbell, R. M. and Kosterlitz, H. W. (1949). *J. Endocr.* 6, 171–183.

Chanutin, A., Hortenstine, J. C., Cole, W. S. and Ludewig, S. (1938). *J. biol. Chem.* **123**, 247–256.
Clerici, E., Cammarano, P. and Mocarelli, P. (1965). *Experientia* **21**, 143–144.
Daniel, P. M. and Prichard, M. M. L. (1951). *J. Physiol.* **114**, 521–537.
Davis, J. M. G. (1962). *Acta radiol.* 58, 17–32.
Doljanski, F. (1960). *Int. Rev. Cytol.* **10**, 217–242.
Doniach, I. and Weinbren, K. (1952). *Brit. J. exp. Path.* **33**, 499–505.
Echave Llanos, J. M. and Bade, E. G. (1963). *Naturwissenschaften* **50**, 97–98.
Echave Llanos, J. M. and Bordín, C. (1963). *Acta physiol. latinoam.* **13**, 337–346.
Echave Llanos, J. M. (1965). This issue.
Emmelot, P. and Benedetti, E. L. (1961). *In* "Protein biosynthesis" (R. J. C. Harris, ed.). Academic Press, London.
Fawcett, D. W. (1955). *J. nat. Cancer Inst.* **15**, 1475–1502.
Ferrari, V. and Harkness, R. D. (1954). *J. Physiol.* **124**, 443–463.
Glinos, A. D. (1958). *In* "The chemical basis of development" (W. D. McElroy and B. Glass, eds.). Johns Hopkins Press, Baltimore.
Grindlay, J. H. and Bollman, J. L. (1952). *Surgery Gynec. Obstet.* **94**, 491–496.
Grisham, J. W. (1960). *J. Histochem. Cytochem.* 8, 330.
Grisham, J. W. (1962). *Cancer Res.* **22**, 842–849.
Halberg, F. (1957). *Experientia* **13**, 502–503.
Halberg, F., Barnum, C. P., Silber, R. H. and Bittner, J. (1958). *Proc. Soc. exp. Biol. Med.* **97**, 897–900.
Harkness, R. D. (1952). *J. Physiol.* **116**, 373–379.
Harkness, R. D. (1952). *J. Physiol.* **117**, 267–277.
Harkness, R. D. (1957). *Brit. med. Bull.* **13**, 87–93.
Howard, A. and Pelc, S. R. (1953). *Heredity*, suppl. Vol. 6, 261–273.
Hübner, G. and Bernhard, W. (1961). *Beitr. path. Anat.* **125**, 1–30.
Jackson, B. (1959). *Anat. Rec.* **134**, 365–377.
Jaffe, J. J. (1954). *Anat. Rec.* **120**, 935–954.
Kennedy, G. C. and Parrott, D. M. V. (1958). *J. Endocr.* **17**, 161–166.
Kennedy, G. C., Pearce, W. M. and Parrott, D. M. V. (1958). *J. Endocr.* **17**, 158–160.
Lafontaine, J. G. and Allard, C. (1964). *J. Cell Biol.* **22**, 143–172.
Lahtiharju, A. and Teir, H. (1964). *Exptl. Cell Res.* **34**, 205–207.
Leong, G. F., Grisham, J. W., Hole, B. V. and Albright, M. L. (1964). *Cancer Res.* **24**, 1496–1501.
Lippman, R. W., Marti, H. U. and Jacobs, E. E. (1952). *Arch. Path.* 169–184.
MacDonald, R. A., Rogers, A. E. and Pechet, G. (1962). *Lab. Invest.* **11**, 544–548.
Marsh, J. B. and Drabkin, D. L. (1958). *J. biol. Chem.* **230**, 1063–1071.
Maurer, W. (1960). *In* "10. Colloquium Ges. Physiol. Chemie". Springer, Berlin.
Mukherjee, T., Gustavson, R. G., Afzelius, B. A. and Arrhenius, E. (1963). *Cancer Res.* **23**, 944–953.
Munro, H. N. and Clark, C. M. (1959). *Biochim. biophys. Acta* **33**, 551–553.
Needham, A. E. (1952). "Regeneration and wound-healing." Methuen, London.
Novikoff, A. B. and Potter, V. R. (1948). *J. biol. Chem.* **173**, 223–238.
Oehlert, W., Hämmerling, W. und Büchner, F. (1962). *Beitr. path. Anat.* **126**, 91–112.
Oudea, P. R. (1963). *Lab. Invest.* **12**, 386–394.
Paschkis, K. E. (1958). *Cancer Res.* **18**, 981–991.
Peters, R. (1962). *Z. Naturforsch.* **17b**, 164–168.
Pichotka, J. (1942). *Beitr. path. Anat.* **107**, 117–155.
Podwyssozki, W. von (1886). *Beitr. path. Anat.* **1**, 259–360.

Rabinovici, N. and Wiener, E. (1963). *J. surg. Res.* 3, 3–8.
Royer, M. and Bogetti, A. (1939). *Rev. Soc. argent. Biol.* 15, 297–301.
Russo, J. and Echave Llanos, J. M. (1964). *Z. Zellforsch.* 61, 824–828.
Saetren, H. (1956). *Exptl. Cell Res.* 11, 229–232.
Smuckler, E. A., Iseri, O. A. and Benditt, E. P. (1962). *J. exp. Med.* 116, 55–72.
Steiner, J. W. (1961). *Am. J. Path.* 38, 411–436.
Steiner, J. W., Carruthers, J. S. and Kalifat, S. P. (1926a). *Z. Zellforsch.* 58, 141–159.
Steiner, J. W., Carruthers, J. S. and Kalifat, S. P. (1926b). *Virchows Arch. path. Anat. Physiol.* 336, 99–114.
Stenram, U. (1954a). *Acta anat.* 21, 386–390.
Stenram, U. (1954b). *Acta physiol. scand.* 30, 226–230.
Stenram, U. (1957). *Acta histochem.* 5, 156–166.
Sternheimer, R. (1936). *Endorinology* 25, 899–908.
Stich, H. F. and Florian, M. L. (1958). *Can. J. Biochem. Physiol.* 36, 855–859.
Teir, H. and Ravanti, K. (1953). *Exptl. Cell Res.* 5, 500–507.
Trotter, N. L. (1961). *Cancer Res.* 21, 778–782.
Trotter, N. L. (1965). *Anat. Rec.* 151, 427 (Abstr.).
Trowell, O. A. (1946). *J. Physiol, Lond.* 105, 268–297.
Virolainen, M. (1964). *Exptl. Cell Res.* 33, 588–591.
Weinbren, K. and Billing, B. H. (1956). *Brit. J. exp. Path.* 37, 199–204.
Weinbren, K. (1959). *Gastroenterology* 37, 657–668.
Weinbren, K. and Tarsh, E. (1964). *Brit. J. exp. Path.* 45, 475–480.
Weiss, P. (1955). *In* "Biological Specificity and growth" (E. G. Butler, ed.). University Press, Princeton.

DISCUSSION SUMMARY

Chairman: Professor W. S. Bullough

In reply to a question from Dr. Glinos regarding the possible standardization of experimental conditions by force-feeding, Professor Echave Llanos considered this technique unsatisfactory as it causes severe stress; his method of standardization was to starve the animals for 12 hours before the operation and before the time of death. This is the minimum starvation period needed to ensure an empty stomach, and it is also probably the maximum that the animals can tolerate without stress. He quoted evidence that the mitotic reaction to partial hepatectomy is decreased after more than 24 hours starvation. Dr Wrba also emphasized the possible inhibitory effects of such simple stressful factors as change of box or change of room.

In answer to Professor Jalavisto, Professor Echave Llanos stated that the diurnal mitotic rhythm in the foetal liver appeared to be different from that in the mother, and the Chairman also pointed out that there is evidently no diurnal mitotic rhythm in foetal epidermis.

Both Professor Iversen and Dr. Weiss emphasized the importance of a thorough understanding of diurnal (circadian) mitotic rhythms since these form the baseline to which the mitotic reaction in liver regeneration must be related. The action of any humoral agent on the process of regeneration has to be judged against this baseline which is evidently in the form of a sine wave.

Following a question from Dr. Weiss, Dr. Wrba explained that he and Dr. Rabes had used embryo extract in their culture medium, but that this extract was made only from embryos from which the livers had been removed. He believed that the humoral factor of adult liver has a relatively small molecule, but he knew nothing of its nature.

In answer to Dr. Glinos, he added that the mitotic response in a hepatectomized parabiont was about 300% that of a normal parabiont, while that of the normal partner of a hepatectomized parabiont lay between these two extremes.

Dr. Glinos expressed his appreciation of the beautiful work described by Dr. Virolainen and argued that it strengthened the theory that liver regeneration is related to the dilution of particular plasma proteins.

Dr. Craddock asked whether the remarkable reduction in ploidy in the early stages of liver regeneration, as described by Professor Grundmann, might be due simply to the splitting of the polyploid cells into diploid cells. Professor Grundmann considered that this, which was a form of amitosis, might be possible, but in spite of detailed observations he had no conclusive evidence.

A series of questions were asked, especially by Dr. Wrba and Dr. Koss, regarding the significance of the liver cell basophilia seen shortly after partial hepatectomy. Dr. Bade considered that this basophilia may have a direct causal relationship with the high mitotic activity that developed about 20 hours later, and he was supported by Dr. Glinos who stressed that such basophilia must be regarded as a specific sign of increased protein synthesis. Professor Iversen asked whether the basophilic cells were indeed those destined to enter mitosis or whether perhaps they were merely showing signs of non-specific damage. Dr. Bade replied that he had found no signs of cell necrosis such as might follow cell damage, and Dr. Wrba added that similar basophilia, which was perfectly reversible, was seen in liver cells after any kind of damage.

The Chairman observed that almost all the speakers, in one way or another, had referred to the possibility that the increased functional activity of a surviving liver fragment might be a possible factor in the induction of the mitotic response. Dr. Goss then examined this suggestion. He pointed out that because the liver is such a versatile organ physiologically it does not really lend itself to studies of function in relation to growth. However, it is already known that in the presence of certain toxic substances the liver cells produce the necessary detoxication enzymes and that this is followed by a mitotic response. It is also known that a high protein diet, leading perhaps to especially active deamination, leads to liver hypertrophy, as also does albuminuria. The basic question may therefore be whether the humoral growth regulators, which evidently exist, control only mitotic activity or whether they are also involved in the regulation of tissue function.

Section VI

Miscellaneous Aspects

Age, A Factor in Growth of the Exorbital Lacrimal Gland*

T. E. BOLDEN†

Department of Oral Pathology and Oral Medicine, School of Dentistry, Meharry Medical College, Nashville, Tennessee, U.S.A.

I. Introduction

WEIGHT and height are the most critical indices for measuring growth in the human. Wet weight is the usual parameter upon which the growth curve of an animal or portions of an animal is based. The wet weight of an animal will vary depending upon its state of health, sex and age. The growth curve of different glands of the rat have been compared with the growth curve of the body weight. All glands do not mimic this curve (Loewenthal 1900, Venable and Grafflin 1940, Sreebny 1954, Boas and Scow 1954).

Parameters other than wet weight can be used to measure growth of the exorbital lacrimal gland. Among these would be embryogenesis, the development of function, the development of functional components of the gland, its histologic density and the level of enzyme activity of the gland.

Each of these parameters can be expected to be modified by age. It is the purpose of this report to summarize a series of investigations in which the effect of age as a factor in the growth of the exorbital lacrimal gland was tested.

II. Materials and Methods

This investigation is based on morphological, histological, histochemical and biochemical studies of the exorbital lacrimal glands of 167 untreated male rats of the Wistar strain. The glands of animals from seven age groups (15, 30, 60, 90, 150, 300 and 500 days) were studied. The glands of 35 rats were used for special histochemical investigations and the glands of 70

* This presentation was supported in part by research grants B–1722–A, B–1722–B and B–1722–C from the Department of Health, Education and Welfare, United States Public Health Service, Council on Neurological Disease and Blindness, Bethesda, Maryland, U.S.A., by a grant from the Patterson-Hettinger-Cary Foundation, Minneapolis, Minnesota, U.S.A. to improve faculty training and by a special travel grant from the Sigrid Jusélius Foundation, Helsinki, Finland.

† Professor of Dentistry, Chairman, Department of Oral Pathology and Oral Medicine. Director of Research for the School of Dentistry.

animals were used for biochemical studies. With the exception of the dry weight, glands from at least 5 animals in each age group were used for each determination.

A. *Experimental procedures*

Animals were killed using ether anesthesia. Body weight and body length (top of nose to tip of tail with the tail extended as the animal lay on its right side) were recorded for each animal. Glands were exposed by skin incision, dissected free and removed. Before histologic or biochemical treatment, the glands were weighed and studied grossly.

B. *Histological examination*

Left glands used for histological examination were fixed in Helly's fluid for 6 to 24 hours, washed in running water for 24 hours, then placed in 80% alcohol, treated routinely and embedded in paraffin. Sections were cut at 6 μ, decrystalized in Lugol's solution and stained with Ehrlich's hematoxylin and eosin. Some sections from each gland were stained for collagen with Mallory's aniline blue.

C. *Quantitative histological examination*

A quantitative estimate was made of the relative areas in sections of the glands which were occupied by acinar and by intralobular duct tissue. A grid divided into 66 squares was placed in the right ocular of the microscope. A total of 1,200 squares were counted. The number of squares in which one-half or more of the area was occupied by acini or ducts was counted in 5 specimens from each age group.

Right glands used for histological examinations were fixed according to the technique of Smith-Atkinson (1956), stained with toluidine blue and counterstained with eosin. Mast cells were counted in one section from each of five glands in each age group. The surface area of the section was measured with a compensating polar planimeter at 18·5 magnification. The ratio of cells to surface area was then computed and recorded as cells per square millimeter of surface area of tissue.

D. *Histochemical examination*

Right glands of 5 rats from each age group were tested for alkaline phosphatase activity. Using the technique of Gomori (1941) as modified by Moog (1946), sections were incubated for 15 minutes. A section inactivated by Lugol's solution was carried on the same slide. One slide from each age group was run simultaneously. Sites of localization of alkaline phosphatase were designated as positive (+) and the intensity of the positive reaction was graded numerically, i.e. 1+ least positive, 4+ strongly positive.

E. *Biochemical studies*

Fresh glands from 70 animals were used for biochemical assay. Alkaline phosphatase activity was assayed spectrophotometrically in right glands using the procedure suggested by Hofstee (1954).

Acid phosphatase activity was assayed spectrophotometrically in left glands using the method described by Holter and Li (1949). For alkaline phosphatase, a homogenate was prepared by grinding the gland with small amounts of glycine buffer at pH 9·2. For acid phosphatase homogenates were prepared using acetate buffer at pH 5·0. Activity was determined by measuring, according to Fiske and Subarrow (1925), the amount of inorganic phosphorus set free from sodium beta glycerophosphate in incubations lasting 20 minutes at 37°C. Duplication determinations were made. Tissue inorganic phosphorus was determined by measuring the optical density of the test solution in the Coleman Junior Spectrophotometer. Each solution contained 4 ml of supernatant (0·5 ml tissue extract, 3·0 ml TCA, 0·1 ml $MgCl_2$, 1·9 ml buffer, 1·0 ml molybdate, 0·4 ml sulfonic acid and 4·6 ml H_2O).

III. Findings

A. *Gross findings*

The exorbital lacrimal glands of the rat are flat racemose glands which lie between the eye and the ear with their long axes perpendicular to the zygomatic arches. The main excretory duct leads from the hilus of the gland toward the posterior corner of the eye. Here the duct joins the excretory duct of the intra-orbital lacrimal gland.

Just before reaching the lid, the common duct opens into the outer part of the superior conjunctival fornix (Venable and Grafflin 1940).

B. *Gland size*

The pattern of growth of right and left exorbital lacrimal glands was quite similar. Right glands were uniformly smaller then left glands from animals at 30, 90, 150 and 500 days (Table I).

The size of right glands increased markedly between 15 and 30 days and

TABLE I. Average size and wet weight of exorbital lacrimal glands of male albino rats.

Number glands	Age days	Right glands					Left glands				
		Length cm	Width cm	Thickness cm	Area cm^2	Wet weight mg	Length cm	Width cm	Thickness cm	Area cm^2	Wet weight mg
8	15	0·64	0·41	0·10	0·26	20·7	0·73	0·40	0·14	0·041	19·7
7	30	0·90	0·59	0·16	0·085	48·6	0·96	0·57	0·17	0·093	47·1
7	60	1·29	0·77	0·23	0·228	114·4	1·20	0·81	0·21	0·204	113·8
5	90	1·28	0·80	0·18	0·184	108·0	1·26	0·78	0·20	0·196	116·8
7	150	1·27	0·84	0·19	0·204	124·2	1·36	0·93	0·19	0·270	136·5
7	300	1·34	0·81	0·17	0·184	119·0	1·39	0·74	0·17	0·175	134·4
7	500	1·30	0·76	0·21	0·207	137·5	1·35	0·81	0·20	0·219	145·0

Reproduced with permission from Bolden, Barnes and Hall, The Meharri-Dent, Nashville, Tennessee, 1963.

between 30 days and 60 days (Figs. 1 and 2). The thickness of the glands remained unchanged so that the increases were primarily in width and length (Table I).

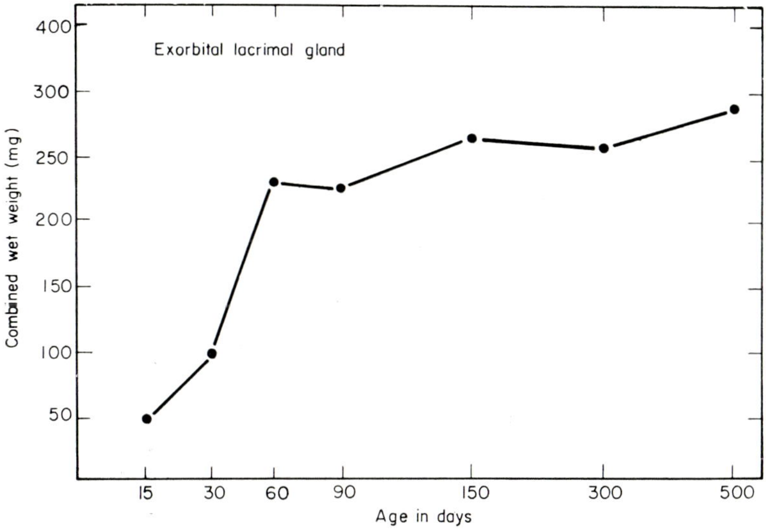

FIG. 1. Wet weights of exorbital lacrimal glands of male albino rats at seven different ages.

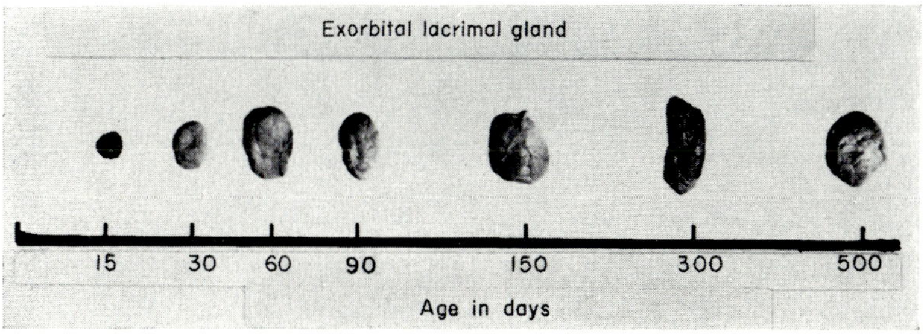

FIG. 2. Photograph or resected fixed left glands from male albino rats at seven different age groups.

C. *Histological findings*

Histologically, the gland is composed of ducts, interstitial tissue and acini and resembles a serous salivary gland. Intralobular ducts are seldom observed (Fig. 3).

The excretory system starts with the lumen of the acini which connect with that of the intercalated ducts. These empty into the intralobular ducts which join the interlobular ducts. The latter join to form the major excretory duct. Intercalated ducts are lined by low cuboidal cells with scanty agranular cytoplasm and an oval nucleus orientated parallel to the lumen of the duct

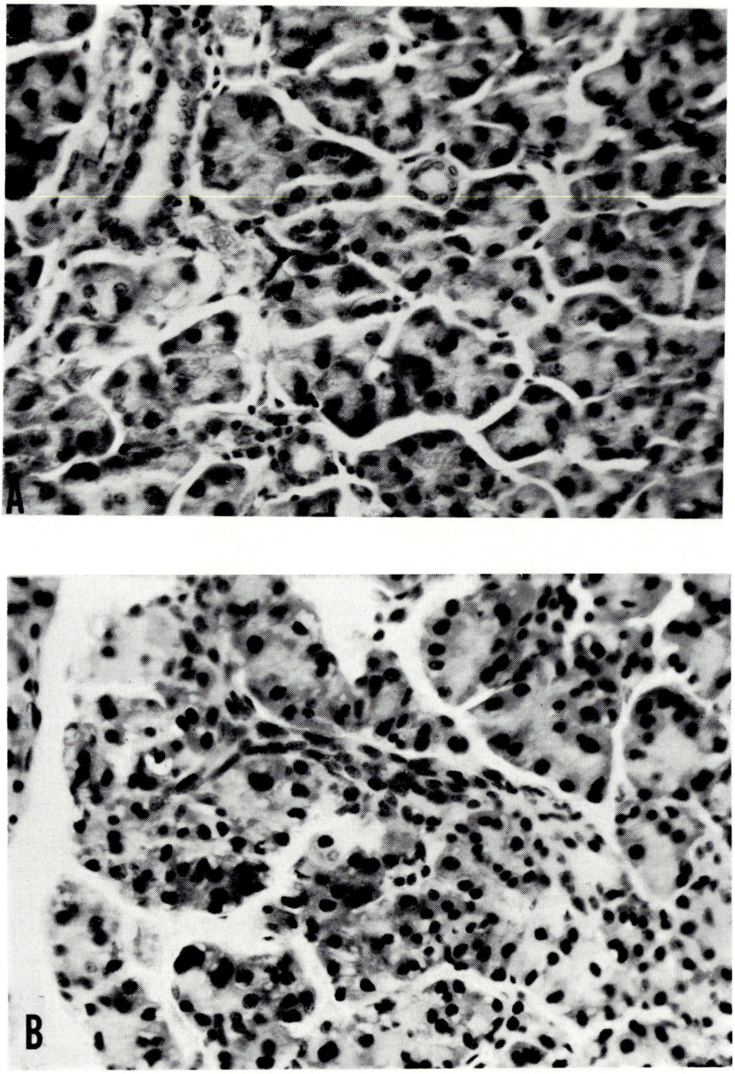

FIG. 3. Photomicrographs of a section of normal exorbital lacrimal gland of a male albino rat.
A—Note that the gland is composed of acini, infrequent intralobular ducts and non-secretory tissue.
B—Note the intercalated duct. Hematoxylin and eosin. Magnification 450×.

(Fig. 3). Intralobular ducts are seen infrequently, but are lined by a single layer of low cuboidal cells with agranular cytoplasm and a central oval nucleus. The interlobular and major excretory ducts are lined by a stratified cuboidal epithelium. The cytoplasm is homogenous.

The gland is invested by a thin, tenacious connective tissue capsule. The acinar and duct elements are surrounded by delicate interstitial connective tissue in which lie nerves, blood and lymph vessels. The acini consist of pyramidal shaped cells with granular vacuolated cytoplasm and a large

TABLE II. Body length, body weight, combined wet weight and percentage contribution of exorbital lacrimal glands to total body weight of male albino rats of seven age groups.

Rats		Body		Exorbital combined wet weight (R. & L.)	Lacrimal gland contribution to body weight
Age in days	Number	Length cm	Weight gm	mg	%
15	8	5·7	27·2	40·4	0·0014
30	7	8·6	64·4	95·7	0·0015
60	7	13·1	236·8	228·2	0·0010
90	5	14·6	250·1	224·8	0·0009
150	7	15·9	293·1	260·7	0·0009
300	7	18·3	399·4	253·4	0·0007
500	7	18·7	440·0	282·5	0·0006

Reproduced, with modifications, with permission from Bolden, Barnes and Hall, The Meharri-Dent, Nashville, Tennessee, 1963.

round or oval nucleus (Fig. 3). The nuclei show great variation in size. Some nuclei contain two or more nucleoli. Lumina of the acini are seen with difficulty but communicate with the short intercalated ducts. Myoepithelial cells are found at the base of the acini. They rest between the basement membrane and the secretory cell. Their processes extend over the outer surface and between acinar cells (Figs. 11, 12, 13).

D. *Quantitative histologic findings*

1. Composition

The normal lacrimal gland consisted of about 3% intralobular duct tissue, 82% acinar tissue and 15% non-secretory or interstitial tissue.

The number of squares occupied by acini and intralobular duct tissue showed only slight fluctuations with age (Table III).

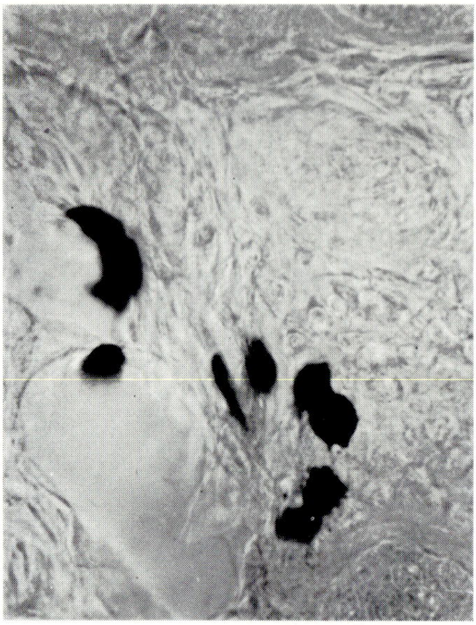

FIG. 4

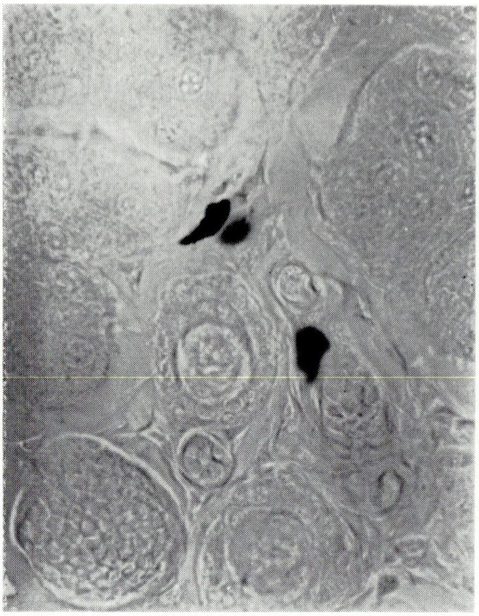

FIG. 5

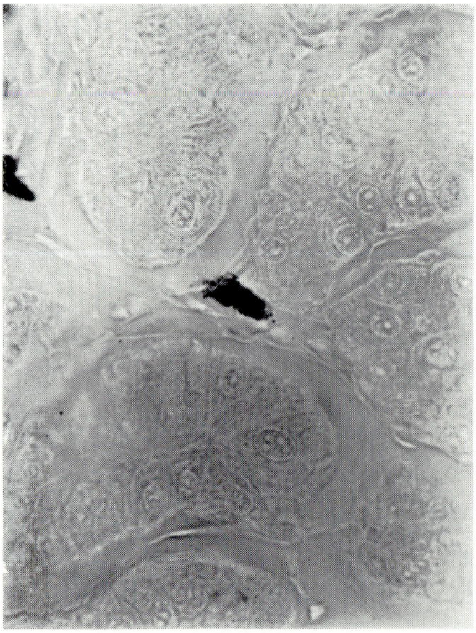

FIG. 6

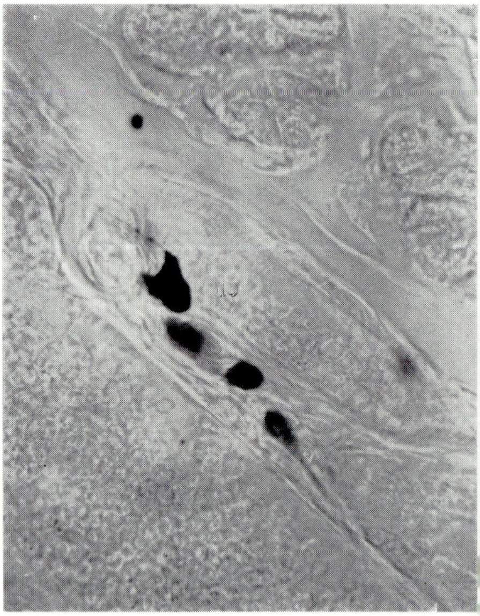

FIG. 7

FIG. 4, 5, 6, 7. Photomicrographs showing the morphology of mast cells and their distribution within sections of the exorbital lacrimal gland of the male albino rat. Note that some mast cells lie near the base of acini and others approximate blood vessels.

Formalin—alcohol—calcium acetate. Eosin. Magnification 450×.

TABLE III. Number of squares in 1,200 squares occupied by and relative proportion of acini and intralobular ducts of exorbital lacrimal gland of rats of seven age groups.

Age of rats in days	Number of squares occupied by Acini	Intralobular ducts	Percentage composition Acini	Intralobular ducts
15	931·4	65·4	77·6	5·5
30	973·8	47·0	81·1	3·9
60	1022·6	45·0	85·2	3·8
90	968·8	44·2	80·7	3·7
150	1009·6	25·8	84·1	2·2
300	928·2	37·8	77·3	3·2
500	924·4	47·2	77·0	3·9

Reproduced with permission from Bolden *et al.*, the Quarterly of the National Dental Association, Tuskegee, Alabama, 1962.

2. *Mast cells*

(*a*) *Distribution.* Mast cells stained metachromatically and selectively by the Smith-Atkinson technique. They were distributed throughout the lobule of the gland, but predominantly within the capsule and the interstitial connective tissue. Many approximated blood vessels. They were quite variable in shape and size (Figs. 4, 5, 6 and 7). Many were in various stages of degranulation (Figs. 8 and 9).

(*b*) *Quantitative analysis.* The concentration of mast cells showed mild fluctuations with increase in age after 30 days (Table IV).

TABLE IV. Mast cell density of exorbital lacrimal gland of the male albino rat.*

Mast cell	15	30	60	90	150	300	500
Average density	277·8	103·9	155·2	134·0	108·0	123·3	107·0
% Reduction from 15 days		64	44	52	60	54	60

* Per square millimeter of surface area.

The highest average concentration was observed in sections of glands from 15 day old animals. The average density at all other periods was nearly 50% of the 15 day figure. The next highest average mast cell density was found at 60 days. Nearly identical densities were found at 30, 150 and 500 and at 90 and 300 days.

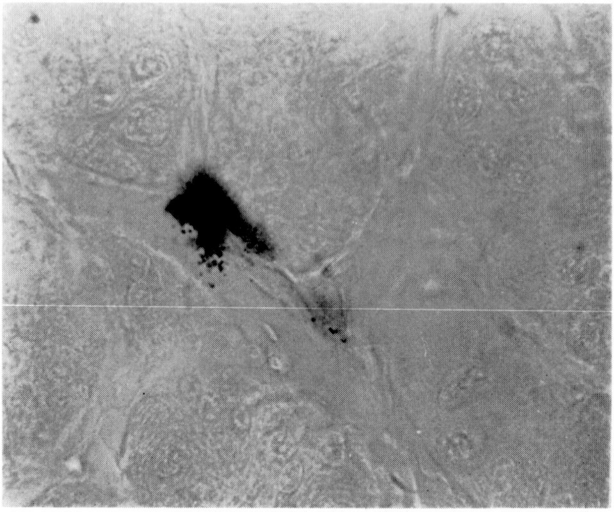

FIG. 8

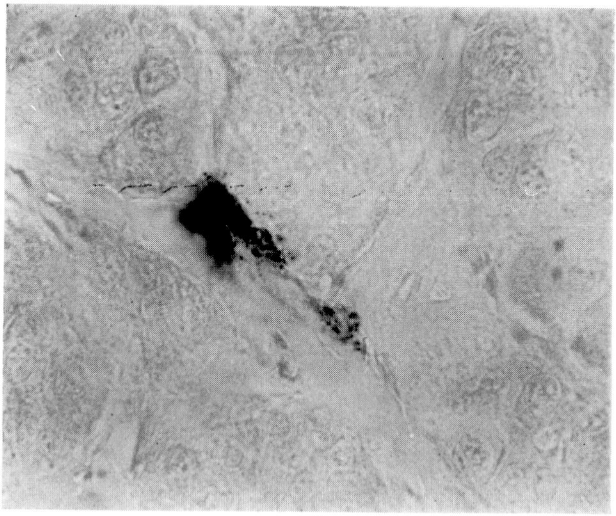

FIG. 9

FIG. 8 and 9. Photomicrographs of the same mast cell at the base of an acinus, taken at different levels of focus. Note the distribution of granules.

Formalin—alcohol—calcium acetate. Eosin. Magnification 450×.

E. *Histochemical findings*

1. Intensity

The intensity of the reaction for alkaline phosphatase activity varied from 1+ to 4+. The least intense reaction was given by the basal end of the acini. The most intense reaction was given by the myoepithelial cells and the walls of the capillaries.

The reaction in sections of glands from 60– to 500–day old animals was of nearly equal intensity. The most marked reaction was observed in 15– and 30–day glands.

2. Site

At all ages, the same sites were positive for alkaline phosphatase activity (Bolden 1963). These were the peripheral edge of the acini, the intercalated duct, the intralobular ducts, walls of the capillaries and myoepithelial cells at the base of the acini (Fig. 10—Figs. 11, 12 and 13). Both the nucleus and the protoplasm of the myoepithelial cell appeared to be active sites.

F. *Chemical findings*

1. Dry weight

Exorbital lacrimal gland dry weight nearly tripled between 30 days and 90 days. From 90 to 500 days, the dry weight remained near the 90–day level (Table V).

TABLE V. Average dry weights (mg) of right exorbital lacrimal glands of male albino rats.

Age of rats in days	Dry weight of lacrimal gland (mg) (No.)
30	8 (4)
60	10 (5)
90	21 (5)
150	22 (5)
300	21 (5)
500	22 (4)

Reproduced with permission from Bolden *et al.*, the Quarterly of the National Dental Association, 1962.

2. Wet weight

Between 15 and 30 days the average combined wet weight of right and left lacrimal glands nearly doubled. It increased from 40·4 to 95·7 milligrams.

By 60 days the weight had nearly tripled. Left glands were lighter than right glands from animals 90 to 500 days of age.

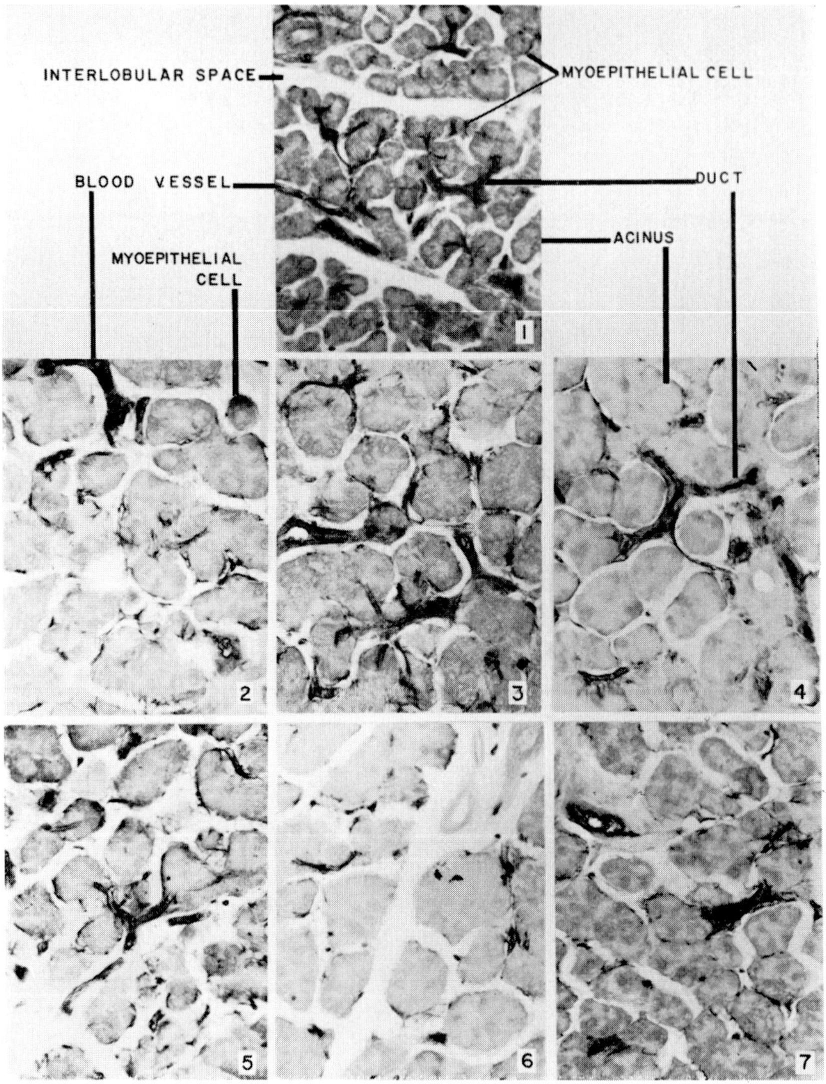

FIG. 10. Photomicrographs showing sites of alkaline phosphatase activity in sections of rat exorbital lacrimal glands incubated for 15 minutes. All photomicrographs and prints were treated in an identical manner. Magnification 120× —reduced ½.

Figure 1—15 days, Figure 2—30 days, Figure 3—60 days, Figure 4—90 days, Figure 5—150 days, Figure 6—300 days, Figure 7—500 days. Reproduced with permission from Bolden, Journal of the Connecticut State Dental Association, 1964.

The greatest percentage contribution of gland weight to total body weight was found in 15- and 30-day old animals. A progressive decrease in percentage contribution was noted for each succeeding period up to 500 days (Table II).

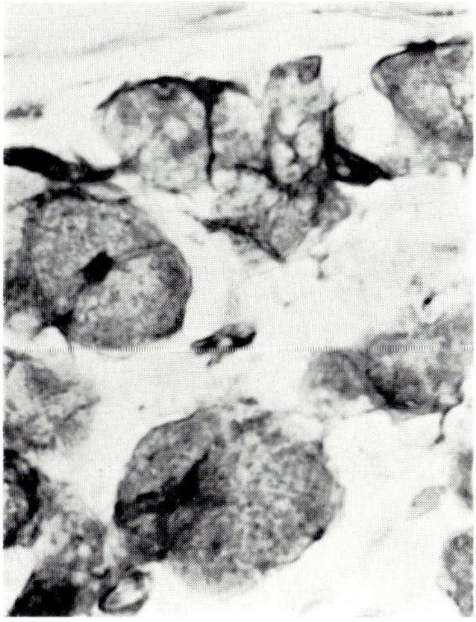

FIG. 11

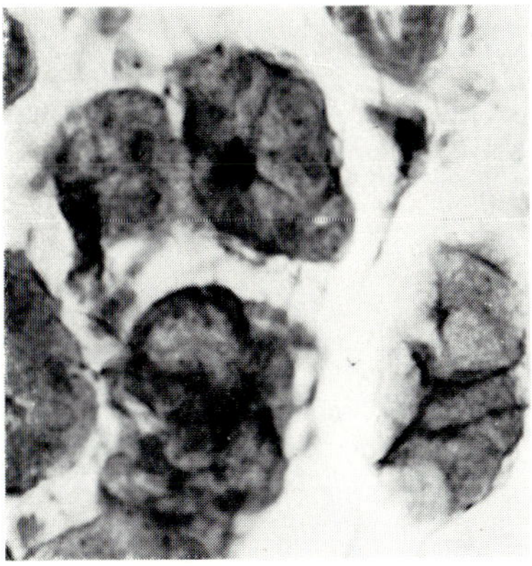

FIG. 12

FIG. 11 and 12. Photomicrographs of a group of acinar cells of the exorbital lacrimal gland of a male albino rat showing localization of alkaline phosphatase activity. The spiderlike processes of the myoepithelial cells are seen radiating across the acini. Magnification 120×. Figure 11—90 days, Figure 12—300 days. Reproduced with permission from Bolden, Journal of the Connecticut State Dental Association, 1964.

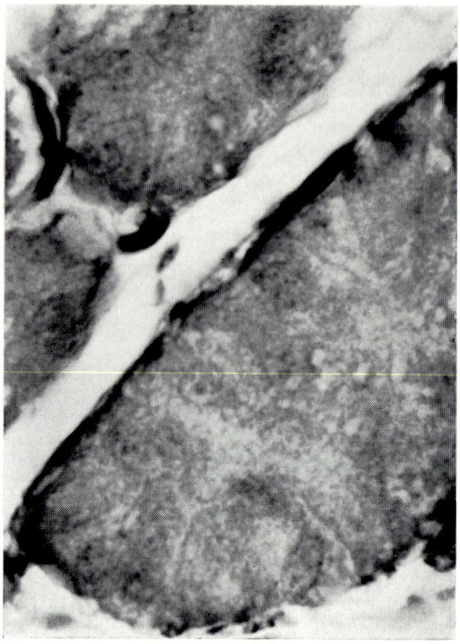

FIG. 13. Photomicrograph of acini of exorbital lacrimal gland of a 30 day old rat. The nucleus of a myoepithelial cell may be observed at the base of the acinus between the basement membrane and the secretory cell. Magnification 450×. Reproduced with permission from Bolden, Journal of the Connecticut State Dental Association, 1964.

TABLE VI. Alkaline phosphatase activity, acid phosphatase activity and inorganic phosphorus levels of exorbital lacrimal gland homogenates with increase in age.

Age of rats in days	No. of rats	Alkaline phosphatase activity*	Acid phosphatase activity†	Tissue inorganic phosphorus‡	
				pH 9·2	pH 5·0
15	5	0·00054	0·000072	0·00049	0·000362
30	5	0·00079	0·000102	0·00054	0·000362
60	5	0·00078	0·000102	0·00044	0·000405
90	5	0·00085	0·000114	0·00058	0·000446
150	5	0·00086	0·000072	0·00052	0·000334
300	5	0·00116	0·000076	0·00040	0·000333
500	5	0·00114	0·000096	0·00055	0·000286

* Alkaline phosphatase milligrams of phosphorus liberated from sodium beta glycerophosphate in 20 minutes enzyme action at pH 9·2/milligrams of wet weight of tissue.

† Acid Phosphatase: milligrams of phosphorus liberated from sodium beta glycerophosphate in 20 minutes enzyme action at pH 5·0/milligrams of wet weight of tissue.

‡ Milligram/milligram wet weight.

Reproduced, with modifications, with permission from Bolden, The Meharri-Dent, Nashville Tennessee, 1963.

3. Alkaline phosphatase activity

Gland homogenates from 15–day old animals had approximately 62% less alkaline phosphatase activity than did glands from senile rats, 500 days of age. The increase in activity was consistent with increase in age up to 300 days (Table VI).

4. Acid phosphatase activity

The level of acid phosphatase activity showed mild fluctuations. The highest levels were found between 30 to 90 days. At all ages the amount of acid phosphatase was less than that of alkaline phosphatase (Table VI).

5. Inorganic phosphorus

Age did not significantly affect the level of tissue inorganic phosphorus. Homogenates prepared with acetate buffer (pH 5·0) had slightly lower inorganic phosphorus levels than homogenates prepared with glycine buffer (pH 9·2) (Table VI).

IV. Discussion

A. *Body weight and body length*

The most rapid period of growth for male Wistar rats occurs within the first 60 days of life (Bolden *et al.* 1963). Animals more than doubled their weight between 15 and 30 days of age, and by 60 days exhibited a ten-fold increase over the 15-day weight. Constant weight gains were observed after 60 days. Increase in body length paralleled increase in body weight (Table II), (Figs. 14, 15 and 16).

These findings substantiate data reported by King (1915), Sreebny (1954), and Walker (1958). Walker reported that the body weight of male albino rats reached a plateau before 300 days, while females grew more slowly and increased in weight up to 900 days. Donaldson (1924) had already shown that the body weight of the albino rat is very rapid during the first 120 days postnatally.

B. *Gland weight*

Glandular growth is influenced by sex, and can be accelerated by specific drug action. Minami and Kamei (1959) found a sexual difference in the weight of the lacrimal gland of the rat. Glands from female rats, 300 to 500 days of age, had a greater relative weight (63·8±2·7) than did exorbital glands from male rats (47·9±2·6) at the same age.

Walker (1958) found that the exorbital lacrimal glands grew only about half as fast as body weight. From 21 days to 100 days, the left gland started at

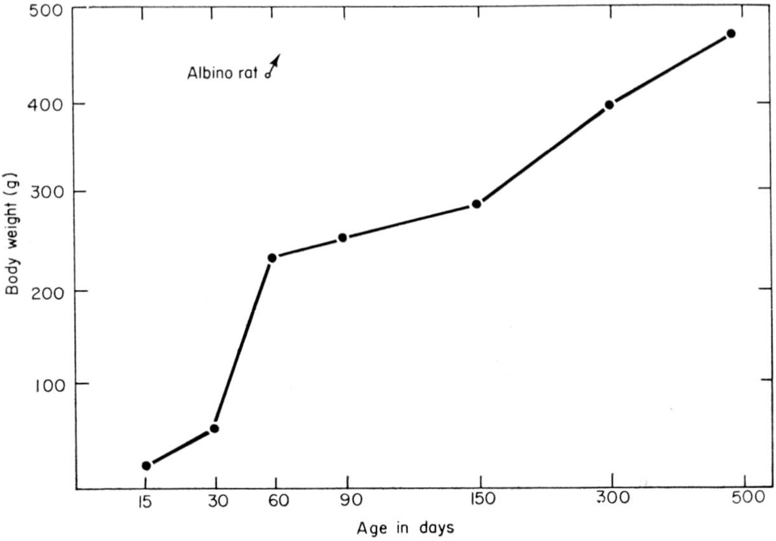

FIG. 14. Body weights of male albino rats at seven different ages.

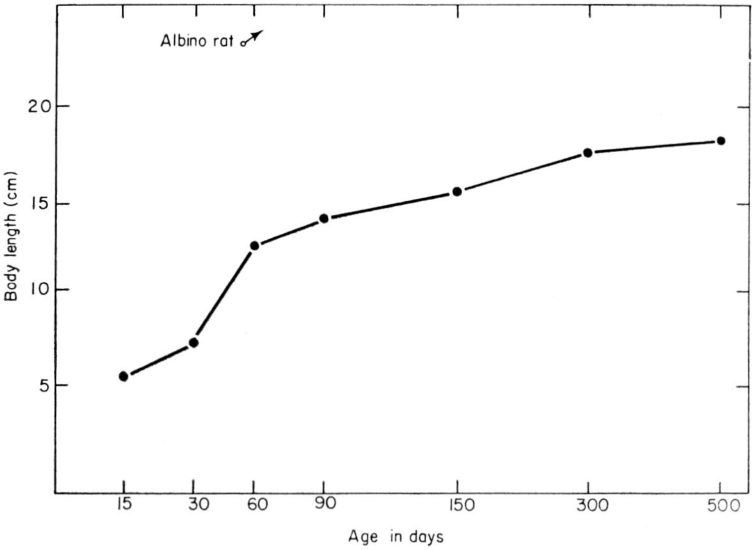

FIG. 15. Body lengths of male albino rats at seven different ages.

about 0·5 mg per gram of body weight ended at about 0·25 mg/mg. Gland weight had nearly reached its plateau by 100 days. Senile albino rats (700–1000 days) experienced a drop in gland weight from 64 mg to about 50 mg.

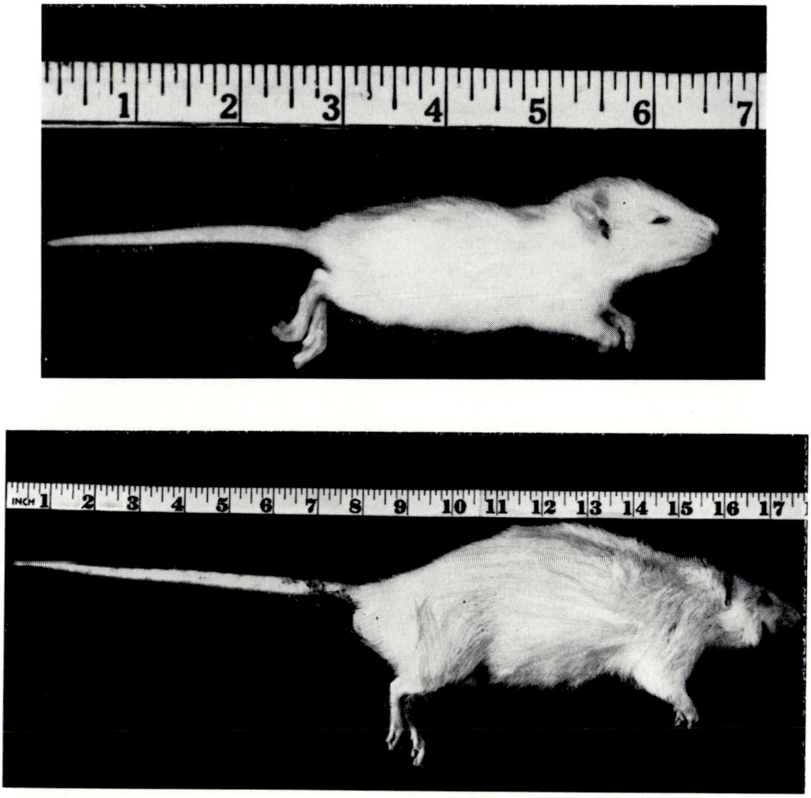

FIG. 16. Photograph of male albino rats showing the plane of measurement of body length—tip of nose to tip of tail. Top 15 days—bottom 500 days of age.

Reproduced, with modifications, with permission from Bolden, Journal of the Connecticut State Dental Association, 1964.

Glandular growth is also influenced by age. In the current series, right and left (combined) exorbital lacrimal glands contributed more to total body weight at 15 and 30 days of age than at any other period. After 30 days, there was a progressive decline in percentage contribution of gland weight to total body weight (Table II).

The weight of individual glands continued to increase and reached a plateau around 150 days of age. The highest value reported by Walker (1958),

64 milligrams at 100 days, was reached by glands in our series between 30 and 60 days (Fig. 1) (Table I).

C. *Age and dry weight*

The relative amount of solids in a given tissue is reflected by the dry weight of the tissue. These data indicate that the maximal accumulation of solids is achieved in the exorbital lacrimal gland from 90–day old animals. The average dry weights of lacrimal glands nearly tripled between 30 and 90 days. This growth was accompanied by proportional increases in non-secretory and parenchymatous elements. After 90 days, and up to 500 days, the average dry weight of lacrimal glands constituted the same fraction of the total body weight (Table V).

D. *Age as a factor in composition*

Acini of exorbital lacrimal glands are assumed to be functionally mature at 15 days. Teir (1951a, 1951b) demonstrated that the growth of the outer orbital glands, as measured by mitotic activity, was quite rapid at birth still high at 15 days, then rapidly diminished. Only a few new cells were seen at 42 to 56 days. At 210 to 270 days, only a rare isolated mitosis was observed.

Benson (1964) induced growth changes in the lacrimal glands of female white rats by giving repeated and alternating doses of DL-ethiomine and N-2 fluorenylacetanide (2 FAA) alone and in combination. Glands from animals 200 to 450 days of age were induced to assume several histologic characteristics normally found in glands from animals at senility. These included increased mitotic activity, increased incidence and degree of variability of cellular and nuclear sizes, a prosoplasia to Harderian type glandular epithelium, and the appearance of acidophilic cells.

Exorbital lacrimal glands in this series showed very slight variations in the relative proportions of acinar to intralobular duct tissue with increase in age. Beyond 15 days of age there was little growth of these two elements and no appreciable difference in the area occupied by intralobular ducts (Table III).

E. *Mast cell density*

1. Effect of age

Wide variability exists in the mast cell populations in tissue from animals of different ages (Carranza and Cabrini 1955, Calonius 1960, Dewar 1955, Dummett *et al.* 1961, Dummett *et al.* 1963, Stella 1955, Vantrease and Bolden 1964).

The fluctuation in mast cell density in sections of normal exorbital lacrimal gland at seven different age groups is in keeping with observations made of mast cell densities in other normal tissues.

Human and rat connective tissue (Higuchi 1930, Bates 1935) have been shown to demonstrate decreased mast cell populations with increase in age. Spicer (1960), studied mast cells in the exorbital lacrimal glands of the mouse and reported that glands from year old animals had nearly twice as many mast cells as glands from month old animals.

Age appeared to have its greatest bearing on the number of mast cells present in the exorbital lacrimal gland of rats in this series between 15 and 30 days of age. The large number at 15 days, had dropped by more than 50% by 30 days. At all other periods, including adulthood and senescence, periods characterized by fluctuations in hormonal levels, there were fluctuations in mast cell density but with no precipitous decreases or spectacular increases.

Riley (1954) has shown that the tissue mast cells disappear in areas of acute tissue injury and reappear when connective tissue fibrils begin to be laid down and the ground substance is shrinking.

Flachmeyer and Valu (1962) studied lacrimal glands histologically and reported an increase in the collagenous and fatty contents of the capsule, with coarser and more irregular argyrophilic fibers. Fatty infiltration and an increase in the numbers of atypical serous cells were also observed in salivary glands of the rat (Andrew 1949) and humans (Bauer 1950) as a result of increase in age.

The maximum relative amount of interstitial tissue had already been acquired by the exorbital lacrimal glands in this series by 15 days post-natally.

2. *Response to drugs*

Factors other than age influence the density of mast cells. The mast cell also shows wide variation in its reponse to drugs which have been shown to affect parenchymal tissue.

Benson (1964) observed that the number of mast cells were increased in sections of lacrimal glands potentiated by DL-ethionine or N-2-fluorenyl-acetamide (2 FAA) alone or in combination. The glands took on the histologic characteristics of glands of senile rats.

The external lacrimal glands of mice given twice daily doses of iso-propylnoradrenalin sulphate for 5 to 8 days (Brown-Grant 1961) showed a variable increase in the size of the acinar cells. However, isoproteronol was shown by Vantrease and Bolden (1965) to have no specific effect on mast cell density.

F. *Chemical changes with age*

These studies have shown that alkaline and acid phosphatase activity is present in the exorbital lacrimal gland of the male rate at 15 days after birth and that this activity is not appreciably altered up to 500 days of age.

1. Localization of alkaline phosphatase

The sites of alkaline phosphatase activity were limited to the capillaries, base of acini and myo-epithelial cells (Bolden 1964). The latter were well developed and prominent at 15 days. This tends to verify observations by Leeson (1960a, 1960b) that this gland is functional 15 days post-natally. Similar sites of localization of alkaline phosphatase were demonstrated in normal and duct ligated submaxillary glands of the rat by Bolden (1957, 1962, 1963) and Bolden and Weinmann (1958) and in the submaxillary gland of the rat and mouse by Hill and Bourne (1954). The nuclei of myoepithelial cells of human lacrimal glands have been shown by Kato (1958) to have marked activity for alkaline phosphatase.

2. Content of phosphatases and phosphorus

Greatly elevated alkaline phosphatase activity is found in tissues during rapid growth (Schlamaowitz 1958) and heightened function (Folley and Greenbaum 1947). The placenta shows increased alkaline phosphatase activity during pregnancy which reaches a maximum just before term (Nagnostopoulos and Matsudana 1958).

The greatest period of growth of the exorbital lacrimal gland, as measured by wet weight, was within the first 150 post-natal days.

The rare mitotic activity at 100 days (Teir 1951a) would tend to show that increase in age is accompanied by a rapid decline in cellular growth.

However, increase in age did not appear to cause a decline in functional activity. The high levels of alkaline and acid phosphatase activity persisting up to 500 days was probably in direct response to a constant need for heightened functional activity of these glands, in this species. Age had no apparent effect on inorganic phosphorus content.

It is of interest to note that ligation atrophy (Bolden and Bolden 1960) is accompanied by a decline in the content of alkaline phosphatase (Junqueira and Rabinovitch 1954, Bolden 1957) and inorganic phosphorus (Bolden 1957) of homogenates of rat submaxillary glands. RNAP (ribonucleic acid, phosphorus and DNAP), (desoxyribonucleic acid phosphorus), (Rabinovitch *et al.* 1952) and creatin phosphate labile phosphorus levels (Fernandes and Junqueria 1953), have also been shown to be reactive to ligation atrophy.

A species and tissue variation in enzymatic activity also exists. Mizukawa and Morisue (1961) found differences in enzymatic activity in homogenates of rabbit lacrimal glands. Acid phosphatase, lactic dehydrogenase and isocitric dehydrogenase were active in the lacrimal gland while alkaline phosphatase, malic dehydrogenase and 6-phospho-gluconic dehydrogenase were active in Harder's gland.

V. Summary and Conclusions

(1) Age, as a factor in the growth of the exorbital lacrimal gland, was determined in male Wistar rats at seven age groups representing youth, maturity and senility. (2) The greatest increase in gland growth, size and weight, occurred within the first 60 days of life. Gland growth mimiced body growth. (3) The exorbital lacrimal gland of the male rat is composed of three basic elements (i) acini, (ii) ducts, and (iii) non-secretory tissue. The approximate proportion of 82% acini, 3% intralobular duct and 15% non-secretory tissue remained fairly constant at each age tested. (4) Maturation of parenchyma occurs at approximately 15 days in the exorbital lacrimal gland. (5) Mast cell density followed the same pattern of variability observed in other normal tissues. The highest concentration was found at 15 days. No trend was shown which could be attributed to age. (6) Maximal accumulation of solids is achieved in the exorbital lacrimal gland by 90 days. (7) The sites of localization of alkaline phosphatase activity are principally the capillaries, the acini and the myoepithelial cells. Age does not cause significant alterations in the intensity of the reaction. (8) Alkaline phosphatase content of gland homogenates showed a consistent increase with increase in age. However, age would appear to have little effect on the content of inorganic phosphorus or acid phosphatase.

Acknowledgment

The author acknowledges the assistance of Mrs. Lola W. Nathan in preparation of the manuscript and Mr. Edwin Wilson for photomicrographs, Fig. 3.

REFERENCES

Andrew, W. (1949). *J. Geront.* 4, 95–104.
Bates, E. (1935). *Anat. Rec.* 61, 231–239.
Bauer, W. H. (1950). *J. dent. Res.* **29**, 686.
Benson W. R. (1964). *Am. J. Path.* 45, 587–597.
Boas, N. F. and Scow, R. O. (1954). *Endrocrinology* 55, 148–155.
Bolden, T. E. (1957). Thesis for the Doctor of Philosophy Degree in Pathology, University of Illinois, Chicago, U.S.A.
Bolden, T. E. (1962). *J. Conn. St. dent. Assoc.* 36, 19–25.
Bolden, T. E. (1963). *J. dent. Res.* 37, 967–968.
Bolden, T. E. (1964). *J. Conn. St. dent. Assoc.* 38, 5–9.
Bolden, T. E. and Bolden, D. M. F. (1960). *Q. natn. dent. Ass.* **22**, 53–55.
Bolden, T. E. and Weinmann, J. P. (1958). *J. dent. Res.* 37, 967–968.
Bolden, T. E., Barnes, S. and Hall, M. L. (1962). *Meharri-Dent.* **21**, 6–12.
Brown-Grant, K. (1961). *Nature* **191**, 1076–1078.
Calonius, P. E. B. (1960). *Dent. Abstr.* 5, 356.
Carranza, F. A. and Cabrini, R. L. (1955). *Oral Surg.* 8, 1093.

Dewar, M. R. (1955). *J. Periodont.* **29**, 67.
Donaldson, H. H. (1924). *Mem. Wistar Inst. Anat.* 6.
Dummett, C. O., Bolden, T. E. and Ashurst, J. C. (1961). *J. dent. Res.* **40**, 921–928.
Dummett, C. O., Bolden, T. E. and Goldsberry, S. (1963). *J. dent. Res.* **40**, 647.
Fernandes, J. F. and Junqueira, L. C. U. (1953). *Exp. Cell Res.* **5**, 329–334.
Fiske, C. H. and Subarrow, Y. (1925). *J. biol. Chem.* 66, 375–400.
Flachmeyer, R. and Valu, I. (1963). *Albrecht v. Graefe Arch. Ophthal.* **165**, 295–303.
Folley, S. J. and Greenbaum, A. L. (1947). *B. J.* **41**, 261–269.
Gomori, G. (1941). *J. cell comp. Physiol.* **17**, 71–84.
Higuchi, K. (1930). *Folia haemat. hpz.* **41**, 401–411.
Hill, R. C. and Bourne, G. H. (1954). *Acta Anat.* **20**, 116–128.
Hofstee, B. H. J. (1954). *Archs Biochem. Biophys.* **51**, 139–146.
Holter, H. and Li, Si-OH. (1949). *Compt. rend. d. Chim.* **27**, 393–407.
Junqueira, L. C. and Rabinovitch, M. (1954). Thesis. *Tex. Rep. Biol. Med.* **12**, 94–97.
Kato, Y. (1958). *Acta Soc. Ophthal. Japan* **62**, 175.
King, H. D. (1915). *Anat. Rec.* **9**, 751–776.
Leeson, C. R. (1960a). *Anat. Rec.* **137**, 45–55.
Leeson, C. R. (1960b). *Acta anat.* **40**, 87–94.
Loewenthal, N. (1900). *Arch mikrosk. Anat.* 56, 535–552.
Minami, A. and Kamei, T. (1959). *C.r. Séanc. Soc. Biol.* CLIII (2), 269–273.
Mizukawa, T. and Morisue, T. (1961). *Folia ophthal.* jap. **12**, 616–622 (6).
Moog, F. (1946). *Biol. Rev.* **21**, 41–59.
Nagnostopoulos, C. A. and Matsudana, H. (1958). *In* "Proceedings of the International Symposium on Enzyme Chemistry", Tokyo-Kyoto (K. Ichihara ed.), Academic Press. New York.
Rabinovitch, M., Rothschild, H. A. and Junqueira, L. C. U. (1952). *J. biol. Chem.* 194, 835–838.
Riley, J. F. (1954). *Lancet* **1**, 841.
Schlamaowitz, M. (1958). *Ann. N.Y. Acad. Sci.* 75, 373–379.
Smith, E. W. and Atkinson, W. B. (1956). *Science* **123**, 941.
Spicer, S. S. (1960). *Am. J. Path*, pp. 457–475.
Sreebny, L. M. (1954). Ph.D. Thesis. University of Illinois, Chicago.
Stella, A. (1955). *Anales facodont. Montevideo* **2**, 125.
Teir, H. (1951a). *Soc. Scient. Fenn. Comm. Biol.* XIII (1), 1–32.
Teir, H. (1951b). *Soc. Scient. Fenn. Comm. Biol.* XIII (6), 1–16.
Vantrease, D. B. and Bolden, T. E. (1964). *Meharri-Dent.* XXII, 16–24.
Vantrease, D. B. and Bolden, T. E. (1965). *J. dent. Res. Suppl.* **44**, 128.
Venable, J. H. and Grafflin, A. L. (1940). *J. Mammal.* **21**, 66–71.
Walker, R. (1958). *Anat. Rec.* **132**, 49–70.

Humoral Influences on Mitotic Activity of Gastric Epithelium

T. RÄSÄNEN

II Department of Pathology, University of Helsinki, Finland, and Central Hospital of Savonlinna, Finland

THE humoral factors which regulate the mitotic activity in the gastrointestinal epithelium may affect the epithelial cells directly or indirectly by liberating from the environment factors which possibly influence the regeneration of epithelium. Furthermore the gastric epithelium has a special position in its secretory function with almost non-existing absorptive capacity, contrary to the intestinal epithelium which is also capable of readsorbing the fragments of nucleic acids.

In the gastric mucosa it may be that the changes in concentrations of hydrogen and other ions cause lability in the mitotic activity.

A single injection of ACTH (15 i.u. Cortrophin Z, Organon) within 5 hours of the application leads to decrease of mitoses in the rat gastric mucosa. In the intestinal mucosa the amount of mitoses does not decrease—in some parts it increases. If the ACTH application is extended to 4 days then we can observe a weak inclination towards an increase in the mitotic activity (Räsänen, 1963). The mitotic activity clearly increases after 10 days application of ACTH (4 i.u. per day). The increase in the count of mitosis in the rat gastric mucosa in three experiments is about treble that of the control value (Räsänen and Teir 1961).

In the same way as after a single injection of ACTH the stress decreases the mitotic activity in the gastric mucosa. It happens in a few hours during physical stress and in the intoxication. In the intestinal mucosa the count of mitoses does not decrease (Räsänen 1963). The thyroid gland possibly does not participate in this phenomenon because the long term application of thyreotropin (1 i.u. per day of Ambinon, Organon) in 2 weeks does not lead any changes in the regeneration of gastric mucosa (Table I).

It is well known that glucocorticoids lead to a drop in the mitotic activity in the gastric epithelium. Also two days application of dexamethasone and prednisolone (1 mg × 6 at 6 hours interval) causes almost total failure in the gastric mitotic activity; the drop after cortisol was also 80%. In the duodenal mucosa the decrease in the count of mitoses was much weaker, for dexamethasone only 60% and for the other glucocorticoids even less (Räsänen

1963). The effect may be partly to the glycogenic activity of glucocorticoids which is associated with their antitumour and anti-inflammatory potencies (Glenn 1964). A single injection of dexamethasone of 0·05 mg in mice caused a distinct decrease in the thymidine index only in the gastric mucosa and not in the duodenum and liver (Lahtiharju *et al.* 1964).

TABLE I. The count of mitoses in the body mucosa per 4000 epithelial cells, and in the abdominal skin per 2000 epidermal cells. The 4 month old male rats were injected intramuscularly during two weeks with 1 unit of thyreotropin per day. Mean ±S.E.

	No.	Body mucosa	Epidermis
Thyreotropin	20	20·3±1·0	13·0±1·1
Controls	10	21·7±2·9	19·7±3·5

Mineralocorticoid, deoxycorticosterone acetate in doses of 1 mg given ×7 at 8-hour intervals caused an increase in the mitotic frequency in the rat gastric mucosa. Neither the smaller doses: 0·1 and 0·5 mg nor the greater doses: 5·0 and 10·0 mg affected the mitotic activity. A little amount of

TABLE II. The count of mitoses in the rat gastric and intestinal epithelium after aldosterone application, and after the removal of water. Mean ±S.E.

	Stomach		Duodenum		Ileum		Colon
Aldosterone							
0·1 mg	60±6		194±12		171±9		37±6
0·1 mg × 3	31±4	P<0·01	196±14		182±17		48±11
0·1 mg × 5	55±8		183±13		171±16		45±7
After removing of water							
4 hours (Control)	64±6		197±14		191±16		47±7
28 hours	37±5	P<0·01	142±9	P<0·01	163±8		40±7
52 hours	40±7	P<0·05	105±13	P<0·001	109±9	P<0·01	46±10

prednisolone, 0·3 mg given simultaneously with 1 mg of deoxycorticosterone acetate prevented the stimulating effect on mitotic activity (Räsänen 1962). Deoxycorticosterones do not either affect the liver glycogen synthesis or the inflammation. It changes the pH of intracellular water to the alkaline direction (Withrow and Woodbury 1964). Possibly it changes the cellular milieu favourable to RNA-DNA-synthesis by increasing the content of cellular sodium and water.

Aldosterone first decreases the count of mitoses in the gastric mucosa but later the inhibition disappears (Table II). Aldosterone was given in doses of 0·1 mg at 12 hour intervals. In the intestinal mucosa it does not cause any

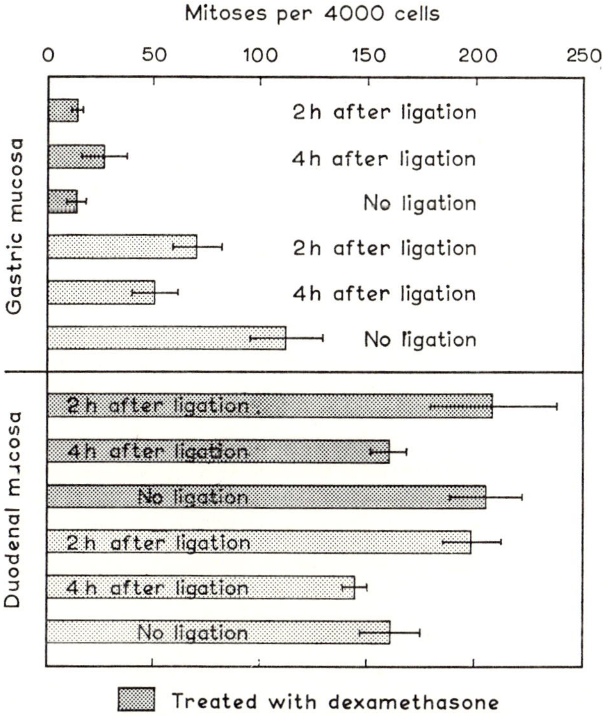

FIG. 1. Mitotic count in gastric and duodenal mucosa 2 and 4 hours after ligation of pylorus in dexamethasone-treated (5 × 1 mg of dexamethasone at 12 h intervals) and in untreated rats. Each column is the number of mitoses per 4000 cells (mean ± S.E.) in five rats.

change in the mitotic activity. It is observed that aldosterone has also a glucocorticoid effect on the metabolism (Schuler *et al.* 1954).

In thirst the decrease of the intra- and extracellular water may partly affect the mitotic activity in the gastro-intestinal canal leading to decrease of mitoses in the epithelium except in the colon (Table II).

A specific feature in the regeneration of gastro-intestinal epithelium is that the mitotic activity of the gastric mucosa is very labile compared with that of intestinal mucosa. The count of mitoses in the gastric mucosa decreases significantly 2 to 4 hours after pyloric ligation by untreated rats. In dexamethasone-treated rats it is decreased *ad minimum* and the pyloric ligation does not change the count of mitoses. In the duodenal mucosa of these rats there are no clearly significant changes in the count of mitoses (Fig. 1). Similarly the filling of an isolated jejunal loop with different solutions (Fig. 2) for one hour does not cause any change in the mitotic activity of epithelium.

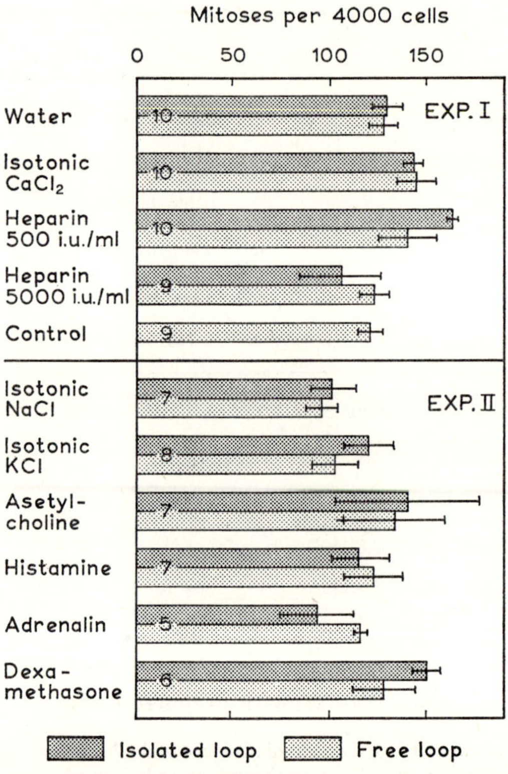

FIG. 2. Mitotic count in jejunal mucosa of isolated and free loop. The loop was filled with 2 ml of test solution 2 hours prior to sampling. Each column is the number of mitoses per 4000 cells (mean ± S.E.); the numbers of rats are indicated in the columns.

Polysaccharide with properties of heparin has been isolated from gastric mucosa (Smith and Gallop 1953). The intestinal polysaccharide is heparinoid differing from heparin of gastric mucosa (Bianchini 1958). According to Roth (1953) heparin inhibits the function of RNase and DNase, more effectively in pH 6 than pH 8. The inhibition of gastric mitotic activity falls until the instant at which the mucosal mast cells are degranulating and probably liberate their heparin. This may be a local, genuine inhibitor of mitotic activity in the gastric mucosa. If we heparinize the test animals the same happens again: gastric mucosa loses its mitotic activity. The decrease in the intestinal mucosa is very weak or unobservable (Räsänen 1963). When during ACTH-treatment in ten days the gastric mucosa has totally lost the granulation of mast cells (Räsänen 1961) then the mitotic activity is increased in the epithelium (Räsänen and Teir 1961).

The mast cell content in the human gastric mucosa is in average 10,000 per mm^3 of tissue (Räsänen 1958) and in the superficial part of rat body mucosa 50,000. This may be the local, mitosis-inhibiting factor which functions in the gastric mucosa during neural and endocrinic stimulation.

A chronic peptic ulcer is rich in mast cells which are shown to secrete polysaccharide to the surrounding tissue. If this is heparin, it may partly prolong the local lesion of gastric mucosa by inhibiting the regeneration of epithelium, especially in acid milieu. It is demonstrated that glucocorticoids decrease the thymidine index of gastric mucosa neighbouring an artificial ulcer (Myhre 1960).

REFERENCES

Bianchini, P. (1958). *Boll. Soc. ital. Biol. sper.* 34, 685–688.

Glenn, E. M. (1964). *In* "Hormonal Steroids" (L. Martini and A. Pecile eds.). Vol. 1, pp. 319–349. Academic Press, New York.

Lahtiharju, A., Räsänen, T. and Teir, H. (1964). *Growth*, **28**, 221–224

Myhre, E. (1960). *Acta path.* **70**, 476–485.

Räsänen, T. (1958). *Acta path. microbiol. scand.* Suppl. **129**.

Räsänen, T. (1961). *Acta physiol. scand.* **52**, 162–166.

Räsänen, T. (1962). *Growth*, **26**, 1–14.

Räsänen, T. (1963). *Acta physiol. scand.* 58, 201–210.

Räsänen, T. and Teir, H. (1961). *Growth*, **25**, 139–149.

Roth, J. S. (1953). *Archs Biochem. Biophys.* **44**, 265–271.

Schuler, W., Desaulles, P. and Meier, R. (1954). *Experientia*, **10**, 142– .

Smith, H. and Gallop, R. C. (1953). *Biochem. J.* 53, 666–671.

Withrow, C. D. and Woodbury, D. M. (1964). *In* "Hormonal Steroids" (L. Martini and A. Pecile eds.), pp. 503–513. Academic Press, New York.

An Active Cell Factor Inhibiting Body Growth in Mice: Its Origin and Mode of Action

A. CLAUDE

Laboratoire de Cytologie et de Cancérologie Expérimentale, Université Libre de Bruxelles, Brussels, Belgium

Two main sublines of the Bagg Albino/c strain of mice have been maintained in our laboratory since 1953. The first subline, referred to as the C^- line, constitutes a direct descent by brother × sister mating of the original Bagg Albino strain (BALB/c) and has retained some of its main characteristics (Staats 1964). The second subline, referred to as the C^+ line, differs from the original BALB/c line (C^-) by a high mammary tumour incidence, established by Andervont (1940–41 and 1944–45) as a result of the introduction in this line of the milk agent by foster-nursing youngs on C3H mothers. The two C^- and C^+ lines (without, and with the milk factor) reached our laboratory through Andervont (Bethesda), Dmochowski (Leeds), and Haddow (London), pairs of both lines being received from Prof. Haddow in December 1953; until then, no kidney tumours had been reported in neither of these lines.

The first case of a renal adenocarcinoma in the C^+ line was observed in our laboratory in 1955 (Claude 1958). Subsequent observations have revealed that the incidence of this adenocarcinoma, of apparently spontaneous occurrence, would vary from a few percent to 40 to 60% of the individuals, depending on age and on differences between selected, inbred families of the same line. This sub-strain of mice, carrying the potentialities for kidney adenocarcinoma has been entered in the "Standardized Nomenclature for Inbred Strains" under the denomination: BALB/c Cd (Mouse News Letter 1961, Staats 1964). Continuous inbreeding and selection, carried out in the past few years has produced a descent presenting 60 to 70% of spontaneous adenocarcinoma, at the age of 9 to 15 months. Characteristically, mammary tumours are no longer observed in this renal tumour sub-strain, suggesting that the milk factor has been lost, or is being suppressed by a kidney tumour influence.

The spontaneous tumours arise in the cortical zone of the kidney, apparently from cells of convoluted tubules. The tumours affect equally both sexes, and both kidneys are involved, each being found occupied by multiple tumours in various stages of evolution, from plain cysts to papillomatous growths and solid carcinomas. Histologically the tumours found in one organ present the same progressive evolution, from normal tubule cells to cysts, plain or papillomatous, and to solid tumours, the latter being represented by compact columns of large, cuboidal cells, characterized by a pronounced polymorphism, a variety of sizes, and a uniformly intense basophilia (Claude 1958, 1962). This particular morphology is illustrated in Fig. 1.

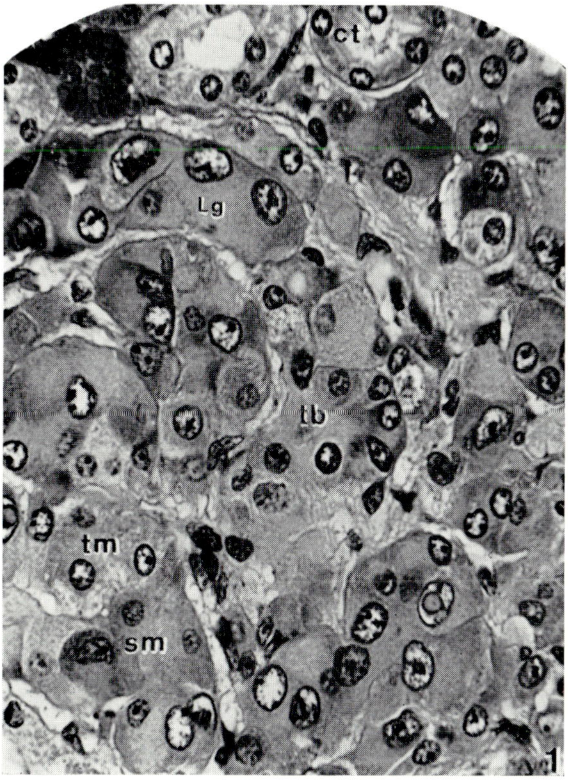

FIG. 1. Spontaneous carcinoma developing in the renal cortex of a BALB/c mouse. The upper part of the micrograph shows a portion of normal kidney, including a section of a convoluted tubule (ct); the rest of the micrograph is occupied mostly by carcinoma cells distinguished (a), by a variety of sizes and shapes, from relatively small cells with small nuclei (sm), to large, bi-nucleated cells, with comparatively large nuclei (Lg) and (b), a relatively homogenous cytoplasm with intense and diffuse basophilia; in this respect, it may be noted a progressive morphological transformation from swollen tubule cells, somewhat granular and moderately basophilic (tm), disorganized and strongly basophilic tubule cells (tb), to enlarged and strongly basophilic carcinoma cells (Lg). It seems that it is the latter cellular types that are serially transplantable. Fixative: Bouin-Hollande; Stain: Hematoxylin Eosin, ×380.

The kidney carcinoma cells have proven to be transplantable in mice of the same strain. Since 1955 a number of strains of transplanted tumour cells have been established, some of them having attained or passed the 70th generation of successive grafting. An outstanding feature of the transplanted kidney tumour cells is their property to inhibit body growth. When 0·02 to 0·05 m of tumour pulp is inoculated subcutaneously to young mice, 10 to 12 g in weight, effect on the weight curve is first noticed 3 to 4 days after

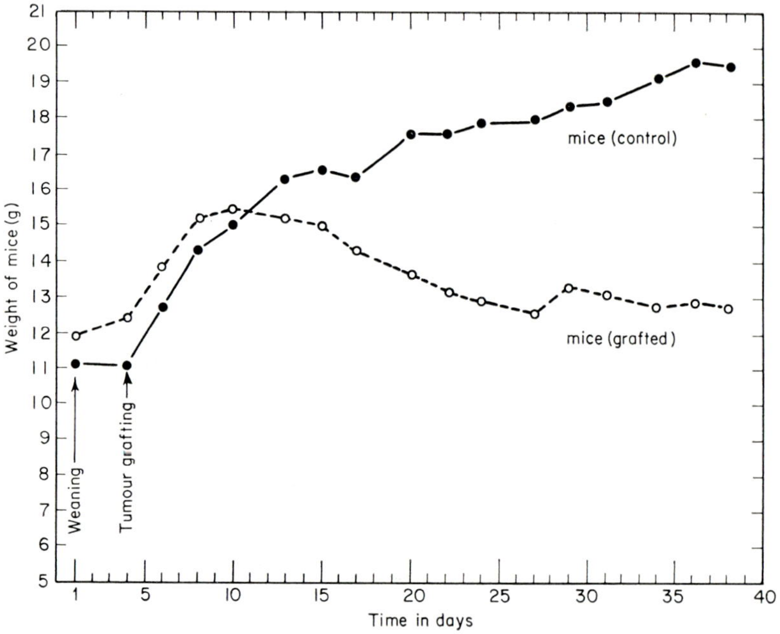

FIG. 2. Diagram showing the inhibitory effect of grafted tumour cells on the body growth of mice. In these experiments, the grafted mice and their controls were siblings from the same litters. Because of the differences of growth rates in males and females, groups of grafted and control mice were always of the same sex. Animals selected for growth tests were young mice shortly after weaning. As shown in the diagram, the inhibitory action on body growth became apparent about four days after grafting, i.e. when the cells came in contact with new capillaries, and began to grow actively. When the tumour cells were removed two or three weeks later (not illustrated in these graphs), body growth was resumed within 24 hours, and maintained until the weight of individual animals approached that of the controls.

injections, at about the time when tumour cells have established contact with the capillary bed and start growing actively; a few days later inhibition is complete, the weight curve showing no gain, or even some loss of weight. This effect on the body weight of grafted animals, as compared to the weight of normal control mice, is illustrated in the graphs shown in Fig. 2. Twenty to thirty days after inoculation the treated animals show no more than their initial weight of 10 to 12 g, in contrast with the young of the same litter used as controls which by then, depending on sex, have nearly twice that weight, i.e. 19 to 22 g. The two mice, the photographs of which appear in Fig. 3, were sisters of the same litter, kept together under the same conditions of feeding. On the left, is the untreated control; on the right, the test animal which received a renal tumour graft when both were 44 days old, and their

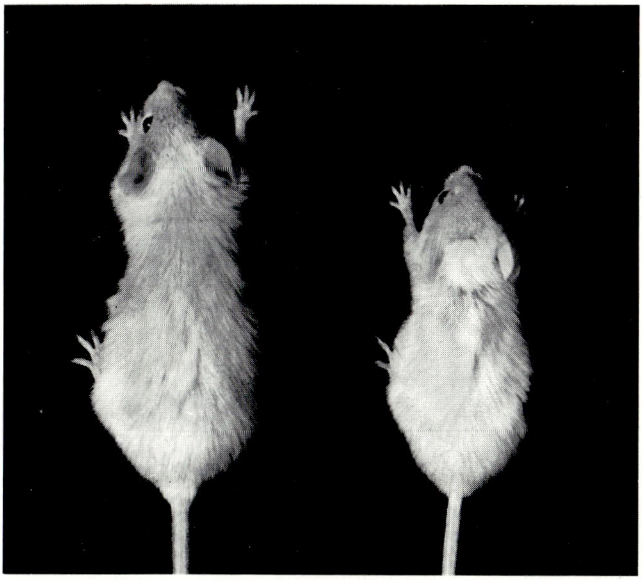

FIG. 3. Growth inhibition test. Two female BALB/c mice, sisters from the same litter. On the left, the practically full grown control; on the right, the test animal which received a subcutaneous graft of tumour cells when 12 g in weight; when the picture was taken, 42 days later, the body weight of the test mouse was 9·8 g, a loss of 2·2 g, against 20·2 g, for the untreated control.

weight was 12·5 and 12 g, respectively. The picture in Fig. 3 was made 42 days later, when their respective weights were 20·2 g for the control, and 9·8 g for the test animal, i.e. a difference of 9·9 g, indicating not only a complete suppression of growth as a result of inhibitory effects, but some loss of body weight. Nevertheless, the test animals continued to appear healthy and normally active. When animals bearing grafts are kept under observation for one year or more, they begin to show signs of premature senility. The inhibitory effect has been found to be entirely reversible: as soon as 24 hours after surgical removal of the graft, recovery of body growth was detectable and the weight curve of the test animals approached rapidly that of the controls. Growth inhibition may reappear in cases of local recurrence, or lung metastasis of the tumour (Claude 1958, 1962).

The property to inhibit growth is retained indefinitely by the transplanted tumour cells, even after 70 or more generations of grafting. Inhibition cannot be attributed to coarse toxic effect since, during the growth test, tumours are small, perhaps because of auto-inhibition, and necrosis is absent. During treatment the test animals appear healthy and as active, if not more than the controls. The inhibition phenomenon seems to affect the general metabolism of the host; its nature and mechanism of action are at present unknown. An electron microscope study of the transplanted tumour cells has been made.

Like the original carcinoma cells which develop spontaneously in the cortex of the kidney (Fig. 1), renal tumour cells maintained by successive transplantation in mice, peritoneally or in the subcutaneous tissue, retain their general characteristics as demonstrated by light microscopy: they are large elements, ten times or more in volume the parent convoluted tubule cells, with a highly basophilic cytoplasm, and a relatively large nucleus.

Areas of such transplanted cells, as seen in the electron microscope, are illustrated in Figs. 4 and 5. A characteristic feature exhibited by these cells is the presence of slender filaments, about 70 A in diameter, which are found in abundance in the cytoplasm: they are usually parallel to each other, although without assuming close contact. The latter situation is illustrated in Fig. 5(f), and especially in (cs), where the filaments are viewed in cross section. Closer examination suggests that the 70 A filaments may be compound structures made up of at least two finer filaments, possibly disposed in helix, with lateral projections which give them a coarse or beaded appearance and suggesting, in some regions, the possibility of cross-binding. Perhaps as a result of this condition, the filaments have a tendency to associate themselves in bundles, occasionally interrupted by a band of different density, as illustrated in b, Fig. 5. A similar fibrillar apparatus has been described previously (Claude 1961, 1965). Its nature and functions have not been elucidated, but it has been tentatively suggested that it may be related to mitotic and amoeboïd movements.

Another morphological feature of the kidney tumour cells has to deal with a particular activity of the ergastoplasm. The electron microscope observation reveals the presence in the cytoplasm of bodies of various sizes, which contain a finely granular material of uniform density (Fig. 6). Under favourable conditions of sectioning, it is found that these bodies are in continuity with the endoplasmic reticulum, and that they actually represent distended segments of the latter (er, Fig. 6). This view is also supported by the fact that the envelope of these bodies is lined by a practically continuous array of ribosomes (r, Fig. 6), and that both the body and the connected profile of an endoplasmic reticulum cisternae (r, Fig. 6), contains a material of similar granularity and density (p, Fig. 5). Because of its

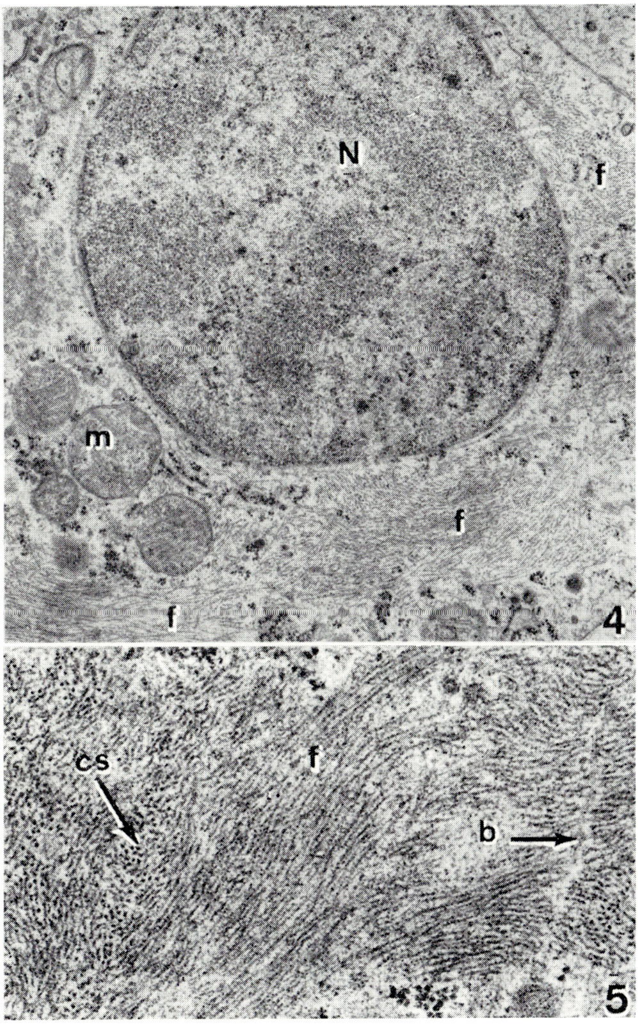

FIGS. 4 and 5. Electron micrograph of portions of typical renal carcinoma cells after serial transplantation in the peritoneal cavity of BALB/c mice.

Figure 4 shows a large nucleus (N), mitochondria (m), and bundles of fine filaments (f) in different regions of the cytoplasm. Figure 5 shows, at higher magnification, similar filaments from another cell. As described in the text, the filaments, about 70 A in diameter, present a finely beaded structure. They follow a course more or less parallel, and can be seen along their length (f), or in cross section (cs). In both cases, they appear to remain at some distance from each other. A band of lesser density interrupts the main bundle (b). Magnifications, for Fig. 4: $\times$ 15,500; Fig. 5: $\times$ 39,000.

particular connection with ribosome lined membranes of the endoplasmic reticulum, it may be inferred that the content of these bodies is protein in nature.

The fact that ribosomes remain associated with the membrane of the

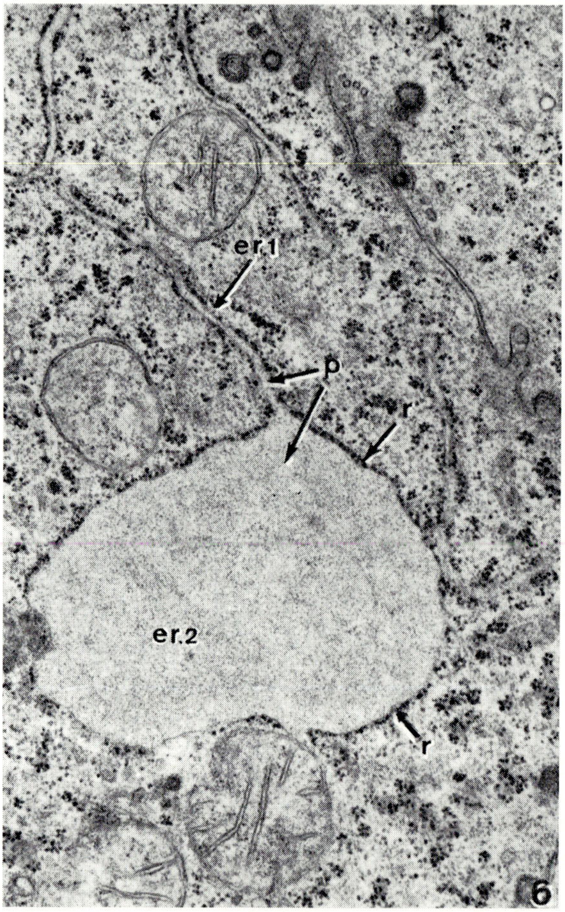

FIG. 6. Electron micrograph illustrating the particular morphology of inclusions found in the cytoplasm of transplanted carcinoma cells. In the present case, orientation of the plane of section permits to demonstrate the relation which exists between these bodies and rough regions of endoplasmic reticulum. The inclusion (er 2) is found to consist of a distended portion of endoplasmic reticulum as shown by its continuity with the profile of a regular cisternae of rough endoplasmic reticulum (er 1). Like the latter, the wall of the inclusion is lined with ribosomes (r). The finely granular contents of the inclusion (p) has the same appearance and density as the material observed within the adjoining endoplasmic reticulum (p, and er 1). The observations suggest that the substance accumulated in the endoplasmic sac is a protein. Magnification: ×23,500.

distended cavity suggests that protein synthesis may still take place at its surface. The production of proteins and their accumulation in endoplasmic reticulum sacs of various sizes and shapes seems to be particular for the renal tumour cells. The nature of the material demonstrated is unknown. The possibility may be considered, however, that it might be related, directly or indirectly, to the growth inhibiting property otherwise exhibited by the transplanted tumour cells.

REFERENCES

Andervont, H. B. (1940–41). *J. natn. Cancer Inst.* 1, 147; (1941–42), 2, 307, and (1944–45), 5, 383.
Claude, A. (1958), *Revue fr. Etud. clin. biol*, 3, 261.
Claude, A. (1961). *C.r. hebd. Séanc. Acad. Sci., Paris* 253, 2251.
Claude, A. (1962), *J. Ultrastruct. Res.* 6, 1.
Claude, A., Naissance de la Biologie Moléculaire, in: *Acquisitions Récentes en Biologie*, Centre International de Synthèse (XXVIe Semaine de Synthèse) Editions Montaigne (Aubier-Montaigne), Paris, 1965, pp. 13–34.
Inbred Strains of Mice No 2 (1961). Companion issue to "Mouse News Letter No 25", *No 2*, p. iv.
Staats, Joan (1964), *Cancer Res.* 24, 147.

Aspects on the Kinetics of the Cell Population in Cestodes

BO-JUNGAR PAUL WIKGREN

Åbo Akademy, Åbo (Turku), Finland

I. Introduction

IT is believed that the intensive research on the growth controlling factors in some organisms, e.g. mammals, leads to a general understanding of the nature of growth control. However, in many invertebrates we meet special growth patterns which are still not properly understood. For instance in cestodes we are confronted with some tricky problems concerning growth pattern and growth control. One complication arises with the occurrence of several stages in the life cycle. The transition from one stage to the next is in general coupled to a change of host animal. Each stage, including the larval ones, apparently has its own mode of growth, and in each stage factors which control the growth must be in operation. In respect to growth control the larvae can thus be considered equivalent to "adult" organisms.

II. Growth Control in Larvae

Regulation of growth by a gradual decrease of the specific growth rate with increasing size, and the attainment of a maximal size, is in cestodes perhaps best known in the *larval* stages (Hutchison 1958, McCaig and Hopkins 1965, a.o.). It can be shown that this regulation also manifests itself in the mitotic activity of the cell population in plerocercoids of diphyllobothrid species. It was found that the mitotic rate (determined by the use of colchicine) is lower in large plerocercoids than in small ones. The mitotic rate at 38°C in small plerocercoids of *Diphyllobothrium latum* was 3·47 cells/10^2 cells/h, but only 0·86 cells/10^2 cells/h in large plerocercoids. The corresponding values for plerocercoids of *D. dendriticum* were 3·70 and 2·82. This may depend on a progressive decrease of the metabolic activity during growth (cf. Walkey and Davies 1964).

Measurements of the DNA-amount in nuclei (by Feulgen microphotometry) show a slight difference in the distribution of DNA-classes in plerocercoids of different size (Fig. 1). It seems as if some cells in large plerocercoids should be arrested in the G_2-stage, being unable to enter into mitosis.

It may be asked whether there are any specific growth stimulating or inhibiting factors. The experiments designed to give an answer to this question are still in progress, and the results are hence quite preliminary.

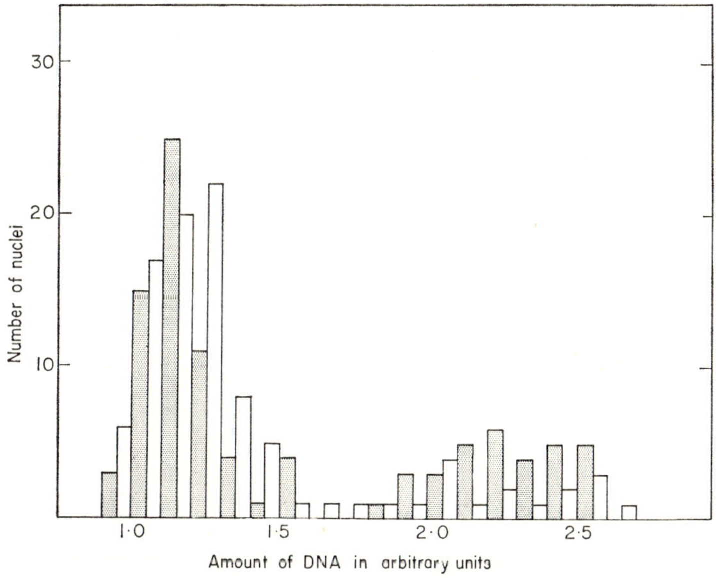

FIG. 1. The distribution of the amount of DNA in interphase nuclei in large plerocercoids (black staples) and small plerocercoids (open staples) of *Diphyllobothrium dendriticum*. The estimated proportions of the G_1, S, and G_2 stages are: in large plerocercoids 63% G_1, 8% S, and 29% G_2. In small plerocercoids 80% G_1, 10% S, and 10% G_2.

By using the rather crude experimental approach of treating large plerocercoids with extracts of small and large plerocercoids it appears, however, as if these extracts have a stimulating effect on the mitotic activity and it is also possible that extracts of small plerocercoids are more powerful than those of the larger variety.

III. The Dynamic Growth Pattern of Adults

In many cestodes we are confronted with the somewhat paradoxical fact that while mitotic homeostasis may occur in larvae, it certainly does not occur in the adults.

During the growth of the young adult there is first a phase of rapid, exponential growth. When sexual maturity has been reached the rate of increase in size decreases until a given size, typical for each species in a particular host, is reached (Wardle and Green 1941, Archer and Hopkins

1958, Hutchison 1959). Although this mimics the growth pattern of mammals, the underlying mechanism is quite different. In many tapeworms there is a continuous formation of new tissue in the neck region and proglottids are shed at the posterior end. Broadly speaking the size is a function

FIG. 2. Diagrammatic representation of a diphyllobothrid cestode. The mitotic indexes of the parenchymal cell populations in different parts of the body are indicated. The values are relative.

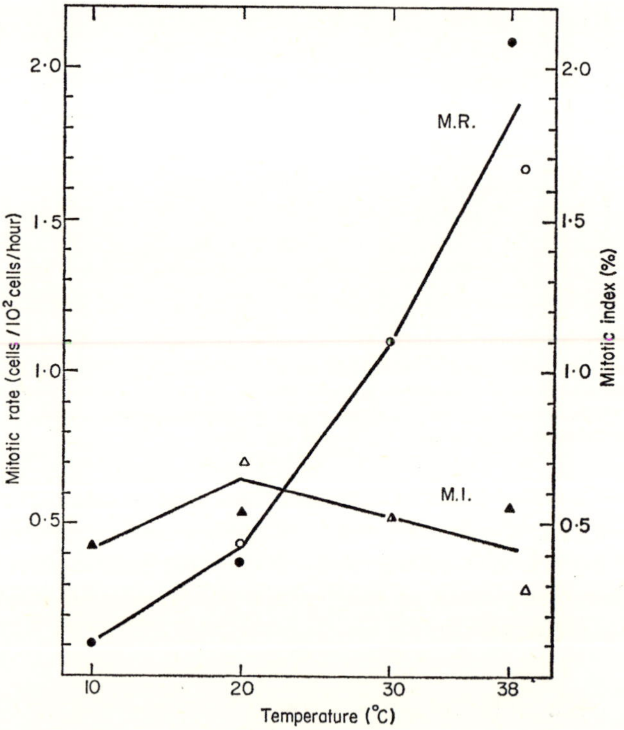

FIG. 3. The relationship between mitotic index, mitotic rate and temperature in plerocercoids of *Diphyllobothrium osmeri*. The closed and open symbols refer to two experimental series.

of the rate of growth and the rate of apolysis. This is also reflected in the kinetics of the cell population of adults. In Fig. 2 some data on the mitotic index of the parenchymal cell populations in different regions of adult *Diphyllobothrium dendriticum* are included. The parenchymatous cells (including the "germinative" cells) seem to give rise to all types of specialized cells in cestodes. It should be noted that while there is no mitotic homeostasis in the worm as such, a marked mitotic gradient exists along the length of the body.

Although it is somewhat difficult to visualize how a feedback regulation works in cestodes with a dynamic growth pattern, such a regulation may

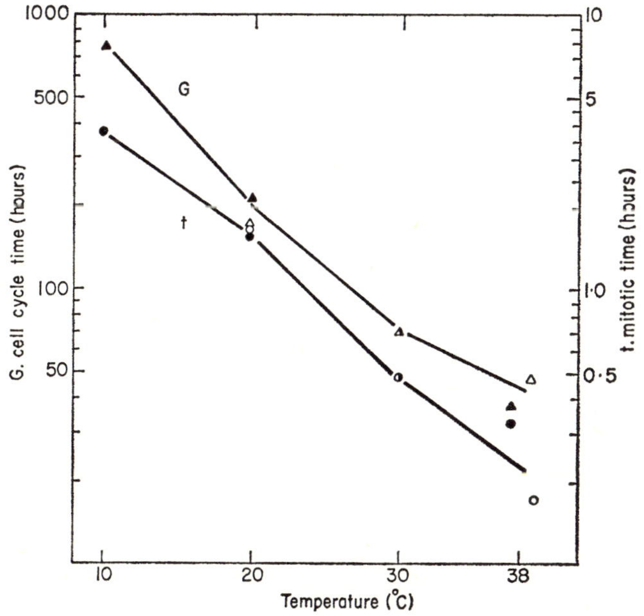

FIG. 4. The effect of temperature on the duration of mitosis (t) and cell cycle (G) in plerocercoids of *Diphyllobothrium osmeri*. The closed and open symbols refer to two experimental series.

exist. The final size may also, however, be highly influenced by the relative length of the different regions of tapeworms, and the metabolic and mitotic activities of these regions. For instance, the observation of Archer and Hopkins (1958) that the size of the adult *D. dendriticum* at maturation is proportional to the size of the plerocercoid, may be due to a correlation between the length of the neck region and the size of the plerocercoid. Host induced variations in size (and shape) may perhaps also be explained in this way.

IV. Cell Population Kinetics and Temperature

Considering the difference in growth pattern of larvae and adults of *Diphyllobothrium*, and the change of host (from fish to birds or mammals) which is associated with the transition from the larval to the adult stage, it may be asked whether high temperatures exert a specific influence (trigger action) on cell proliferation. This question may be answered in the light of the results of an investigation on the effect of temperature on the cell division cycle of diphyllobothrid plerocercoids (Wikgren 1966). In these experiments the plerocercoids were kept in Hanks' solution. The main results were obtained during the first few hours of incubation, during which the plerocercoids are thought to be able to rely on their stores of carbohydrates and other materials. Colchicine was used to determine the rate of entry of cells into mitosis. The main results are summarized in Fig. 3 and 4.

It was found that the time parameters of the cell division cycle approximate those of vertebrate tissues with a high mitotic activity. The duration of mitosis was remarkably short. It must be noted that experimental errors tend to result in too short prophase times. With regard to the influence of temperature it was observed that the mitotic index is only slightly affected while the rate of entry of cells into mitosis is very temperature sensitive. The durations of both mitosis and total cell cycle, when plotted against temperature on a semilogarithmic paper, form an almost straight line. The temperature response of these events thus closely follows the pattern observed in other biological material. It is thus concluded that high temperatures alone do not provoke a special response of the cell proliferation activity in cestodes. The transition from the larval to the adult growth pattern is apparently due to other factors than temperature.

REFERENCES

Archer, D. M. and Hopkins, C. A. (1958). *Expl Parasit.* 7, 125–144.
Hutchison, W. M. (1958). *J. Parasit.* 44, 574–582.
Hutchison, W. M. (1959). *Expl Parasit.* 8, 557–567.
McCaig, M. L. O. and Hopkins, C. A. (1965). *Parasitology* 55, 257–268.
Walkey, M. and Davies, P. S. (1964). *Parasitology* 54 (Proc.), 6–7.
Wardle, R. A. and Green, N. K. (1941). *Can. J. Res.*, Sect. D. 19, 245–251.
Wikgren, B.-J. P. (1966). *Acta zool. fenn.* 117, 1–27.

DISCUSSION SUMMARY

Chairman: Professor Dr. Dr. H. Wrba

Replying to a question by the chairman, Dr. Bolden explained that there were no studies on the dependence of the growth of the exorbital lacrimal gland on other organs.

Dr. Teir replied to Dr. Alexander that the action of ACTH shown in the adrenalectomized animal by Dr. Räsänen must be a direct one. Even for intact animals ACTH acted on the mitotic activity of the mucous membrane of the stomach. Dr. Weiss pointed out that there are various pharmacological influences on mitotic activity which had nothing to do with growth control, as exemplified by the action of heparin on the division of the sea urchin egg (change of colloidal state).

During the discussion of the paper by Dr. Claude the chairman pointed out certain parallels to the infection of mice with polyoma virus (kidney tumours and loss of weight = runting disease). Dr. Claude had never succeeded in effecting a cell-free transplantation of his kidney tumour. But implanted subcutaneously or intraperitoneally the tumour showed a measure of inhibitory action on the kidney. Dr. Alexander referred to a considered, though unconfirmed, endocrine dependence of these tumours. Dr. Claude explained that at Bethesda kidney tumours appeared, after application of human tumour extracts, only in animals of the particular stock used by him. Apparently the tumour has some inhibitory effect on growing mammary carcinoma. Dr Koss mentioned the endocrine activity of the kidney and pointed out that tumours, derived from the kidney, frequently show a different behaviour.

The discussion of Dr. Wikgren's paper offered an opportunity of debating once more the relationship of mitotic rate and mitotic index, with Drs. Iversen, Rytömaa and Weiss speaking. The relation is obviously exponential rather than linear. With regard to the duration of mitosis, temperature is the determinant factor in the present system. Dr. Weiss remarked in conclusion that this relationship must become more complex in non-homogeneous systems. The linear relation must be upset in any case where cells are eliminated from the line of proliferation by differentiation.

Section VII

Aspects in Malignancy

Differential Sensitivity of Growth-Controlling Systems in Normal and Neoplastic Tissues to the Toxic Action of Chemical Carcinogens

J. M. VASILIEV AND V. I. GUELSTEIN

Institute of Experimental and Clinical Oncology, Academy of Medical Sciences, Moscow, USSR

I. Characteristics of the Phenomenon

TOXIC growth inhibiting properties of chemical carcinogens have been known for many years. Haddow (1938) was first to suggest almost 30 years ago that neoplastic cells may be more resistant to the toxic effects of carcinogens than their normal ancestors. Occasional observations of the pathologists published in the following years (Vasiliev and Guelstein 1963) indicated that manifestations of the toxic effects of various carcinogens upon the normal target tissues gradually decreased in the course of carcinogenesis. However, only recently differential sensitivity of normal and neoplastic cells to carcinogens had been demonstrated in this laboratory in direct tests *in vitro*. Up to the present time differential sensitivity to carcinogens had been observed in two *in vitro* systems: (a) normal rat, mouse and hamster fibroblasts cultivated as monolayers or on the plasma clots were much more sensitive to the toxic action of carcinogenic hydrocarbons [7,12-dimethylbenzo/a/anthracene (DMBA), benzo/a/pyrene (BP), and 20-methylcholanthrene (MC)], than homologous sarcomatous cells (Starikova and Vasiliev 1962, 1963, Starikova 1964). First experiments demonstrating differential sensitivity of normal and neoplastic fibroblasts to these hydrocarbons had been confirmed and extended by other investigators (Berwald and Sachs 1963, Alfred *et al.* 1964). (b) Normal epithelial cells of mouse liver cultivated on the collagen gel were more sensitive *in vitro* to the toxic effects of two hepaticarcinogenic azodyes [o-aminoazotoluene (AAT) and N.N.-dimethylaminoazobenzene (DAB)], than the cells of several strains of transplantable mouse hepatomas cultivated in a similar way (Guelstein and Yagorskaja 1964, Yagorskaja, unpublished data).

Experimental studies of differential sensitivity to carcinogens in both systems had revealed two essential characteristics of this phenomenon: lack of etiological specificity and chemical selectivity. The former had been demonstrated in experiments with connective tissue cells. Increased resistance to the growth-inhibiting action of carcinogenic hydrocarbons was shown to be characteristic not only for sarcomas induced with these substances but also for sarcomas induced by plastic films (Starikova and Vasiliev 1962) and by polyoma virus (Alfred *et al.* 1964). Thus, decrease in the sensitivity to carcinogenic hydrocarbons regularly accompanied neoplastic transformation of fibroblasts, even if transformed cells or their ancestors were never treated with these agents. In our experiments all primary sarcomas tested *in vitro* had an increased resistance to hydrocarbons; similar resistance was characteristic for long-transplantable strain of rat sarcoma 45 and for L strain of neoplastic mouse fibroblasts. Obviously, this alteration took place at some early stage of tumour progression and was not lost at the following stages. All these facts suggest that decreased sensitivity to carcinogens is not a secondary cell characteristic typical only for neoplasms of certain etiology and/or for neoplasms having certain degrees of malignancy.

Therefore an alteration of the sensitivity to carcinogens seems to be linked with some basic cell change which makes the cells neoplastic.

One can hardly doubt that most essential biological characteristics of neoplastic tissue is a deficiency (or deficiencies) of growth-controlling systems. We may assume, as working hypothesis, that altered cell sensitivity to carcinogens is correlated in some way with the deficiencies of growth-controlling systems responsible for neoplastic cell properties.

Experimental studies on differential sensitivity to carcinogens had revealed also chemical selectivity of this phenomenon. To get reliable estimates of the relative toxicity of certain compounds for normal and neoplastic cells *in vitro* one has to test several doses of this compound and to repeat each experiment several times: therefore these tests are rather time-consuming. Up to the present time we had tested about a dozen substances; among them few were tested both in the cultures of fibroblasts and in those of hepatic epithelium (see Table I). These experiments had clearly shown that selective toxicity for normal cells in the *in vitro* systems is not characteristic for all types of cell-damaging and growth-inhibiting agents. Within the same chemical group one could find certain substances which were more toxic for normal cells than for their neoplastic homologues, other compounds were equally toxic for both cell types and at last, some substances were more toxic for neoplastic cells than for normal ones. Thus alterations of the cell sensitivity accompanying neoplastic transformation are rather specific: the cells become more resistant only to certain compounds but not to other chemical related substances.

Selective toxicity of certain compounds to normal vs. neoplastic tissue *in vitro* seems to be correlated with the carcinogenicity of the same compound for the same tissue *in vivo*. E.g., as mentioned already, very active carcinogenic hydrocarbons (MC, BP and especially DMBA) were much more toxic for the normal mouse fibroblasts than for neoplastic ones. A somewhat lower degree of selectivity was observed in similar experiments with dibenzo (a,h) anthracene which was less active carcinogen *in vivo*. Dibenzo (a,c) anthracene which was inactive *in vivo* was also less selective for normal fibroblasts *in vitro* than its structural analogue dibenzo (a,h) anthracene. Non-carcinogenic anthracene had no selective toxicity for normal cells; on the contrary, this hydrocarbon was slightly more toxic for neoplastic fibroblasts than for normal ones.

In the experiments with hepatic epithelium active hepatocarcinogens DAB and AAT were more toxic for normal cells than for neoplastic ones.

TABLE I. Differential toxicity of various compounds for normal and neoplastic cells of two rodent tissues

Compound	Fibroblasts		Hepatic epithelium	
	Differential toxicity for normal cells *in vitro* †‡	Carcinogenic activity in *in vivo* *	Differential toxicity for normal cells *in vitro* †‡	Carcinogenic activity in *in vivo* *
DMBA	++++	+++	.	o
BP	+++	++	—	o
MC	+++	++	.	o
Dibenzo/a,h/-anthracene	++	+	.	o
Dibenzo/a c/-anthracene	+	o	.	o
Anthracene	—	o	.	o
Chrysene	o	o	.	o
AAT	o	o	+	++
DAB	.	o	+	++
DEAB	o	o	—	o
AAF	.	o	o	+

† Summarized data from the experiments of Starikova and Vasiliev (1962, 1963), Starikova and Solina (1966), Guelstein and Yagorskaja (1964), and Vasiliev *et al.* (1966). Data on chrysene are taken from the experiments of Berwald and Sachs (1963). See text for explanations of abbreviations.

‡ Designations: ++++, +++, ++, + compounds are more toxic for normal cells than for homologous neoplastic cells. Various degrees of selective toxicity for normal cells are designated by various numbers of crosses. O—compound is equally toxic for normal and for neoplastic cells. — compound is more toxic for neoplastic cells than for homologous normal elements. . compound was not tested.

* Designations: +++, ++, + various degrees of carcinogenic activity for mouse and rat connective tissue (respectively mouse liver) are expressed by various numbers of crosses. O—compound is not carcinogenic for given tissue.

Non-carcinogenic azocompound, N,N diethylaminoazobenzene (DEAB) was equally toxic for normal hepatic cells and for the cells of one of the hepatoma stains (#46): cells of other hepatoma strain (#22) were more sensitive to DEAB than normal epithelium. The only substance in this series with the exceptional behaviour was fluorenylacetamide (AAF). This hepatocarcinogen was equally toxic for normal and neoplastic liver cells.

It is known that *in vivo* AAF must undergo certain metabolic transformations before carcinogenic metabolite is formed (Miller *et al.* 1961). Obviously there may be striking differences between the state of cells and metabolism of carcinogens *in vivo* and *in vitro*. Therefore one can hardly be surprised to find certain compounds that are carcinogenic *in vivo* but have no selective toxicity *in vitro*. In fact it is more striking that for other compounds it had been possible to find some correlation between the results obtained *in vitro* and *in vivo* in spite of all the differences of the test systems.

It is necessary to stress that conclusion postulating the existence of this correlation is a preliminary one because it is based on the experiments with limited number of substances. In future it is essential to test many other compounds in various systems.

Results of our experiments indicate that certain compounds which can induce tumours of a given tissue, have selective toxicity for the same tissue *in vitro*, but the same compound may not have selective toxicity for other tissues for which it is not carcinogenic. This was demonstrated in the experiments with BP and AAT. In the experiments with liver and hepatoma explants one could observe differential sensitivity of normal cells to hepatocarcinogenic AAT but not to BP, which is not carcinogenic for the mouse liver. The reverse was true for mouse fibroblasts; they were more sensitive to BP than sarcomatous cells but both cell types had equal sensitivity to AAT which was not carcinogenic for connective tissue. These results agree well with the suggestion made above: increased resistance to carcinogens is linked in some way with the deficiency of growth controlling systems. These systems in the adult organism are tissue-specific (Bullough and Rytömaa 1965) therefore different systems may become deficient in the course of neoplastic transformation of various tissues. Thus, tissue specificity of differential toxicity to carcinogens may be a manifestation of the tissue-specific alterations of growth-controlling systems during carcinogenesis.

II. Targets for the Toxic Effects of Carcinogens in Sensitive Normal Cells

What is the possible cause of the observed correlation between altered sensitivity to carcinogens and neoplastic transformation? The easiest possible explanation is as follows: Carcinogens as well as physiological growth-controlling factors (hormones, "chalones", "antitemplates", etc.) may have the same target ("template") within the sensitive normal cells. The normal state of this target is essential for physiological effects of these agents. If we assume that this hypothetical common target is altered or disappears from neoplastic cells, this may explain deficient growth control as well as resistance to the toxic effects of carcinogens. Then by studying reactions of various components of normal and neoplastic cells with carcinogens we may hope to find something about the nature of these targets. Carcinogens may react with various cell components including proteins and nucleic acids. There are no facts and no theoretical considerations to suggest that reactions of carcinogens with nucleic acids of normal and neoplastic cells may be different in any way. Of greater interest in this connection are reactions with proteins. Several investigators (see Heidelberger 1965) had shown that carcinogenic hydrocarbons and azodyes are bound in normal cell by a particular protein fraction which is absent from homologous neoplastic cells. Miller and Miller (1953) suggested that carcinogen-binding protein is involved in growth control. Later it was suggested (Vasiliev and Guelstein 1963, Prehn 1964) that reaction of carcinogens with this protein might be essential for their toxic action on normal cells. However, in a recent experiment of Alfred (1964) labelled BP was accumulated in protein-bound form in neoplastic fibroblasts at a much greater rate compared with normal cells. Further experiments are obviously needed to find the role of protein binding in the toxic effects produced by carcinogens.

The same problem of the mechanisms of differential sensitivity to carcinogens may be approached in a somewhat different way: that is, by studying early alterations produced by these compounds in various cell organelles and in the cell behaviour. We cannot discuss here the vast literature on cytological alterations produced by various carcinogens and will review only some of the experiments performed recently in our laboratory. In these experiments we have studied various alterations produced *in vitro* by carcinogenic hydrocarbons (DMBA and BP) in normal and neoplatic mouse fibroblasts, that is in the system in which differential toxicity of these compounds is most obvious. The most typical manifestation of the toxic action of carcinogens on normal fibroblasts was growth inhibition, therefore we tried to find out by autoradiographic methods what phase of mitotic cycle was sensitive to these hydrocarbons. (Erofeejeva and Vasiliev 1966, Yaskovetz 1966). Experiments made with tritiated thymidine (H3T) had shown that inhibition of the DNA synthesis was observed 6 hours after the addition of DMBA suspension to culture medium; the rate of DNA synthesis continued to decrease in the following days. Analysis of the distribution of the numbers of grains over labelled nuclei in control and in the DMBA—treated cultures had shown that degree of inhibition at any given time was similar for all the cells that were in the S-phase. Inhibition of DNA synthesis by carcinogenic hydrocarbons had been observed also in various normal tissues *in vivo* (McCarter and Quastler 1962, Iversen and Evensen 1962).

In the culture treated for 2 to 3 days with DMBA and then exposed for 1 hour to H3T per cent of labelled cells was not significantly lower as compared with control cultures: from this one had to conclude that DMBA increased the duration of GI phase to the same degree as that of the S phase. Another group of experiments had shown that DMBA proportionally increased also the duration of G_2 phase. Thus, DMBA added to the medium progressively slowed down the passage of normal fibroblasts through all three main phases of the mitotic cycle (G_1, S and G_2). The same and greater doses of DMBA in similar experiments had no effects on any stage of mitotic cycle of neoplastic cells (explants of rat sarcoma 45 and L strain of mouse fibroblasts).

On the basis of these facts one may suggest that DMBA inhibits not a single "critical" stage of the mitotic cycle but some general process limiting the passage through all stages of mitotic cycle in normal cells. One of the possibilities is that this is some energy-yielding process and that the action of carcinogens upon the mitotic cycle is a consequence of their toxic action on the mitochondria. Damage to mitochondria produced by carcinogenic hydrocarbons in various normal tissues *in vivo* had been observed by many authors (Graffi and Bielka 1959, Arcos *et al.* 1960, Iversen and Evensen 1962, Vasiliev *et al.* 1962). In our laboratory using histochemical methods Olshevskaja and Raikhlin (1966) also revealed alterations of mitochondria in normal fibroblasts treated *in vitro* with DMBA, but it is not yet clear whether these alterations are casually related to the inhibition of cell proliferation.

The use of other experimental methods had shown that early toxic effects of DMBA on normal fibroblasts were rather selective. DMBA-treated fibroblasts stained vitally with neutral red dye segregate this dye in the cytoplasmic granules in the same way as control cells, although the cells reversibly damaged by a number of other agents (e.g. by heating) are stained diffusely with this dye (Slavnaja 1965). Recent studies (Koenig 1965) indicate that cytoplasmic granules accumulating neutral red are identical to lysosomes. Thus, experiments with neutral red suggest that lysosomes of fibroblasts are not likely to be the primary targets of the toxic action of DMBA. Time-lapse cinematography (Ivanova and Komm, unpublished results) had shown that DMBA (10^{-4} mg/ml) in the first days after addition to the culture medium did not change significantly the pattern of the movements of the cell surface; at later time (at 4 to 5 days) disorganization of the movements of undulating membranes had been observed. Normal fibroblasts were characterized by a formation of one fan-like pseudopodium which moved in a co-ordinated fashion while neoplastic cells (L strain) had numerous small independently moving villi on their surfaces (see Abercrombie and Ambrose 1962). 4 to 5 days after addition of DMBA many normal fibroblasts became somewhat similar in their locomotory behaviour with the neoplastic cells; they did not form stable contacts with their neighbours, single small villi or small groups of these villi were formed on their surfaces. One had often observed also local non-co-ordinated contractions of small parts of cell surfaces. The form of these cells was also changed; many fibroblasts became fusiform and had very long cytoplasmic processes. When L cells were treated with DMBA no apparent changes in the cell movements or cell form were observed.

Obviously, all these alterations produced by the carcinogen in the locomotory behaviour of fibroblasts indicate that DMBA interferes in some way with the processes controlling the movements of cell surface as well as the formation of contacts with neighbour cells and with the substrate. Thus toxic effect of DMBA may lead to the appearance of cells somewhat similar in their behaviour with neoplastic elements. We do not know at present whether these effects are specific for the carcinogenic substances and whether these alterations may become stable, that is whether they are the first manifestations of the neoplastic transformation produced by hydrocarbon (Berwald and Sachs 1963).

One of the interesting effects produced by DMBA and by BP in the cultures of fibroblasts was stimulation of the synthesis of acid polysaccharides. Histochemical tests as well as chemical analysis had revealed large amounts of acid mucopolysaccharides digestible by testicular hyaluronidase at the bottom of the culture flasks after DMBA treatment (Ivanova 1964). Synthesis of mucopolysaccharides is one of the specialized functions of fibroblasts. Thus, action of DMBA is somewhat similar to that of growth-controlling factors: it switches the cells from participation in the mitotic cycle to the tissue-specific activity. Results of the studies of these first early cell alterations induced by carcinogenic hydrocarbons *in vitro* are not sufficient to make any conclusions, but at least, these results are in good agreement with the hypothesis that carcinogen-binding target play an essential role in the control of cell behaviour. Most carcinogens are lipid-soluble and it is reasonable to suggest that cell membrane as well as internal membraneous organelles should be regarded as the most probable candidates for the role of cell structures responsible for the toxic effects of carcinogens (Abercrombie and Ambrose 1962, Pitot 1962). One cannot discuss here the facts confirming the leading role of membraneous structures in the control of cell activities. It is sufficient to remind striking experiments of Nomura (1964) who had shown that absorption of a single molecule of colicine on the surface of bacterial cell may switch off selectively synthesis of DNA and of RNA.

It is important to stress that the pattern of toxic effects may be different when the same carcinogen re-acts with the cells of different tissues, e.g. BP had no selective growth-inhibiting effect on the cultures of normal or neoplastic hepatic epithelium; it killed some of the explants of liver tissue but mitotic activity in the surviving explants was not inhibited as compared with controls (Yagorskaja—unpublished results). These facts confirm the tissue-specificity of the cell targets reacting with carcinogens. It is reasonable to suggest that certain compounds will be carcinogenic for the cell of given tissue only if (among other conditions) it will affect selectively the same cell structures that are essential for the growth control in this tissue; these structures may be different in various tissues.

III. Alterations of Cell Sensitivity to Carcinogens at Various Stages of Neoplastic Progression

In most cases malignancy is not achieved in a single step but is a result of a series of independent alterations of the cell characters: this process is usually designated as neoplastic progression. In the first part of this paper we mentioned certain facts indicating that alteration of the sensitivity to carcinogen occurs at some early stage of progression. Now we will discuss some additional data confirming this conclusion. Starikova in collaboration with Irlin (laboratory of Professor L. A. Zilber at the Gamaleya Institute, Moscow) had compared the sensitivity to DMBA of three types of hamster fibroblasts: first subculture of normal embryonic fibroblasts. (b) Line #917 transformed with polyoma virus; this line is characterized by loss of contact inhibition *in vitro* and by low cell number needed for positive take *in vivo*. (c) Subline #917 K 3 retaining contact inhibition; this line gives tumours *in vivo* only if large numbers of cells are inoculated. It was shown that normal fibroblasts are more sensitive to the growth-inhibiting action of DMBA than both cell lines. Cells of 917 K3 subline were not more resistant to DMBA than morphologically transformed cells (line 917). When cloning efficiencies in DMBA-containing medium were compared transformed cells proved to be slightly more sensitive to this hydrocarbon than non-transformed cells. Thus, initial transformation of normal cells into neoplastic cell line was accompanied by the alteration of the cell sensitivity to carcinogen, but additional increase of malignancy induced by virus and associated with the loss of contact inhibition was not correlated with further augmentation of resistance. These results confirm that various stages of malignancy can be achieved by qualitatively different cell alterations.

Another example of the early alterations of the cell sensitivity to carcinogen in the course of tumour development is provided by the *in vivo* studies of the reaction of azodye-treated liver to partial hepatectomy. Short-term treatment of animals with hepatocarcinogens inhibits mitotic response of the hepatic tissue to partial hepatectomy. At the later stage of carcinogenesis hyperplastic nodules appear in the liver; these nodules may be regarded as earliest foci of neoplastic growth. If hepatectomy is performed at this stage, nodules react by an increased mitotic activity while reaction of surrounding hepatic tissue remains inhibited (Laws 1959, Guelstein 1963a, 1963b). Obviously cells of the nodules are more resistant to the growth-inhibiting action of carcinogens than other cells.

Besides giving another example of the early alteration of cell sensitivity to carcinogens these experiments are also important with regard to the other problem. We briefly discussed here already deletion hypothesis postulating that neoplastic transformation is accompanied by the loss of cell component essential for carcinogen binding and for cell reaction to growth-controlling factors. If this component completely disappeared from neoplastic cells, development of the resistance to carcinogens should have been simultaneous with the complete loss of the cell sensitivity to growth-controlling factors. However, cells of hyperplastic nodules as well as the cells of certain strains of transplanted hepatomas are resistant to carcinogen but retain some sensitivity to hepatectomy, that is certain degree of reactivity to the action of organ-specific growth-controlling factors. Therefore one has to suggest that alteration of the sensitivity to carcinogens is linked with some transformation of the hypothetical cell system reacting with growth-controlling agents, may be with the deletion of some parts of this system but not with its complete disappearance and/or inactivation. Resistance to carcinogens may not be the only manifestation of this transformation of growth-controlling mechanisms at the early stages of progession. At these stages cells are still sensitive to the tissue-specific growth-controlling factors but the threshold level of these factors inducing proliferative response in neoplastic tissue and in normal tissue may be different. This was clearly shown in the experiments of De Ome (1963): proliferation of the mammary gland epithelium in the mice infected with mammary tumour virus was induced by much lower doses of hormones as compared with the proliferation of non-infected epithelium. Cutaneous epithelium of the rabbit ear after 30 paintings with DMBA became more sensitive to the growth-stimulating action of scarlet red dye (Vasiliev and Au Bow Cheung 1962). Even cells of very anaplastic tumours are sensitive to the growth-stimulating factors provided by damaged normal tissues (Vasiliev 1962). Heightened sensitivity to growth-stimulating factors as well as heightened resistance to carcinogens are possibly characteristic for many types of early neoplastic lesions, produced by various agents (chemical compounds, viruses, etc.). In certain types of chemical carcinogenesis toxic carcinogen is probably one of the factors favouring the selective multiplication of neoplastic cells. It is, however, beyond the scope of this paper to discuss mechanisms of formation of first neoplastic cells and of neoplastic foci in various types of carcinogenesis.

IV. Conclusions

Summarizing the facts discussed above we may conclude that alterations of the sensitivity to the toxic effects of carcinogen often develop at the early stages of neoplastic progression. These alterations are not dependent of tumour etiology and seem to be tissue-specific. There are reasons to suggest that alterations of the cell sensitivity to carcinogens are intrinsically linked to the transformation of the tissue-specific growth-controlling mechanisms essential for neoplastic change of the cells. Therefore, by studying mechanisms of early toxic effects of carcinogens one may hope to find more about the nature of cell components involved in cell reaction to growth-controlling factors. However, these investigations are still at their beginning and the main aim of this communication has been not to present any final conclusions but to justify a somewhat new approach to the studies of growth-controlling mechanisms in normal and neoplastic cells.

REFERENCES

Abercrombie, M. and Ambrose, E. I. (1962). *Cancer Res.* **22**, 525–548.
Alfred, L. J. (1964). *Br. J. Cancer* **18**, 564–574.
Alfred, L. J., Globerson, A., Berwald, Y. and Prehn, K. T. (1964). *Br. J. Cancer* **18**, 159–164.
Arcos, J. C., Griffith, G. W. and Cunningham, R. W. (1960). *J. biophys. biochem. Cytology* **7**, 49–60.
Berwald, J. and Sachs, L. (1963). *Nature* **200**, 1182–1184.
Bullough, W. S. and Rytömaa, T. (1965). *Nature* **206**, 573–578.
De Ome, K. B. (1963). *In* "Viruses, nucleic acids and cancer". Williams and Wilkins, Houston.
Erofeejeva, L. V. and Vasiliev, J. M. (1966). *Cytologia* (in Russian). (In press).
Graffi, A. and Bielka, H. (1959). "Probleme der experimentellen Krebs. forschung." Geest and Portig, Leipzig.
Guelstein, V. I. (1963a). *Acta Univ. int. Cancr.* **19**, 549–551.
Guelstein, V. I. (1963b). *Vop. Onkol.* **9**(4), 61–68 (in Russian).
Guelstein, V. I. and Yagorskaja, T. A. (1964). *Vop. Onkol.* **10**(9), 58–65 (in Russian).
Haddow, A. (1938). *Acta Univ. int. Cancr.* **3**, 342–352.
Heidelberger, C. (1965). *Usp. sovrem. Biol.* **59**, 101–103 (in Russian).
Ivanova, O. J. (1964). *Vest. akad. med. Nauk SSSR* **11**, 28–30 (in Russian).
Iversen, O. H. and Evensen, A. (1962). *Acta path. microbial. scand.* Suppl. 165.
Koenig, H. (1965). *J. Histochem. Cytochem.* 13–20.
Laws, J. O. (1959). *Br. J. Cancer* **13**, 669–674.
McCarter, J. A. and Quastler, H. (1962). *Biochim. biophys. Acta* **55**, 552–553.
Miller, J. A. and Miller, E. C. (1953). *Adv. Cancer Res.* **1**, 339–431.

Miller, E. C., Miller, J. A. and Hartmann, H. A. (1961). *Cancer Res.* 21, 815–824.

Nomura, M. (1964). *Proc. natn. Acad. Sci., USA*, 52, 1414–1521.

Olshevskaja, L. V. and Raikhlin, N. T. (1967). *Proceedings of the Symposium on Cytochemistry and histochemistry of tumours in Tbilisi* (in Russian). (In press.)

Pitot, H. C. (1962). *Fedn Proc. Fedn. Am. Socs exp. Biol.* 21, 1124–1129.

Prehn, R. T. (1964). *J. natn. Cancer Inst.* 32, 1–17.

Slavnaja, I. (1965). *Cytologia*, 7, 749–752 (in Russian).

Starikova, V. B. (1964). *Vop. Onkol.* 10(5), 55–59 (in Russian).

Starikova, V. B. and Solina, R. M. (1967). *Vest. akad. med. nauk. SSSR*, (in Russian). (In press.)

Starikova, V. B. and Vasiliev, J. M. (1962). *Nature* 195, 42–43.

Starikova, V. B. and Vasiliev, J. M. (1963). *Acta Un. int. Cancr* 19, 620–623.

Vasiliev, J. M. (1962). *In* "Biological interactions in normal and neoplastic growth". Henry Ford Hospital International Symposium (M. J. Brennan and W. L. Simpson eds) pp. 299–309. Little Brown, Boston.

Vasiliev, J. M. and Guelstein, V. I. (1963). *J. natn. Cancer Inst.* 31, 1123–1151.

Vasiliev, J. M. and Guelstein, V. I. (1965). *Cytologia* 7, 302–311 (in Russian).

Vasiliev, J. M., Olshevskaja, L. V., Raikhlin, N. T. and Ivanova, O. J. (1962). *J. natn. Cancer Inst.* 28, 515–559.

Vasiliev, J. M. and AuBow Cheung (1962). *Br. J. Cancer* 16, 238–245.

Vasiliev, J. M., Guelstein, V. I., Starikova, V. B. and Yagorskaja, T. A. (1967). *Neoplasma.* (In press.)

Yaskovetz, A. A. (1967). *Vop. Onkol.* (in Russian). (In press.)

The Failure of Physiological Defence Reactions Directed Against Malignant Cells as the Rate Determining Step in the Occurrence of Cancer

PETER ALEXANDER

Chester Beatty Research Institute, Institute of Cancer Research, Royal Cancer Hospital, London, England

I. Introduction

REVIEWING the biological basis of cancer, Burnet (1957) started with the assumption: "The failure in cancer is due not to any weakness of the organism but to a change in the character of the cells rendering them in one way or another insusceptible to the normal control." His conclusions were summarized by the sentence: "There is little ground for optimism about cancer." I believe that a number of observations made in the last ten years allow us to question the universal validity of the first statement and, as a consequence, I hope to make out a case that we may adopt a more hopeful attitude than that taken by Burnet to the medical aspects of cancer. To support these views I will first bring together some concepts which individually have been discussed before and any originality, which this presentation may have, derives from their juxtapositioning. I will then present some experiments with primary tumours in rats, which suggest that these tumours grow and eventually kill while being opposed by an active reaction of the host and that therapeutic benefit can be obtained by measures, which enhance or augment the natural defence reactions.

II. The Malignant Transformation—a Frequent Occurrence

The biological properties of cells obtained from cancers require the occurrence of a permanent and heritable transformation which has all the properties to be expected from a mutation in normal (i.e. non-neoplastic) somatic cells. Since the somatic mutation hypothesis was first examined (Strong 1926, 1958, Boveri 1929, Lockhart-Mummery 1934, Haddow 1938) evidence has been forthcoming that the rate of appearance in mammals of somatic mutations—or rather of the rate with which they give rise to phenotypic expression in cell populations—is of the same order of magnitude as spontaneous mutations in germ cells. Attwood and Scheinberg (1958) found this in humans by determining antigenic markers in erythrocytes and Fraser and Short (1958) by measuring the appearance of fleece mosaicism in sheep. The rate of induction by ionizing radiations of mutations in somatic cells is also of the same order of magnitude as in germ cells (Russell and Major 1957). An estimate for the spontaneous mutability of genes in somatic cells in man of between one in 10^{-5} to 10^{-6} per cell division can thus be made. In man some 10^{11} cell divisions occur daily and in that time more than 10^{4} cells that have undergone a malignant transformation would be expected if only one mutational step is required. Such potential cancer cells would, of course, be concentrated in those organs where cell division is most frequent.

In the light of these quantitative data on somatic mutation rates and impressed with the carcinogenicity of plastic films, Alexander and Horning (1959) proposed that many carcinogenic agents acted in an indirect way by making it possible for malignant cells of spontaneous origin to establish themselves. Two lines of experimental evidence have since appeared which support the contention that the occurrence of malignant cells is a relatively common event in populations of dividing mammalian cells (cf. Sanford 1965).

1. Normal (e.g. embryonic cells) when grown in tissue culture do not multiply indefinitely; either the cell line dies out or the culture has turned malignant and its cells will give rise to cancer when injected into suitable hosts (cf. Hayflick and Moorhead 1961). This phenomenon has been particularly well documented by Sanford (cf. 1965) and by Fogh and Hok (1958) who used human amnion cells which, after culturing, cause cancer on injection into rats that have been rendered immunologically unresponsive.

2. Embryonic mouse tissue that has been allowed to grow in Millipore chambers placed within syngeneic hosts will cause cancer when removed from the chamber and transplanted subcutaneously (Shelton *et al.* 1963).

The most plausible interpretation is that populations of normal cells when growing under conditions where they are inaccessible to host defences, as is the case in tissue culture or within millipore chambers, become heterogeneous and some readily acquire malignant potential. The corollary is that during the evolution of multicellular life, mechanisms must have been developed to ensure the elimination of cells that have undergone malignant transformations. This is probably the only kind of somatic mutation against which it is essential for an animal to guard, since others will be lost by normal selection pressure or accumulate in a harmless fashion as do erythrocytes with heterogeneous antigenic markers or pigmented hairs in fleece mosaics.

We are thus led to the concept that while the basic requirement for cancer is the malignant transformation of a cell, this process is not in fact the rate determining step in carcinogenesis. Cancer arises when for one reason or another the cells with malignant potential, which are arising constantly, elude the physiological processes responsible for their containment. In a practical sense the genesis of a malignant tumour is not determined by changes in the character of individual cells, but is due to a failure of the organism.

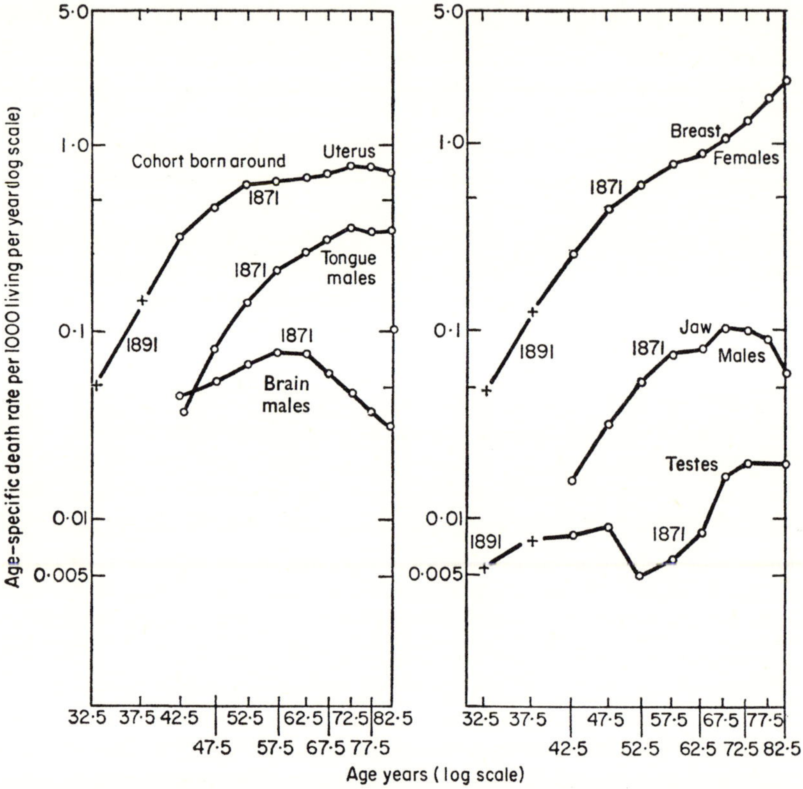

FIG. 1. The relationship between the cancer mortality rate and age. Sites are shown where the increase of log rate with log age is not even approximately linear. (Data for England and Wales.)

Two lines of arguments have frequently been advanced in support of the opposite view: namely that the occurrence of the cellular change (i.e. the transformation to malignancy) is the critical step which decides if and when a cancer will appear.

1. The parallelism between mutagenic and carcinogenic agents. Twenty years ago when very few chemical substances were known to be mutagenic the concordance was much more impressive than it is now when hundreds of diverse compounds have been found to have this property (see Burdette 1955). Thus, evidence for the mutagenicity of the carcinogenic hydrocarbons

and the carcinogenic azo dyes is not very strong and the cancer producing plastic films clearly cannot act in this way. Several powerful mutagens were shown by Trainin *et al.* (1964) to be incapable of initiating skin carcinogenesis and the most potent mutagenic agent so far discovered, ethyl methane sulphonate, produces tumours only after a very long period even when given systemically in repeated doses (Alexander and Connell 1963). For

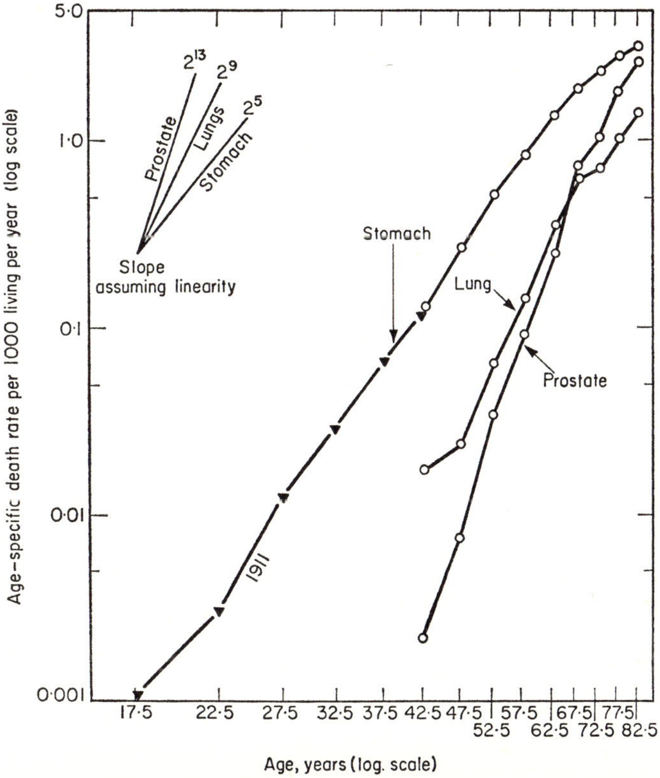

FIG. 2. Cancer mortality rates where tumours at different sites show different relationships with age. The subjects were of the same sex (males), same birth date (cohort born around 1871) and from the same country (England and Wales).

the induction of tumours and leukaemia by ionizing radiations time dose relationships do not fit a mutagenic process (Mole 1958) and suggest that the cause is one of the many other biological changes which radiations bring about.

2. The existence of a constant and significant mathematical relationship between cancer and age. In man, the cancer death rate is said to increase with age at a rate corresponding to a sixth lower law. This curve has impressed Burnet (1957) (see also Armitage and Doll 1954) who speculates that six

control mechanisms have to be abrogated by successive somatic mutations to give eventually a cell which is capable of "the initiation of frank cancer". A detailed study of the mortality statistics shows that the sixth power relationship is a mathematical artefact resulting from the pooling of data. If the mortality rates are analyzed on a cohort basis for individual classes of cancers then they deviate significantly from logarithmic growth rate. But even if on a

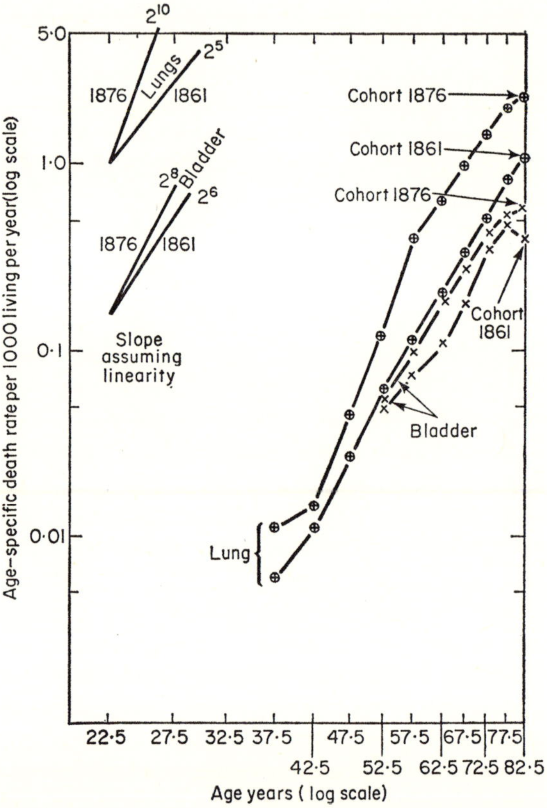

FIG. 3. The difference in relationship of cancer mortality rates with age for tumours of population of different birth dates, but of the same sex (males), at the same site and from the same country (England and Wales).

log-log plot the best straight line is drawn it is apparent that the exponent (i.e. the slope of this curve) varies widely from one situation to another and the average value of six is obviously without significance. This is brought out clearly from data (illustrated in part by Figs. 1 to 3) prepared by my colleague, Dr. R. A. M. Case in presentation to a meeting to the British Society for

Research on Ageing in 1962 and which show that there are not only great variations between different cancers, but that even for one site the exponent varies from country to country, between sexes, and between different dates of birth for the cohorts studied. There would appear to be no justification for giving a simple biological interpretation to the age dependence of cancer mortality (cf. Armitage *et al.* 1965) or for linking this complex relationship to the induction of somatic mutations.

Indeed, the facts would seem to be explained much more satisfactorily by an increase in the likelihood of a failure of a defence mechanism with age as this could be expected to depend on such variables as sex and site. Indications that old animals have a decreased capacity to mount certain immunological reactions are accumulating (cf. Aoki *et al.* 1965) and such changes could be an important factor in the relationship between age and cancer.

III. The Role of the Homograft Reaction in the Elimination of Malignant Cells

In looking for physiological mechanisms for the elimination of malignant cells a clue is provided by the following two observations:

1. The homograft reaction is the most sensitive and the most specific mechanism for the recognition of foreign cells that has been developed by vertebrates. That this system has evolved merely to guard against the unlikely contingency of a foreign tissue graft, seems improbable. The physiological function of the transplantation antigen which determines this exquisitely sensitive immune reaction is unknown.

2. The widespread and possibly universal, presence in tumours of specific antigens having many of the properties of transplantation antigens. In the pioneer investigations of Foley and of Prehn (see review by Old and Boyse 1964), only a small fraction of all chemically induced tumours could be shown to possess antigenic determinants which were absent in the host and which had the property of being able to induce a homograft-like reaction specifically directed against tumours. Their presence seemed to be unrelated to tumour induction and could be explained as a random mutational event produced by the carcinogens as an action independent of carcinogenesis. Refinements in the technique of looking for such antigens have resulted in their being recognized in all tumours whenever the proper tests were used whether the carcinogen was physical, chemical or viral; the latter category includes tumours which occur spontaneously in the sense that no exogeneous stimulus was applied. These findings which have been made in many laboratories (cf. Old and Boyse 1964, Alexander 1965) can be illustrated by results of our own (see Table I) with chemically induced fibrosarcoma in the

TABLE I. Tumour specific antigens in benzpyrene induced fibrosarcoma detected by the presence of cytotoxic spleen cells in syngeneic animals immunized with the tumour.

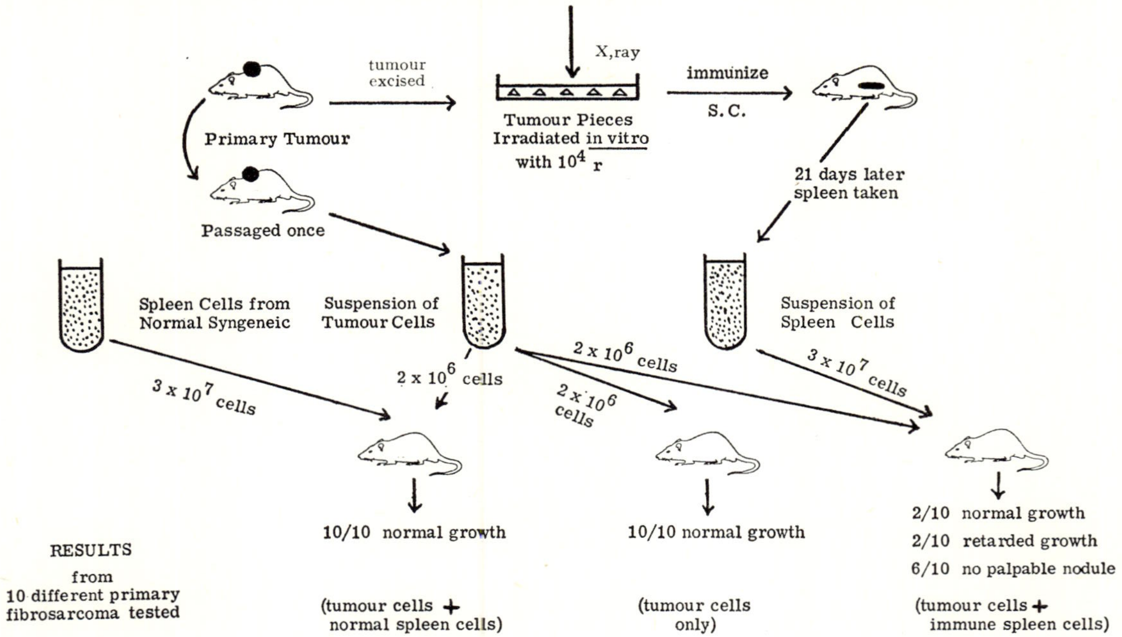

homozygous "hooded" strain of rats. Eight out of ten different primary tumours could be shown to elicit, when injected into syngeneic rats, lymphocytes that were specifically cytotoxic against the cells of the tumours used for immunization. There was no cross resistance between the different tumours.

The hypothesis I would like to put forward is that the antigenic properties of tumours are a necessary step in freeing a cell from restraints exercised by neighbouring cells and in enabling it to achieve autonomy (i.e. an essential phenotypic expression of those somatic mutations which give rise to a transformation to malignancy involves the appearance of a new antigen capable of evoking a homograft reaction). It follows that the physiological role of the homograft reaction is to eliminate cells that have undergone a malignant transformation (cf. Thomas 1959, Prehn 1963, Burnet 1964), and the need to evolve a system of the greatest delicacy is due to the fact that the effective elimination of the malignant cells which are appearing constantly is an essential requisite for the existence of higher forms of metazoan life.

The homograft reaction is, according to this view, linked to the part of the recognition system between cells which is mediated by reactions through the cell surface. The existence and importance of such a system in growth, development, maintanance and repair has been experimentally demonstrated —notably by Paul Weiss and his colleagues—and the need for cells to possess the very many different transplantation antigens could be understood if their physiological role were to perform the functions of recognition and control. The limited experimental evidence about the location of the transplantation antigens is in agreement with this concept as it suggests that they are situated on the cell membrane. The role of the homograft reaction is to eliminate cells which, for one reason or another—including somatic mutations—have undergone a change in their recognition system such that they are no longer susceptible to the normal processes of control. Cells that have undergone a transformation to malignancy fall by definition into this category. Alterations in the substances which make up the system responsible for the recognition are seen by experimenters as changes in transplantation antigens (or as the appearance of new tumour specific antigens) and the mechanism which eliminates cells with a defective recognition system is called the homograft reaction. Experimentally this defence process is observed by the non-physiological test of graft rejection.

If the recognition system (or in the language of homograft reaction the transplantation antigens) is a mosiac made up of a number of macromolecules, then the loss of a component may have the appearance of a new antigen. A cell, that fails to make a component which is required for the functioning of the recognition system, is seen by the homograft reaction as a new pattern in the surface against which it reacts by a process akin to the delayed hypersensitivity type. Transplantation techniques could thus record the loss of a molecule as the appearance of a new antigen.* Serological test may see the same change as a loss of an antigen. This would resolve the paradox that by serological tests the cancer cell is typified as having undergone antigenic simplification (cf. Haddow 1965) while from graft rejection experiments it appears to have acquired new antigens of the transplantation type. The somatic mutations that cause a malignant transformation may therefore be a deficiency; the failure to make an essential part in the complex recognition system. If a part of a structure is missing this constitutes a new pattern; if the structure is complex then many new patterns can arise by loss of different components. Each may fail to act as an effective recognition system and evoke a reaction directed specifically against it. In this way the individuality of specific antigens in tumours induced by chemicals and by physical means would be explained. Attempts to prepare transplantation antigens in a cell-free and soluble form have so far only given preparations which show low activity in producing transplantation immunity. This might be expected from a system that is designed to recognize a pattern and which accordingly responds poorly when presented with parts only.

IV. How Does the Malignant Cell By-pass the Defence Mechanism?

Whatever the relation between malignancy and the tumour specific antigens may prove to be, the fact that a homograft type of reaction is evoked by tumour cells in syngeneic animals poses the question why cells of this type give rise to progressively growing tumours in the autochthonous host or in unimmunized syngeneic animals (the growth of the trophoblast may constitute a physiological model for the avoidance of homograft rejection). Inability to mount a homograft reaction due to disease or senile impairment may occur in some clinical cases. But the indications are that a general immune paralysis only occurs when the disease is already advanced and this situation certainly does not apply generally to induced tumours in experimental animals (the possibility for interference with the immune reaction by the carcinogen will be considered later).

The malignant cell may originate in an immunologically privileged site where the growth of homografts can take place. In the autochthonous host prolonged exposure to an antigenic stimulus may, during the latent period

* Green's (1957) theory on the immunological origin of cancer postulates that an essential lesion is the loss of antigens which act as identity marker and that as a consequence cancer cells produce antibodies against normal cells which have retained these antigens. The hypothesis that autologous cancer cells mount an attack against the host is in direct contradiction with the experimentally well documented existence of an active host reaction directed against syngeneic and primary tumour cells (see p. 335).

of carcinogenesis, produce a state of limited immunological paralysis which evokes a degree of acceptance; the induction of tolerance to allogeneic tissue in adult animals by repeated exposure to cells carrying the antigen (Brent and Gowland 1962) might be considered a parallel. Recently it has been shown Morton *et al.* (1965) that those mouse strains which carry the mammary factor have become tolerant to the tumour specific antigens produced by this virus in the neoplasms it induces. I am inclined to the view that this is rather a special case and for chemically induced fibrosarcoma we were able to show that the animals were not tolerant to the specific antigens present in their own primary tumours.

TABLE II. Failure of tumour autograft to grow in animals with primary chemically induced fibrosarcoma.

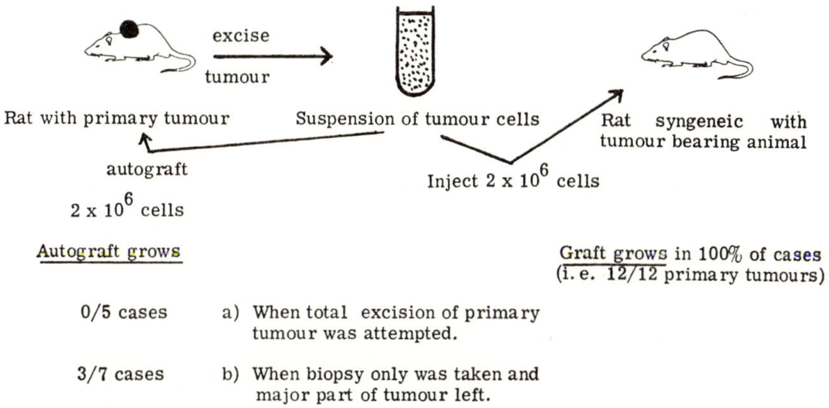

There is a host defence which can reject 2×10^6 cells from the autologous primary tumour but in the presence of large primary the defence is overwhelmed.

Klein *et al.* (1960) detected resistance to transplants of sarcoma cells into the autochthonous host after this had been repeatedly immunized with irradiated tumour cells which had been preserved during the time needed for immunization by passage in isogeneic animals. This important investigation proved that the host can be made to react against its own tumour, but left the question unsettled whether this occurred in the absence of immunization. Experimenting with fibrosarcoma induced with 3:4-benzpyrene in rats we (Mikulska *et al.* 1966) found that inocula of 2×10^6 cells, prepared from a biopsy of the primary tumour, failed to grow as an autograft when injected back into the autochthonous host from which the tumour had been excised, although the cells grew when transplanted into syngeneic animals.

Autografts of tumour cells occasionally grew in those animals where the major part of the primary tumour had been left behind after biopsy (see Table II). These experiments indicate that the autochthonous host is capable

TABLE III. Anti-tumour activity of autologous spleen against primary fibrosarcoma.

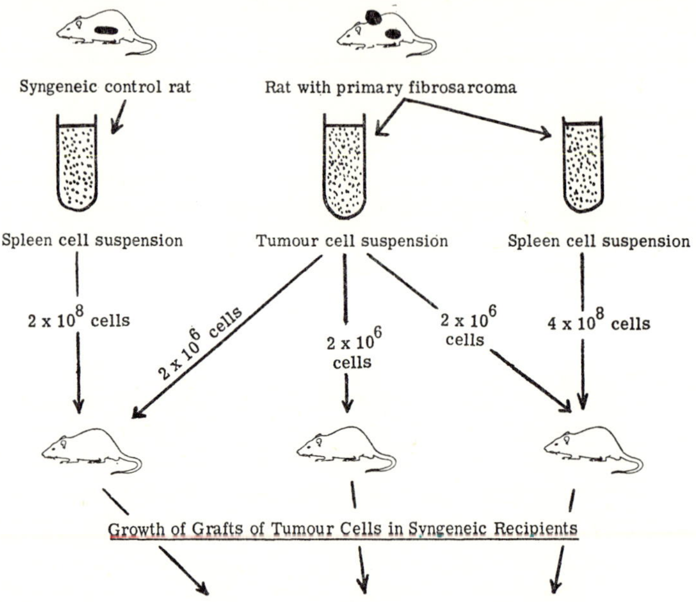

a) In experiments where autologous spleen cells were removed at the same time as the primary tumour.	9/9 normal growth	9/9 normal growth	6/9 normal growth 2/9 retarded growth 1/9 no growth
b) In experiments where autologous spleen cells were removed three weeks after excision of primary tumour.	10/10 normal growth	10/10 normal growth	2/10 normal growth 2/10 retarded growth 6/10 no growth

The spleen cells in a tumour bearing animal react against the cancer cells but the activity is exhausted in the presence of the tumour.

of rejecting the cells of its own primary tumour, but that the magnitude of the reaction is not sufficient to deal with all the cells that are present. Another experiment (see Table III) showed that the reaction against these primary tumours was mediated by lymphocytes, but that not enough of these were produced in the autochthonous host to reject all the tumour cells. The total supply of lymphocytes with anti-tumour activity is exhausted by the rapidly proliferating primary tumour and therefore spleen cells taken from an animal with a tumour are unable to neutralize autochthonous tumour cells. Three weeks after the primary tumour has been removed cells from the spleen of the rat are able to "neutralize" autochthonous tumour cells presumably because in the absence of the primary tumour the immunologically active cells continue to be produced, but are not eliminated. In this way the concentration of lymphocytes capable of reacting against the tumour builds up and the spleen after three weeks contains cells which are effective in preventing tumour cells from growing in syngeneic animals.

I believe that an important mechanism which allows tumours to arise in the face of a potential host resistance is inadequate antigen release in the early stages of tumour development possibly because insufficient cells detach themselves. Careful experimental work, notably by Old *et al.* (1962) and by Klein *et al.* (1960) has shown that the maximum response which it is possible to evoke against cells with tumour specific antigens by repeated immunization of syngeneic animals can only reject a challenge of a relatively small number of cells. The number varies from tumour to tumour and in mice range from 10^3 to 10^6. In the chemically induced fibrosarcoma of the rat with which we have been dealing, an immune response capable of rejecting 10^6 cells has always been obtained but an inoculum of 10^7 cells has usually overwhelmed the resistance. It is clear therefore that if a tumour can reach a size of a few million cells it will defeat any reaction that can be mounted against it and proceeds to kill the animal. The problem of the occurrence of an antigenic tumour is then how to attain a certain minimum size befor the immune reaction is fully established.

Very rapid growth is one possibility; we have worked with a transplanted lymphoma which gives a lethal tumour even when only one cell is injected into syngeneic animals. Yet the tumour is antigenic and resistance to a challenge of up to 10^5 cells can be induced by immunization with irradiated cells (see Table IV). The reason why in the unimmunized animal one cell kills is that these cells divide every 11 hours so that the critical size has been passed before the immune reaction has had time to become established. Clearly if a mutation to give a cell with these properties occurred then it would beat the defence mechanism. Fortunately cells with this kind of growth potential only arise after prolonged selection during transplantation and primary tumours do not grow at these very fast rates. From a number of different experiments (e.g. Boyse 1963) it would appear that an effective antigenic stimulus is only provided when the tumour has reached dimensions which are beyond the control of the maximum host response. The radiosensitizing action of autografts (Haddow and Alexander 1964) suggests that the amount of antigenic stimulation (i.e. the number of antigen-containing cells that find their way to the immunological system of a rat) may not be maximal for fibrosarcoma even when these have reached 1·5 cm in diameter.

The question of accessibility of residual tumour to the host response may be one of the limiting factors in successful surgery. This is illustrated by the autografting experiments already described (see Table II). In animals in which complete surgical excision of the tumour was attempted they were able

TABLE IV. Protection of DBA/2 mice against L5178 Y lymphoma cells by immunization with irradiated tumour cells

Number of L5178 Y cells injected	Number of animals with tumours
1. Non-immunized animals	
1	78%
10	92%
10^3	100%
2. Immunized with 10^7 L5178 Y cells irradiated *in vitro* with 3000r of X-rays.	
10^3	0%
10^4	0%
10^5	20%
10^6	80%

to reject inocula of 2×10^6 of their own tumour cells given subcutaneously, yet in almost every case there was a recurrence of the primary tumour. The defence mechanism of the host against autologous tumour cells was clearly much less effective against residues of tumour tissue than against isolated cells. It is interesting that the magnitude of typical delayed hypersensitivity reactions such as those produced by tuberculin also depend on the site of injection, the intradermal route being the most effective.

V. Indirect Mechanisms of Carcinogenesis

The hypothesis that the critical step in the appearance of cancers is avoidance of a defence reaction and not the induction of somatic mutations requires that carcinogenic agents do not act directly on the cells that eventually give rise to the tumour, but on their environment. At a Ciba meeting in 1958 (Alexander and Horning 1959) we wrote "the induction of tumours by plastic films is indirect and physiological processes are responsible for the malignant transformation. The carcinogen merely provides a favourable opportunity for these to occur." Bates and Prehn (1965) have shown that the fibrous pocket which is formed around the plastic film constitutes a site which is screened from the homograft reaction. The induction of tumours by plastic films is therefore most satisfactorily explained by creating a situation which permits spontaneously occurring cancer cells to establish themselves.

With other carcinogens the situation is less clear. Tissue damage leading to "sheltered sites" may well explain the carcinogenic action of localized irradiation, especially as there is evidence that following exposure to the skin of β-rays (Glücksmann 1963) and of bone to α-ray emitters (Lisco 1956) the tumours arise at a point beyond the range which the particles attained in tissues. The role of generalized tissue damage in chemical carcinogenesis is difficult to assess, but Roe (private communication) has found that rather typical lesions involving the formation of dense connective tissue precede the formation of tumours by the injection of "Imferon".

Apart from the procedures mentioned above any attempt to interpret carcinogenesis as an avoidance of host reaction is pure conjecture. The effect of the carcinogenic hydrocarbons of the immune response has been discussed in relation to the occurrence of antigenic tumours by Prehn (1963) and Rubin (1964). The depression in the immune response seen after low doses which are, however, carcinogenic is small and in our experiments the presence of a pellet of 3 : 4-benzpyrene did not prevent the rats from showing a homograft-type reaction against their own tumours. At best, direct interference with immunity can only be a contributory factor and not the primary cause for the carcinogenic action of the hydrocarbons, though it may play a more important role in the carcinogenic action of whole body radiation and of the radiomimetic alkylating agents which have a pronounced effect on the lymphoid organs. Haddow (1938) stressed that a characteristic feature of carcinogenic situations was a period during which there was continuous interference with normal growth which allowed the emergence of new cell variants. This concept can readily be modified to include the prevention of a homograft reaction which requires the participation of the local lymph nodes draining the site of application of the carcinogen. The carcinogenic agents which are most difficult to reconcile with an indirect action are viruses.

VI. Implications for Therapy

The slow growth, which is a feature of many human tumours, may be the result of a near equilibrium between the number of new tumour cells appearing and the capacity of the host to eliminate them. On biological grounds this would seem to be a more likely explanation for slow growth of tumours than that their cells have an intermitotic period of months or even years. Boosting the host defence may, in such cases, lead to a cure. This concept may be of relevance to surgery and radiotherapy of such tumours as it suggests that the prime consideration of therapy should be to maintain host resistance and this may involve the functioning of the local nodes. Immunological methods of treatment, even when the cancer is far from equilibrium, may be of value as an adjunct to conventional methods of treatment which have removed the bulk of the cancer to assist the elimination of residual and disseminated foci. An example is the effect of injecting immune lymphocytes which has given some promising results in this connection (Delorme and Alexander 1964, Alexander *et al.* 1966) with the primary fibrosarcoma of the rat, if this is not too large. These sarcomata are extremely resistant to radio- and to chemotherapy, yet they respond well to the injection of lymphocytes obtained from allogeneic rats and even from goats and sheep that have been immunized with the tumour to be treated.

In considering practical applications the critical question which it is very difficult to answer experimentally, is whether there are human tumours which have the same immunological properties as have those induced by benzpyrene in the rat.

VII. Summary

It is commonly accepted that cells with a malignant potential arise as the result of somatic mutations and that somatic mutations, as a group, occur spontaneously at a rate comparable to germ cell mutations. Since the occurrence of tumours is a relatively rare event the body must have a defence mechanism which eliminates the large majority of cells that have undergone a spontaneous malignant transformation. Carcinogenesis, in many cases, does not consist of converting normal cells into malignant cells, but in creating the condition by which cells that have undergone a spontaneous mutation to malignancy can establish themselves.

The hypothesis is examined that the role of the homograft reaction is the elimination of cells with a defective recognition system. Since such cells present the most serious threat to the existence of multicellular organisms it is not surprising on evolutionary grounds that an immune mechanism of exquisite sensitivity has been developed to ensure their removal. If the function of the homograft reaction were merely to deal with the unlikely event of accidental introduction of foreign tissue then it would appear to be unduly refined for its purpose. It is suggested that the physiological role of the homograft reaction is to recognize and eliminate cells which by transplantation tests are said to contain tumour specific antigens. If the recognition system is a mosaic made up of a number of macromolecules, then the loss of a component may be seen as the appearance of a new antigen. A somatic mutation resulting in the failure to make a component which is required for the functioning of the recognition system, is then seen by the homograft reaction as a new pattern in the surface against which it reacts by a reaction of the delayed hypersensitivity type. This would resolve the paradox that by serological tests the cancer cell is typified as having undergone antigenic simplification while from graft rejection experiments it appears to have acquired new antigens of the transplantation type.

In a group of chemically induced sarcoma it was shown that tolerance to the tumour specific antigen does not occur and that animals react actively against their own primary tumours, but that the defence reaction is overwhelmed.

Acknowledgments

Professor Haddow's interest and advice is gratefully acknowledged. This investigation has been supported by grants to the Chester Beatty Research Institute (Institute of Cancer Research: Royal Cancer Hospital) from the Medical Research Council and the British Empire Cancer Campaign for Research, and by the Public Health Service Research Grant No. CA-03188-08 from the National Cancer Institute, U.S. Public Health Service.

REFERENCES

Alexander, P. (1965). *In* "Scientific Basis of Surgery" (W. T. Irvine, ed.), pp. 478-495. J. & A. Churchill, London.

Alexander, P. and Connell, D. I. (1963). *In* "Cellular Basis and Aetiology of Late Somatic Effects of Ionizing Radiation" (R. J. C. Harris ed.), pp. 259–265. Academic Press, London.

Alexander, P., Connell, D. I. and Mikulska, Z. B. (1966). *Cancer Res.* 26, 1508–1575.

Alexander, P. and Horning, E. S. (1959). *In* Ciba Symposium on "Carcinogenesis and Mechanisms of Action", pp. 12–23. J. & A. Churchill, London.

Aoki, T., Teller, M. N. and Robitaille, M. N. (1965). *J. natn. Cancer Inst.* 34, 255–264.

Armitage, P. and Doll, R. (1954). *Br. J. Cancer* 8, 1.

Armitage, P., Doll, R. and Pike, M. C. (1965). *Br. med. J.* p. 723.

Atwood, K. C. and Scheinberg, S. L. (1958). *J. cell. comp. Physiol.* 52 (supplement 1) 97–123.

Bates, R. R. and Prehn, R. T. (1965). *Nature* **205**, 303.

Boveri, T. (1929). "The Origin of Malignant Tumours." Williams and Wilkins.

Boyse, E. A. (1963). *Guy's Hos. Rep.* **112**, 433.

Brent, L. and Gowland, G. (1962). *Nature* **196**, 1298.

Burdette, W. J. (1955). *Cancer Res.* **15**, 201.

Burnet, F. M. (1957). *Br. med. J.* 779–786 and 841–847.

Burnet, F. M. (1964). *Br. med. Bull.* **20**, 154–158.

Delorme, E. J. and Alexander, P. (1964). *Lancet* 117–120.

Fogh, J. and Hok, K. A. (1958). *Cancer Res.* 18, 692–697.
Fraser, A. S. and Short, B. F. (1958). *Aust. J. biol. Sci.* **11**, 200.
Glucksmann, A. (1963). *In* "Cellular Basis and Aetiology of Late Somatic Effects of Ionizing Radiation" (R. J. C. Harris ed.), pp. 121–134. Academic Press, London.
Green, H. N. (1957). *Ann. N.Y. Acad. Sci.* 68, 268.
Haddow, A. (1938). *Acta Un. int. Cancr.* **3**, 342–353.
Haddow, A. (1965). *Br. med. Bull.* **21**, 133–139.
Haddow, A. and Alexander, P. (1964). *Lancet* 452–457.
Hayflick, L. and Moorhead, P. S. (1961). *Expl. Cell Res.* **25**, 585–621.
Klein, G., Sjögren, H. O., Klein, E. and Hellström, K. E. (1960). *Cancer Res.* **20**, 1561.
Lisco, H. (1956). *In* Ciba Symposium on "Bone Structure and Metabolism", p. 272. J. & A. Churchill, London.
Lockhart-Mummery, J. P. (1934). "The Origin of Cancer." J. & A. Churchill, London.
Mikulska, Z. B., Smith, C. and Alexander, P. (1966). *J. natn. Cancer Inst.* **36**, 29–35.
Mole, R. H. (1958). *Br. med. Bull.* **14**, 184.
Morton, D. L., Goldman, L. and Wood, D. A. (1965). *Proc. Am. Ass. Cancer Res.* **6**, 47.
Old, L. J. and Boyse, E. A. (1964). *A. Rev. Med.* **15**, 167.
Old, L. J., Boyse, E. A., Clarke, D. A. and Carswell, E. A. (1962). *Ann. N.Y. Acad. Sci.* **101**, 80.
Prehn, R. T. (1963). *In* "Conceptual Advances in Immunology and Oncology" 11th Ann. Symp. Fundamental Cancer Research, at the Univ. of Texas, pp. 470–485.
Prehn, R. T. (1963). *J. natn. Cancer Inst.* **31**, 791.
Rubin, L. (1964). *In* "Progress in Experimental Tumour Research". Vol. 9, p. 100 (F. Homburger ed.). S. Karger. Basel/New York.
Russell, L. B. and Major, M. H. (1957). *Genetics* **42**, 161.
Sandford K. K. (1965). *Int. Rev. Cytol.* 18, 249–311.
Shelton, E., Evans, V. J. and Parker, G. A. (1963). *J. natn. Cancer Inst.* **30**, 377.
Strong, L. C. (1926). *Genetics* **11**, 294–303.
Strong, L. C. (1958). *Ann. N.Y. Acad. Sci.* **71**, 810.
Thomas, L. (1959). Discussion remarks, pp. 529–532. *In* "Cellular and Humoral Aspects of the Hypersensitive State" (H. S. Lawrence ed.). Hoeber-Harper, New York.
Trainin, N., Kaye, A. M. and Berenblum, I. (1964). *Biochem. Pharmacol.* **13**, 263–267.

Malignant Tumours and the Host's Reactivity*

K. O. RENKONEN

Department of Serobacteriology, University of Helsinki, Helsinki, Finland

A STUDY of malignant tumours in man and tolerance against non-self tissue is hardly possible without some statistical guidelines. This paper attempts to relate tolerance in mothers with the incidence of boys and twins born to them.

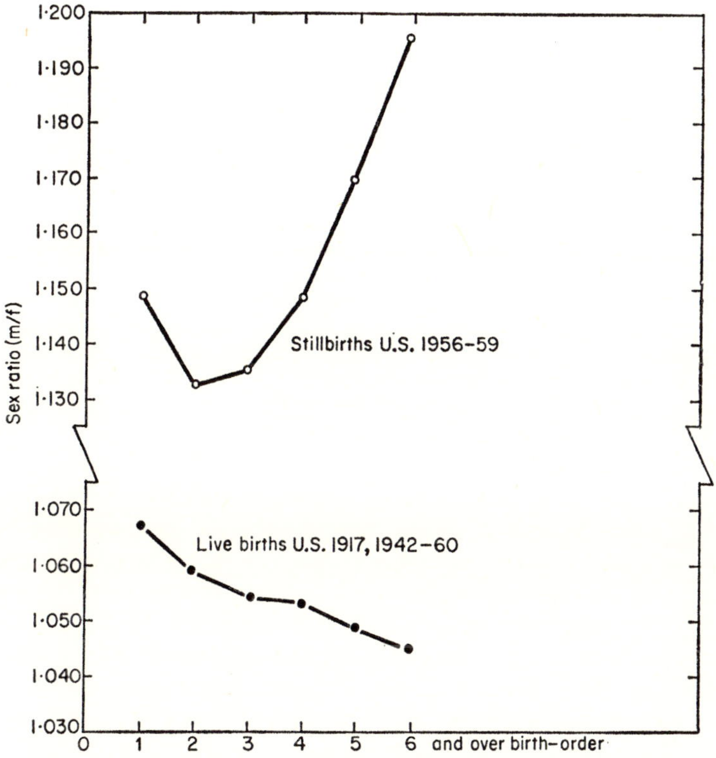

FIG. 1. Birth order and sex ratio among live births and stillbirths (USA).

* This study has been supported by Sigrid Jusélius Foundation, Helsinki.

The well known correlation between the birth order and the sex ratio m/f is seen in Fig. 1 (Renkonen 1964a).

The relative disappearance of boys with increasing birth order could be due to an alteration in the primary sex ratio. However, a corresponding drop in the sex ratio among the stillbirths would be implied, but this is not

TABLE I. Sex ratio of propositus and number of boys among pre-existing sibs.

The composition of the pre-existing families		The sex ratio male/female among the next children: Geissler's data	Finnish data
Three children families			
Boys	Girls		
0	3	1·014±0·015	0·913±0·038
1	2	1·073±0·009	1·047±0·022
2	1	1·029±0·008	0·999±0·022
3	0	1·074±0·014	1·053±0·037
Four children families			
Boys	Girls		
0	4	1·043±0·024	
1	3	1·117±0·012	
2	2	1·053±0·009	
3	1	1·012±0·011	
4	0	1·070±0·022	

apparent. The sex ratio is rising. Accordingly, an acquired reactivity of mothers against her male foetuses is a more reasonable assumption. Such a reactivity may be associated with immunization phenomena. So, the number of boys born before the propositus should be important. In Table I families with an equal number of children are compared (Renkonen *et al.* 1961).

In families, where there are already children of both sexes, the number of males to be born next depends on the number of boys already existing. The more there are, the less are likely to be born.

A complete picture indicating how the sex ratio among the next children depends upon the composition of the pre-existing sibship is given in Fig. 2 (Renkonen 1964b). The picture is based on the data of Geissler. Here each intersection point represents a definite type of family. From the intersection points a black or white arrow is drawn. Black means that among the next children born there were more boys than girls; white indicates the reverse. The lengths of the arrows are proportionate to the deviation of the sex ratio from 1·0.

Several deductions from Fig. 2 are possible. The differentiation of families begins roughly after two children: Then the marginal unisexual families start to contrast with their neighbours. Among families with children of both sexes a big black arrow area becomes differentiated. This area with more girls than boys in the past shows a greater expectancy of boys with increasing birth order. Simultaneously, a corresponding white arrow area is evolved.

TABLE II. The sex ratios among children born to families with two boys and two girls. Role of the sequence

Type of family	Sex ratio
mmff	$0{\cdot}94 \pm 0{\cdot}06$
mfmf	$1{\cdot}02 \pm 0{\cdot}06$
mffm	$1{\cdot}03 \pm 0{\cdot}06$
fmmf	$1{\cdot}04 \pm 0{\cdot}06$
fmfm	$1{\cdot}14 \pm 0{\cdot}06$
ffmm	$1{\cdot}19 \pm 0{\cdot}06$

This area with more boys than girls in the past, does not show the advance of the relative number of males with increasing birth order, which was visible in the black arrow area. Between the black and white arrow areas a narrow intermediate line is noticeable. It consists of families with an equal number of boys and girls.

On this basis, I should like to stress that unisexual families are sui generis and should be treated and analyzed with special interest. The families in the white arrow area do not produce the expected amount of boys among children to be born. It seems acceptable that several of those mothers might turn against their male foetuses.

The number of boys and girls in the already existing family is not the only factor governing the sex ratio among the expected children. For instance, there is quite a difference between the arrows attached to the seven-children families 3m+4f and 4m+3f. Here the order in which the first children are born may play an important role. This factor is treated in Table II (Renkonen 1964b).

In the intermediate type of family consisting of two boys and two girls the males born later suggest a tolerance of the mother because further males are likely to be born. Females born later might be an indication of an already existing reactivity. So, the pre-existing boys in the black arrow area in Fig. 2 might be born later and those in the white arrow families earlier.

The assumption of an acquired reactivity of mothers implies that the male foetuses are killed more frequently by strongly re-acting mothers than by the

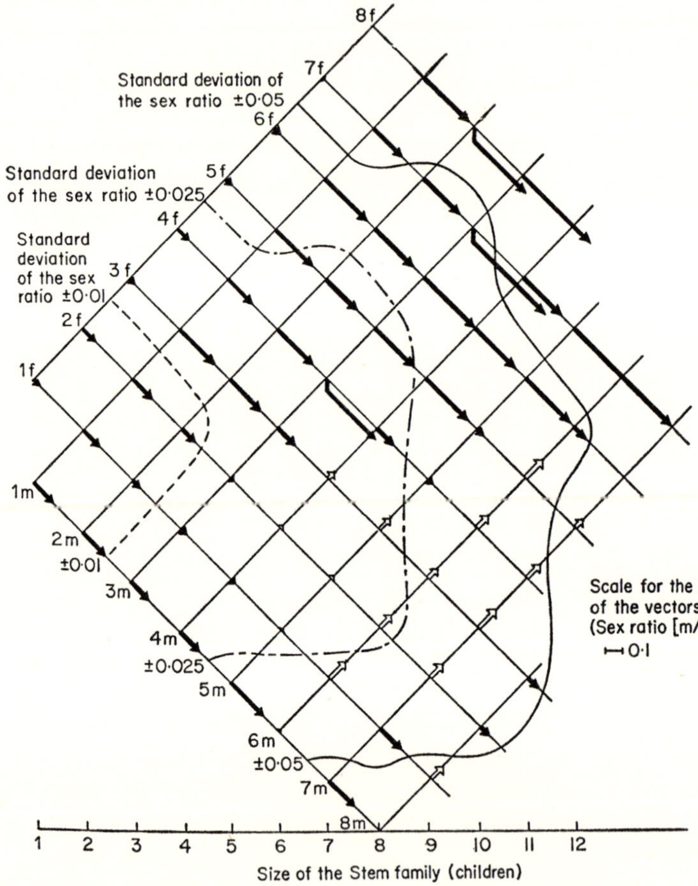

FIG. 2. Graphical presentation of Geissler's data.

weaker ones. If some male foetuses have been killed then a difference in the time interval to the next child may be noticed. We have recorded the time interval from the third to the fourth child in families composed of one boy and of two boys out of three children (Table III). (Renkonen 1964a)

There is the expected difference between the one and two boy families, but this difference is due, solely, to families obtaining a girl as their fourth child. If the difference had been observed when a boy was born fourth it could hardly be explained by an incompatibility of the mother with her male foetuses.

Summarizing the statistical data available I think they should be considered as evidence for the existence of an acquired reactivity of some mothers against their male foetuses. This reactivity is strengthened by the birth of boys. Many mothers (Fig. 2, black arrow area) became more tolerant with increasing birth order. Males born later might indicate a state of tolerance.

TABLE III. Time interval between the third and the fourth child.

Stemfamilies with three children	Size of the group	Birth interval in years between the third and the fourth children	Difference of the birth intervals
		fourth child, boy	
1 boy and 2 girls	2·537	2·3495±0·0176	0·0139±0·0244
2 boys and 1 girl	2·700	2·3634±0·0169	
		fourth child, girl	
1 boy and 2 girls	2·451	2·3355±0·0175	0·0633±0·0245
2 boys and 1 girl	2·637	2·3988±0·0171	

Considering the birth of a boy as an indication of a relative tolerance of the mother, we might find signs of tolerance among mothers known to give birth to an unusual number of boys. Older Japanese mothers are a good example of this (Fig. III). (Takahashi, 1954)

The preponderance of boys may be caused either by a tolerance of older mothers or because the primary sex ratio has been altered. The former assumption is more likely because the sex ratio among the stillbirths seems to drop with the increasing age of the mothers (Renkonen 1964a).

Something similar to the Takahashi phenomenon is visible also in the vital statistics of the U.S.A., if the *primiparae* are considered (Fig. 4). (Renkonen 1964a)

The mongoloids provide perhaps further examples of tolerance. Their cytological picture differs from that of their mothers. They are known to be

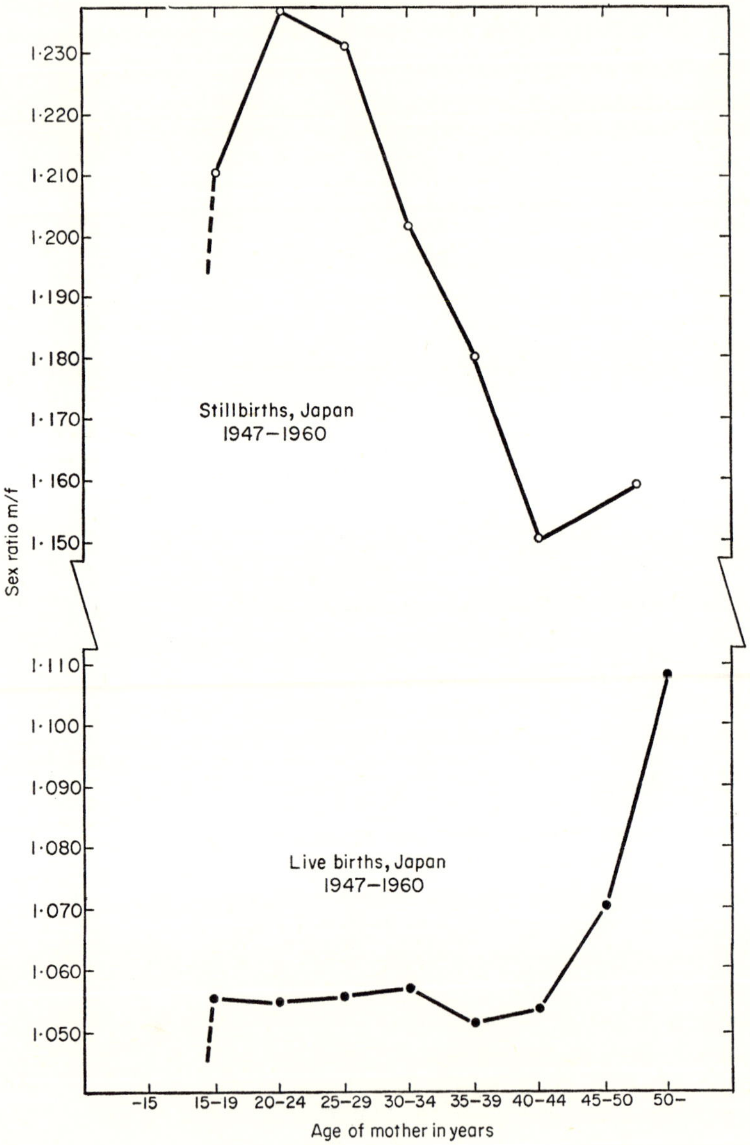

FIG. 3. Age of the mother and sex ratio among live births and stillbirths (Japan).

born especially to elderly mothers. There are three times more boys than girls among the mongoloids born to elderly *primiparae* (Renkonen 1966). If the mothers are tolerant for one type of non-self children they seem to be tolerant also to other types. Similarly families with a unisexual male sibship seem likely to get more mongoloids and more males than families with a unisexual sibship consisting of females (Renkonen 1965b). The mongoloids

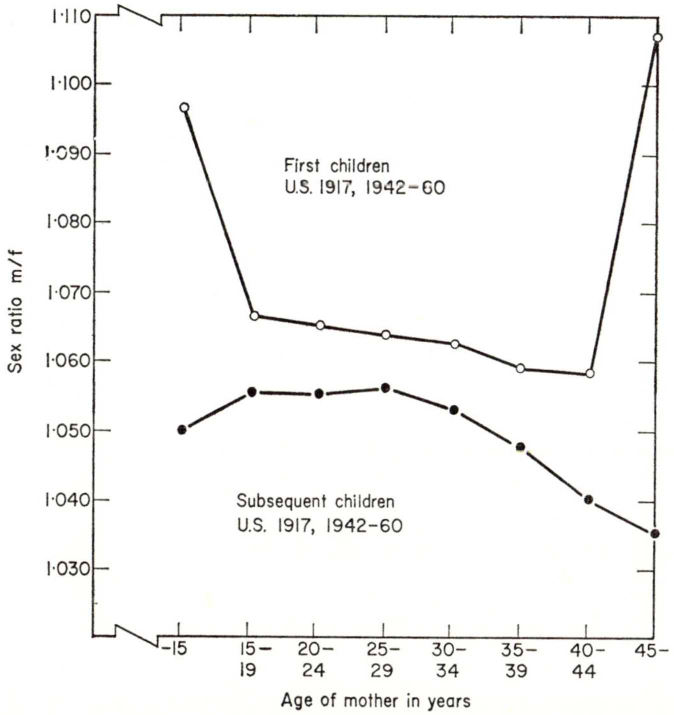

FIG. 4. Age of mothers and sex ratio among first and subsequent children (USA).

have an unexpectedly high frequency of bloodgroups A and B (Shaw and Gershowitz 1962). Could not that also reflect tolerance on the part of their mothers?

Twins are equally interesting. The incidence of dizygotic twins is also increasing with the age of mothers. These mothers should be exceptionally tolerant. In accepting dizygotic twins, they accept two sets of paternally governed antigens of foetuses. The number of twins of unlike sex is roughly 36% in families with one pair of twins only. But, if a further pair is born the first pair of twins includes significantly more dizygotic twins than the normal frequency (Wyshak and White, 1965, and our observation) (Table IV).

A mother of several pairs of twins is likely to be more tolerant than a mother giving birth to only one pair of twins. The state of tolerance of the mother might be reflected also in the following Table.

TABLE IV. Unlike-sexed pairs among the first sets of twins.

	Families with only one set of twins	Families with several sets of twins
Series of Wyshak and White:		
Unlike-sexed twins	2460 (35·6%)	223 (43·9±2·5%)
All twins	6912	508
Our Series:		
Unlike-sexed twins	980 (37·8%)	37 (42·5%)
All twins	2595	87

In families with several pairs of twins there is a trend for them to appear close to each other, as they should do if the mother is in a state of tolerance at the birth of the first pair of twins (Table V). The same holds true in families with several mongoloids. In families of twins and mongoloids a co-incident

TABLE V. Families with multiple sets of twins

	Birth orders after the first set of twins						
	+1	+2	+3	+4	+5	+6	+7–10
Number of twin births	39	20	9	8	4	3	9
Number of all births	87	65	50	39	29	20	34
Percentage of twins	44 (±5·4)	31	18	20	14	15	26

birth is frequent (Renkonen and Donner 1965). Alison McDonald (1964) found among 67 pairs of twins with one or both twins mongoloids, two pairs of male/female mongoloid twins. Mothers bearing twins, including mongoloids, must be rather tolerant.

Summarizing the present position it seems possible that the birth of males and twins (and mongoloids) could be used as statistical markers for the mother's tolerance. A tolerance of aged women against non-self components of tumours has been discussed by several authors. Against this background the family histories of 708 mothers, who have died with different forms of cancer, are worth examining (Renkonen 1965a).

Four observations emerge from the study:

(1) The sex ratio among the last children was high 1·17±0·075. The sex ratio among all the other children was 1·03. The difference is interesting but not significant. (2) The risk for cancer seems to be significantly ($P = 0{\cdot}05$) higher in mothers who have given birth to a unisexual sibship consisting of more than two boys. A comparison of both unisexual sibships is presented in

TABLE VI. The mothers with unisexual progenies deceased from malignant tumours

The type of progeny	Number of children				
	1	2	3	4	5
Males only	110	50	20	7	0
Females only	102	42	11	2	1

Table VI. The same category of families suspected of having a high risk for cancer are, as mentioned previously, likely to get more boys and more mongoloids than the families with a lower risk.

(3) The eventuality of mothers dying before 50 years of age was low (6.7± 2·3%) among mothers who belong to the big white arrow area in Fig. 2. The occurrence of early deaths among all the other types of families was 13·8±1·6%.

(4) Among the very last children of mothers bound to die in cancer there were 10 pairs of twins. The twins seem to be more frequent (4/89) when the mother had died early, compared to the 6 twins born to the 617 mothers dying at 50 years or later. Three out of the former four pairs, but only one out of the six latter ones, were dizygotic.

The question of tolerance of the host seems to merit further studies also in cases of malignant tumours.

REFERENCES

McDonald, A. D. (1964). *J. med. Genet.* **1**, 39–41.
Renkonen, K. O. (1964a). *Acta genet. Statist. med.* **14**, 177–185.

Renkonen, K. O. (1964b). *Ann Med. exp. Biol. Fenn.* 42, 52–54.
Renkonen, K. O. (1965a). *Ann Med. exp. Biol. Fenn.* 43, 155–158.
Renkonen, K. O. (1965b). *Lancet* i, 608.
Renkonen, K. O. (1966). *Acta genet. Statist. med.* 14, 276–282.
Renkonen, K. O. and Donner, M. (1967). *Ann Med. exp. Biol. Fenn.* (in press).
Renkonen, K. O., Mäkelä, O. and Lehtovaara, R. (1961). *Ann Med. exp. Biol. Fenn.* 39, 173–184.
Shaw, M. W. and Gershowitz, H. (1962). *Am. J. hum. Genet.* 14, 317–334.
Takahashi, E. (1954). *Ann. N.Y. Acad. Sci.* 57, 531–550.
Wyshak, G. and White, C. (1965). *Ann. hum. Genet.* 28, 357–359, and a private letter.

The Neoplastic Response: Some Clinical Evidence for the View that Tumour Formation Commonly Results from Disturbances of Normal Tissue-Regulating Mechanisms

D. W. SMITHERS

Radiotherapy Department,
Royal Marsden Hospital and Institute of Cancer Research,
Royal Cancer Hospital, London, England

WE seem to have been passing through a period where appreciation of the importance for experimental tumour research of continual reference to clinical medicine was in decline. Fortunately there are signs of a return to the old regard. If cancer research really is the study of the formation and behaviour of human tumours and not some esoteric discipline devised for an elect, then clinical medicine is its foundation, purpose, inspiration, reason for existence, and final arbiter of relevance. A cancer research worker following an abstraction of his own choice owes no duty to clinicians, but he has to remember that if an understanding of human tumours is his aim, then the responsibility for establishing the relevance of the work he does rests with him. We in clinical medicine should invite him from time to time to look at the facts revealed by the observation of human tumours and should expect him to be concerned that the views he holds about neoplastic disease are compatible with those facts.

Many of the clinical observations which have to be accounted for have been on record for generations, most have been well known during the whole of our lifetimes. They are now better arranged, authenticated and documented then ever before, and there are some relatively recent additions, mostly in the form of re-discovery or the seeing of new significance in old observations (such as the hormone dependence of some tumours of target tissues or the maturation of some tumours of embryonal rests) which have helped to clarify the position.

The clinical picture has been on view long enough for there to be no reason for lack of familiarity. What has to be remembered, however, is that if a man spends most of his working life in a laboratory doing experiments involving, let us say, single-cell tumour transplants, his view of the cancer problem is likely to differ from that of one who spends most of his working time in a consulting room observing the many and varied clinical manifestations of each of such different conditions, for example, as acute leukaemia, neuroblastoma, multiple myeloma, teratoma, chorionepithelioma, adenocarcinoma of the kidney, squamous-cell carcinoma of the vocal cord, cerebral glioma, osteogenic sarcoma, or Hodgkin's disease.

My purpose is to show some examples of observed behaviour patterns in neoplastic disorders in man and to make certain suggestions about their implications. These observations are compatible with the view that most tumour development and behaviour is related to a breakdown in the particular normal mechanisms of control of cellular growth operating in each of the tissues affected.

I. Frequency of Tumour Development

The range of human tumour type seen in clinical practice is as wide as the range of cells in the body capable of division, and the range of tumour behaviour observed must be almost as wide as the complete range of tissue potential for growth and differentiation. Every tissue capable of cell division can produce an invasive tumour, but for most of them this is a rare event. The common tumours which create what is known as "the cancer problem" develop at few sites.

TABLE I. South Metropolitan Cancer Registry, 1960–62 Relative incidence of commonest malignant tumours by site and sex, based on 38,697 males and 38,809 females

Males	%	Females	%
1. Lung	30·5	1. Breast	24·9
2. Digestive Tract	22·3	2. Digestive Tract	21·4
		3. Uterus	10·8
	52·8		57·1
3. Skin	11·7		
4. Urinary Tract	7·1	4. Skin	9·6
5. Prostate	7·1	5. Lung	5·8
	78·7		72·5

The records of our Cancer Registry in the south of England (Table I) show that, in this region, over half the tumours serious enough to endanger life in men arise from only two sites, the linings of the bronchi and of the digestive tract; the skin, the urinary tract, and the prostate follow as the next most common sites of tumour development. Over 75% of all tumours in males arise at only five sites. The exposed surfaces of the lung, the digestive tract, the skin and the urinary tract bear the brunt of an inimical environment; in them the demands for repair are most frequent and multifocal tumour development is most common. In women, in our region, more than half the malignant tumours arise from only three sites, the breast, the digestive tract, and the uterus; tumours of the breast and tumours of the digestive tract are twice as common as tumours at any other site. In women, the skin is fourth in order of frequency of involvement, and the lung fifth, reflecting the diminished risk of damage to these surfaces in comparison with men. The cervix uteri provides an excellent example of a site at which repeated reaction to the environment leads to tumour formation, since carcinoma of the cervix is virtually unknown in the celibate, and has been found to increase with promiscuity (Boyd and Doll 1964, Elliott 1964). Carcinoma of the breast is a hundred times more common in women than in men where there is no cyclical demand for mammary growth and function; breast-feeding, however, provides women with some protection against these tumours. It seems that lack of demand or function fulfilled both incur less tumour liability than fruitless response.

The rise of tumour incidence with age in industrial societies is chiefly due to length of exposure to a harsh environment; it is this which determines the predominance of tumour development at a limited number of sites. In communities living under conditions different from ours, where other hazards to life are more serious, where a high proportion of the population is in younger age groups, where there are clean air and natural foods, and where skin pigment is a protection against the sun, the tumour pattern differs markedly from ours. It is not only much more varied but, unlike ours, may show a fall in tumour incidence in older age groups (Davies *et al.* 1962). Transferred to city life, such people tend to acquire a similar type of tumour pattern to our own.

Tumour morbidity records show that neoplastic disease occurs most frequently at sites where there is the greatest demand for repair, that the longer such repeated demands are made the more likely are tumours to develop, and that reduction in such demand reduces the tumour risk. Further, that tumour development bears no relation to tissue bulk, or to the total number of cells at risk, but does bear a relation to the degree of exposure to a damaging external environment or to a regularly repeated—and especially a fruitless—internal demand for function. Recurrent rather than continuous demand for normal growth is by far the commonest precursor of a disorganized growth pattern in man.

II. Multiplicity of Tumour Origin

Multiple tumours of different types occur in the same patient but no more frequently than chance would allow. Multiple tumours of one tissue, organ, or pair of organs are, however, commonly seen. Bilateral tumours of the breast and testicle, for example, are commonplace. Tumours of the gastro-intestinal tract are frequently multicentric. Simultaneous involvement of lymphnodes by lymphomas at widely separated sites has become even more obvious with the advent of lymphangiography. Multiple tumours of the renal tract present one of its greatest therapeutic problems. The force of multicentric origin can best be seen in the skin; a fair-skinned person living at a low latitude may have many primary tumours. Even in England I have seen more than one patient with over 20 basal-cell carcinomas of the skin, and dozens of intra-epidermal tumours in a woman who was given arsenic in childhood.

III. Flexibility in Tumour Behaviour

Tumours change in structure and behaviour during their life cycle. In the early stages of some human tumour development, the process is easily reversed; other tumours become static, encapsulated, harmless swellings; many progress in disordered growth; a few regress; a number show irregular differentiation with occasional marked metaplasia; and even maturation to a post-mitotic state is seen in some. A few examples will give substance to this important quality of flexibility—or of a disposition in tumours to yield to persuasion.

A. *Early-stage reversal*

Table II shows 8 published cases of carcinoma of the bladder, all seen cystoscopically and confirmed microscopically, all of which regressed spontaneously after diversion of the urine. In all of them cystectomy was performed after ureteric transplantation; in three patients a few residual tumours cells infiltrating the bladder wall were all that could be found, while the other five patients were clear of disease.

B. *Limitation of growth and progression*

Many benign tumours grow to a certain size and then remain static.

Some benign tumours progress to malignant behaviour. Many malignant tumours become in the course of time more anaplastic and more aggressive in their behaviour. Figure 1 shows a giant fibroadenoma of the breast, in which the adenomatous part of the tumour progressed to adenocarcinomatous invasion and eventually to metastasis. A fibrosarcomatous change, or

TABLE II. Reported cases of spontaneous regression of carcinoma of the bladder after pre-cystectomy transplantation of ureters (Smithers 1962)

Case No.	Age of patient	Period of observation	Author
1	63	2 months	Pearse 1943
2	56	3 months	Davis 1948
3	60	14 months	Davis 1948
4	—	6 months	Trabucco 1948
5	55	3 weeks	Goldberg 1950
6	55	6 weeks	Fort, Harlin and Atkinson 1951
7	34	4 months	Creevy 1953
8	62	6 weeks	Creevy 1953

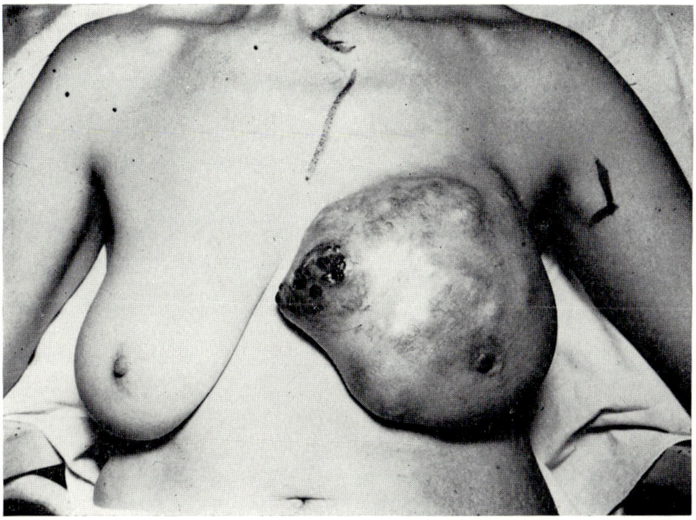

FIG. 1. M.P. (000779), age 48, had a tumour of the left breast which had been observed to be slowly and steadily growing for 8 years before treatment in 1945. The histological report was giant fibroadenoma with adenocarcinomatous change. She died in July 1952 with metastases in the cerebellum.

both fibrosarcoma and adenocarcinoma formation in the same tumour, can also occur. Progression both in the clinical behaviour of many tumours and in their histological structure is well known.

C. *Regression*

There are a number of different reasons other than treatment why a tumour may regress (Table III). Some tumours are conditional on the persistence of an exciting factor, others are hormone-dependent. Table III lists the most often reported cases of spontaneous regression of tumours. The point I want to make here is that regression without treatment does occur. Let me give you two examples in patients with osteogenic sarcomas. Doctors Webster, Phelan and Watne have kindly allowed me to quote a case of theirs, a boy of 16 with an osteogenic sarcoma of the femur and multiple metastases in his lungs. In August 1957 a thoracotomy was performed and one of the metastases was removed; microscopic examination confirmed it as an osteogenic sarcoma. In September 1957 the primary tumour in the femur was irradiated and its size reduced; the lung metastases regressed spontaneously at this time. In November 1960 the primary was enlarging again and amputation of the leg was performed in December. During 1961 the patient was clinically free of disease, but in May 1962 another lung lesion was seen and examination of biopsy material taken at a second thoracotomy again proved it to be a metastatic osteogenic sarcoma. Chemotherapy and irradiation were then tried, but without effect, and the boy died in May 1963. The untreated lung metastases had regressed for a period of four and a half years.

My second example of regression of metastases from an osteogenic sarcoma was a patient of Sir Stanford Cade's, a girl of 12 years of age, with a sarcoma of the mid-shaft of the right femur. The tumour was irradiated in February 1953 and lung metastases were demonstrated for the first time three months later, increasing in size during the course of another three months; then the lung metastases began to regress spontaneously, and by December 1953 none could be seen. The leg fractured at the site of the primary tumour in March 1954 and again 9 months later. Two years after radiotherapy, in March 1955, disarticulation was performed at the hip-joint. This patient was last seen in June of this year, still perfectly well; she is married and has a baby of 15 months. It is now 12 years since the lung metastases disappeared.

D. *Irregular differentiation*

Metaplasia is commonly seen in tumours; a carcinoma of the bladder may, for example, show squamous metaplasia in one part and mucus-secreting adenocarcinoma in another.

E. *Maturation*

Occasionally a tumour may become less malignant. Known examples of this are mostly teratomas or embryonal sarcomas in children. I have published an example of a teratoma of the testicle in a man of 26 who had a right orchidectomy in 1955 with post-operative X-ray treatment, and a left upper lobectomy for a solitary metastasis in the lung in 1957; he is still alive and well 10 years after treatment to the primary and 8 years after removal of the metastasis. The primary was an anaplastic teratoma, the metastasis showed well-differentiated epidermal structure of uniformly benign appearance with abundant keratinization and numerous sebaceous glands (Smithers and Wallace 1962).

There has recently been some experimental support for this clinically observed maturation in embryonal tumours. Goldstein *et al.* (1964) have shown that immature neuroblasts obtained from children with neuroblastomas when cultured *in vitro* have differentiated into more mature neural elements within 5 months.

Clinical observations show that many tumours change their behaviour patterns during their life cycle, a few so drastically as to disappear altogether, others so as to pass from benign behaviour to the most malignant, and some from malignant behaviour to the most benign. Tumour behaviour in man is certainly not an invariable, one-directional, fixed performance, continuous from some presumed sudden starting point, conforming to a pre-determined pattern and beyond all control. A general tendency towards progressive escape from control, with varying degrees of flux in tumour growth, is in fact the usual picture presented.

IV. Influence of Isolation on Tumour Development

If tumour development is associated with failure of normal growth control mechanisms, then it should follow not only on breakdowns due to the stress of repeated demands made on such controls but also on breakdowns due to the weakening of the controls themselves. Bullough (1964) has wondered whether a raised mitotic rate in tumours may be related to a weakening of regulating mechanisms and in particular to a fall in the chalone concentration. The most obvious method of weakening control, however, would be the blocking of the mechanism responsible by the isolation of a cell group from its normal restraining influence. Evidence of tumour formation in isolated cell groups has been known for so long that Cohnheim put forward a mistaken theory in 1877 that all tumours must arise from embryonic cell-rests. Nevertheless, developmental vestiges do produce a fine crop of tumours: the paradental remains of the dental lamina develop cysts, odontomas and adamantinomas; the parapituitary remains of Rathke's pouch, the lateral cervical vestiges, the thyroglossal duct, and the lingual thyroid may all develop cysts and tumours; Meckel's diverticulum contributes its adenomas, lipomas, leiomyomas, and argentaffinomas; the urachus its cystic papillary adenocarcinomas; the notochordal remnants their chordomas; the cordal vestiges their postcoccygeal gliomas; the Wolffian vestiges their rare papillary adenomas and the Müllerian vestiges their vacuolated tumours in the epididymis. Experimentally, first Turner (1941), then the Oppenheimers and Stout (1948) with Danishefsky and Willhite (1959), Mohr and Nothdurft (1958), and Alexander and Horning (1959), all demonstrated tumour production by insertion into tissues of inert plastic barriers, ineffective in grid or powdered form. Willmer (1945) demonstrated the way in which isolated conditions of tissue culture favouring unorganized growth reduced cells to a limited number of types present in the early stages of development. Others have shown (Sanford *et al.* 1959) that normal cells grown in culture may produce tumours when re-implanted. Isolation—which is only a special case of change in environment—is a well-known initiator of physiological new growth; there is some evidence that it can also be an initiator of disorganized new growth as well.

V. Influence of Total Tumour Volume on Tumour Growth

It is interesting to speculate about the possibility that increasing total tumour volume, localized or disseminated, might at times influence tumour growth rate. If a breakdown in the autoregulation of growth plays a part in tumour formation, the system of communication which keeps some cells informed about the size of their community is likely to be to blame. It is not therefore very probable that any feed-back control through function of the total mass will operate efficiently in tumours and, if it does operate at all, it is likely to be difficult to detect. Progressive relaxation of control due to a diminution in availability, amount, or potency of diffusible tissue-specific mitotic inhibitors in a vicious circle of de-differentiation, might, however, provide occasional instances of partial control returning through an over-compensating increase in poorly-functioning volume. Tumour growth rate does indeed have a tendency to slow down as tumours become larger (Laird 1964). This is usually attributed to the debilitating effect which such enlargement has on the whole organism being reflected back on the tumour itself. It may also be due to the tumour outgrowing its blood supply, with resulting central necrosis. That these explanations do not cover all cases is certain however (Brennan *et al.* 1962).

Variable and in some cases clearly diminishing tumour growth rates occur. This is seen in Fig. 2 where curves are plotted of the growth rates of metastases in the lung of a young woman with a chondrosarcoma of the calcaneus treated by amputation in 1963; she is at present well and working, quite unaware that she has any lung metastases.

Experimentally, Schatten (1958) showed that amputation of the hind

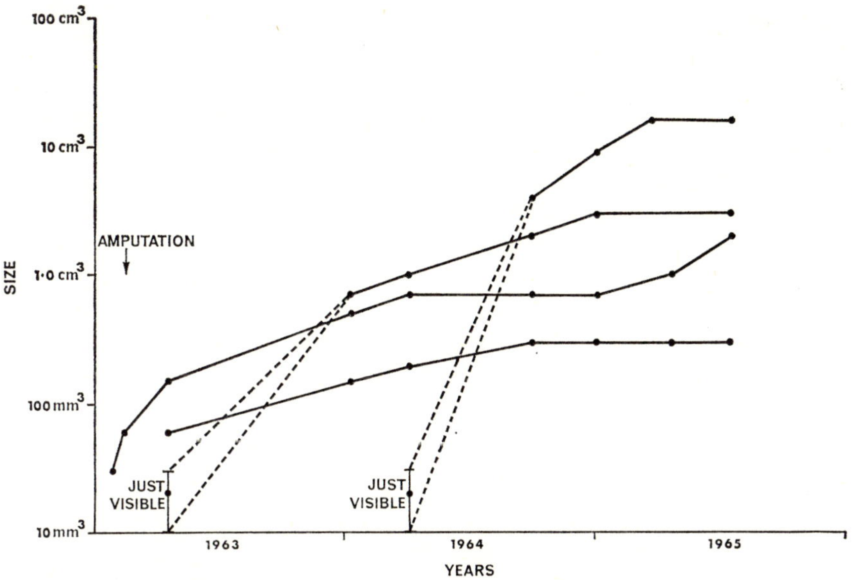

FIG. 2. M.E.B. (051655), a girl aged 22, had a chondrosarcoma of the right calcaneus and lung metastases. The varying growth rates of her lung metastases, some showing a tendency to deceleration after amputation of the limb, are seen in this chart.

TABLE III. Spontaneous regression of tumours

Regression most often reported in	Regression of Primary (P) or Metastases (M)	Possible reasons for regression
Embryonal Tumours	P+M	Maturation
Chorionepithelioma	P+M	Homograft reaction (hormone effect)
Adenocarcinoma of Kidney	M	Hormone effect
Malignant Melanoma	P+M	Hormone effect
Soft-tissue Sarcoma	P	Incomplete treatment restored balance
Carcinoma of Bladder	P	Conditional tumours, urine diverted
Carcinoma of Breast	P+M	Hormone dependent, menopause effect
Carcinoma of Ovary	M	Metastases dependent
Carcinoma of Skin	P	Early stage of imbalance visible

limbs of mice bearing implanted tumours, if done after metastases had appeared in the lungs, might lead to more rapid growth of these metastases. Goodman (1957) has shown that implanted mammary tumour growth rate in mice may be decelerated when more than one tumour is present and that surgical removal of one tumour may accelerate growth in another.

The relation of primary tumours to the growth of their secondary deposits, and the influence of total tumour volume on primaries, secondaries, or both, is a complicated one. For example, there have been a number of reports of regression after nephrectomy of lung metastases from primary carcinomas of the kidney in man (Smithers 1962, Riches 1964).

My point is that we are in no position to dismiss the influence of total volume on tumour growth rate as unimportant while there are so many suggestive observations still to be explained.

VI. Gradualness of Tumour Change

Tumour development is usually slow, tumour doubling times being often less than the turnover times of normal tissues. Gradual change can be observed taking place continuously in tumour structure, function, and behaviour. A continuous gradation of intracellular change as malignancy increases has been shown by electron microscopy (Battifora *et al.* 1964). Change in tumour behaviour may be in a direction either further away from control or towards its re-imposition.

Essentially, the clinical picture of tumour growth is a variable one of gradual change in normal mechanisms. Tumour growth as observed in man bears little resemblance to the popular view of cancer and even seems to be at variance with the concept current in some research departments where emphasis is so often placed on what is called "malignant transformation". If by this term is meant a gradual series of changes in several aspects of tissue behaviour, no harm is done, because the terminal stages of many examples of growth disorganization are much alike and the behaviour of the cell groups involved is indeed transformed. If, however, this term encourages a belief in a cancer-specific intracellular change which transforms a normal cell once for all into a malignant cell and directs work towards such a discovery, it is quite another matter. To the clinician, the transformation concerned is usually to a slowly-acquired terminal phase of tissue behaviour, but to the research worker the transformation may seem to spring from one sudden, irreversible, intracellular event.

One general explanation of the origin and behaviour of tumours is that they usually follow interference with a normal tissue control mechanism, partial or complete, reversible or irreversible, being induced by many factors, especially those leading to repeated demands for normal tissue responses, the weakening or failure of restraining mechanisms, or the isolation of cell groups from their normal restraints. Such an explanation is compatible with the facts observed in human tumour studies. But there are many ways of upsetting a delicate balance in a dynamic system, and a wide variety of tumour-inducing mechanisms is to be expected. A few of these may indeed be the result of the much sought-after single intracellular change producing a wild clone of cells which rapidly overwhelms the system concerned, but clinical observation suggests that a far more prosaic and graded process of disorganization is usually responsible for tumour formation.

Acknowledgments

I should like to record my thanks to Mr. P. M. Payne, Director of the South Metropolitan Cancer Registry, Belmont, Surrey, for the figures shown in Table I, and also to Sir Stanford Cade and to Dr. J. H. Webster, Dr. J. T. Phelan, and Dr. A. L. Watne for allowing me to report histories of their patients.

REFERENCES

Alexander, P. and Horning, E. S. (1959). *In* "Ciba Foundation Symposium on Carcinogenesis: Mechanisms of Action" (G. E. W. Wolstenholme and M. O'Connor eds.), pp. 12–25. J. & A. Churchill, London.

Battifora, H. A., Eisenstein, R., Sky-Peck, H. H. and McDonald, J. H. (1964). *Trans. Am. Ass. genito-urin. Surg.* 56, 31–37.

Brennan, M. J., Prychodko, W. and Horeglad, S. (1962). *In* "Biological Interactions in Normal and Neoplastic Growth" Henry Ford Hospital International Symposium (M. J. Brennan and W. L. Simpson eds.), pp. 739–748. Little, Brown, Boston.

Boyd, J. T. and Doll, R. (1964). *Br. J. Cancer* 18, 419–434.

Bullough, W. S. (1964). *In* "Cellular Control Mechanisms and Cancer" (P. Emmelot and O. Mühlbock eds.), pp. 123–145. Elsevier Publishing, Amsterdam.

Davies, J. N. P., Wilson, B. A. and Knowelden, J. (1962). *Lancet* 2, 328–330.

Elliott, R. I. K. (1964). *Lancet* 1, 231–235.

Goldstein, M. N., Burdman, J. A. and Journey, L. J. (1964). *J. natn. Cancer Inst.* 32, 165–199.

Goodman, G. J. (1957). *Proc. Am. Ass. Cancer Res.* 2, 207.

Laird, A. K. (1964). *Br. J. Cancer* 18, 490–502.

Mohr, H. J. and Nothdurft, H. (1958). *Klin. Wschr.* 36, 493–494.

Oppenheimer, B. S., Oppenheimer, E. T. and Stout, A. P. (1948). *Proc. Soc. exp. Biol. Med.* 67, 33–34.

Oppenheimer, B. S., Oppenheimer, E. T., Stout, A. P., Danishefsky, I. and Willhite, M. (1959). *Acta Unio int. Cancr.* 15, 659–663.

Riches, E. (1964). *In* "Tumours of the Kidney and Ureter" (E. Riches ed.), pp. 128–129. E. & S. Livingstone, Edinburgh and London.

Sanford, K. K., Merwin, R. M., Hobbs, G. L., Young, J. M. and Earle, R. R. (1959). *J. natn. Cancer Inst.* 23, 1035–1059.

Schatten, W. E. (1958). *Cancer* **11**, 455–459.
Smithers, D. W. (1962). *Clin. Radiol.* **13**, 132–137.
Smithers, D. W. and Wallace, E. N. K. (1962). *Br. J. Radiol.* **34**, 422–435.
Turner, F. C. (1941). *J. natn. Cancer Inst.* **2**, 81–83.
Webster, J. H., Phelan, J. T. and Watne, A. L. (1965). Personal communication.
Willmer, E. N. (1945). *In* "Essays on Growth and Form presented to D'Arcy Wentworth Thompson" (W. E. LeGros Clark and P. B. Medawar eds.), p. 284. Clarendon Press, Oxford.

Clinical Studies on the Growth Rate of Human Tumours

L. R. HOLSTI, M. W. BRENNER, P. HOLSTI
and Y. PERTTALA

Radiotherapy Clinic, University Central Hospital, Helsinki, Finland

I. Introduction

SINCE the introduction of the concept of exponential growth pattern for tumours by Collins *et al.* (1956), there has been several justified arguments against its validity. Unfortunately measurements on human tumours are still scarce. This is true especially if tumours and metastases of the lungs are excepted. One can assume that the growth of a tumour is restricted by the geometric boundaries of the available volume of the host tissue. Inhomogenities and anisotropy as well, are expected to influence the tumour growth. So is the shape and size of, for example, a primary tumour of the colon and rectum dependent on the host tissue (Welin *et al.* 1963), which restricts the direction and the rate of growth generally.

It is certainly true that some surroundings offer a higher degree of freedom for a tumour to grow than others. An environment, which does not imply any geometrical bounds with respect to the growth, would offer a suitable "experimental space" for tumour growth studies.

There are reasons to suppose that lung tissue would be unique from this point of view. It is fairly homogenous and provides good conditions for nutrition. The liver tissue as well may offer the same advantages, but the lungs facilitate a quantitative study because of the opportunity to perform exact radiometry. The determination of the size of liver metastases is not as straightforward.

II. Exponentiality of Growth

There are evidences for exponential growth of tumours and metastases of the lung. The series of measurements on lung metastases published by Collins *et al.* (1956) shows a beautiful exponential growth pattern. Spratt and Spratt (1964) found the tumour growth to be predominantly exponential in a material of pulmonary metastases in 118 patients. Schwartz's (1961) results show that deviations from exponentiality are rare for metastases and primary tumours of the lung, even when the tumours have obtained large proportions. Our team has shown exponentiality, within the limits of the error, for primary tumours of the lung and metastases of different origin, among them especially hypernephromas (Brenner *et al.*) (Fig. 1).

On the contrary, the study of experimental tumours of animals yield convex upward growth curves on a semi-logarithmic plot (Mottram 1935). The animal tumours studied are superficial compared with the human lung tumours. The technique of acquiring circular tumours of uniform growth

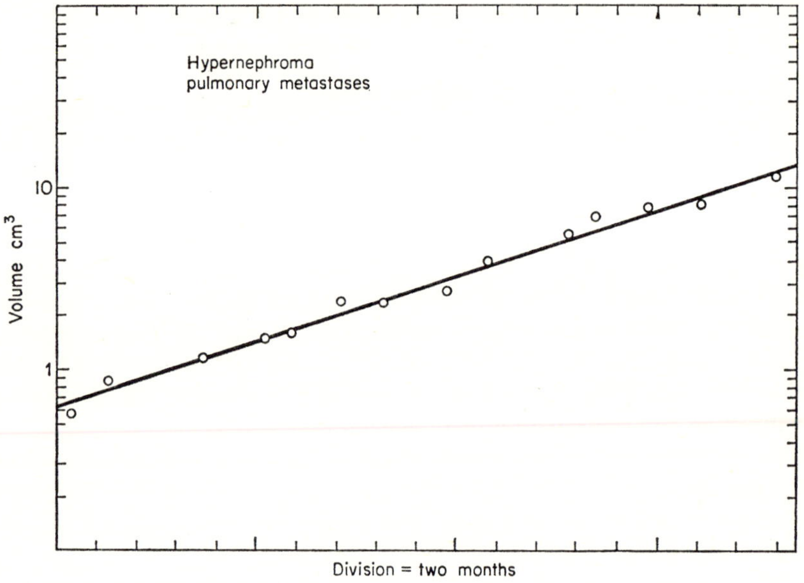

FIG. 1. Exponential growth of a pulmonary metastases of hypernephroma in a semi-logarithmic plot. Fourteen measurements during an observation period of 36 months.

and easily measurable has not been applied in all investigations. Many lung metastases, however, are quite spherical or the structure of them can be divided into spherical nodules (Breur 1965).

As a result of the growth of experimental tumours, as observed in animals, theories tend to find a mathematical expression, which implies a convex growth pattern on a semi-logarithmic graph. A Gompertzian function, which in principle is an exponential function with a time dependent exponent, has been applied by Laird (1964, 1965). A growth fraction, which is proportional to the b : th power of the tumour volume has been applied by Mendelsohn (1963). For b = 1 the growth is pure exponential and for b = 0 linear.

If pure exponential growth is a fundamental growth pattern, which still is an open alternative, then the search for complicated mathematical patterns is misleading. The tendency to use non-exponential approaches to the problem, may mean that disturbing factors, such as the influence of partial necroses are incorporated in a theory aimed at a description of tissue, which is unafflicted by necroses.

Exponential growth should be looked for (a) in a homogenous environment such as lung- and liver-tissue. (b) In cases where necroses do not affect the growth pattern. (If the rate of necrotic to proliferative tissue remains constant during the growth, exponentiality is still expected.) (c) When host factors, such as feed-back and immunity remain constant.

III. What can be Studied from the Growth Curves?

The study of tumours and metastases, which meet the above-mentioned conditions partially or fully, will have relevance to the following problems and concepts. (1) The determination of the volume or cell population doubling time by means of a semi-logarithmic plot. This time is well defined only for exponentially growing tumours. (2) The doubling time together with some kind of mitotic index or data regarding cell cycle time, will give information on the growth of cell populations and especially the ratio of proliferative and non-proliferative cells. (3) By extrapolation backwards it is possible to estimate the time of induction of the malignancy. This hypothetical possibility is useful as a basis for discussions in clinical practice, although it may be misleading, if approached uncritically. It must be pointed out, that our observations on the exponential growth rate of tumours are restricted to the clinical period of growth. We do not know anything about the growth pattern during the silent clinical period of human tumours. (4) If the conditions (a) and (b) above are met, deviations from exponentiality or deviations from linearity in a semi-logarithmic plot, may be considered an indication of the effect of a host factor.

The material of untreated primary lung cancers and lung metastases studied by us did not show any deviations from the exponential law of growth except at a later phase, when a levelling off sometimes follows. Among 38 patients we have noted only one case with a clear effect of a host factor. It was a case with hypernephroma and the primary tumour was surgically removed. After a slight increase or a period of constant volume, all of the four pulmonary metastases disappeared completely within six months (Fig. 2).

The determination of the growth rate or volume doubling time of human tumours is important not only from the point of view of studying the growth rates of different tumours, but also from the viewpoint of radiotherapy and chemotherapy. It seems reasonable to regard the growth rate of a tumour as a factor that must be taken into consideration in the treatment besides the localization, spread, differentiation and oxygen condition. Several human tumours grow quite slowly, having a doubling time of several weeks or months. The renewal time of many proliferating normal tissues is on the other hand only a few days. Rapidly proliferating tissues repopulate quickly

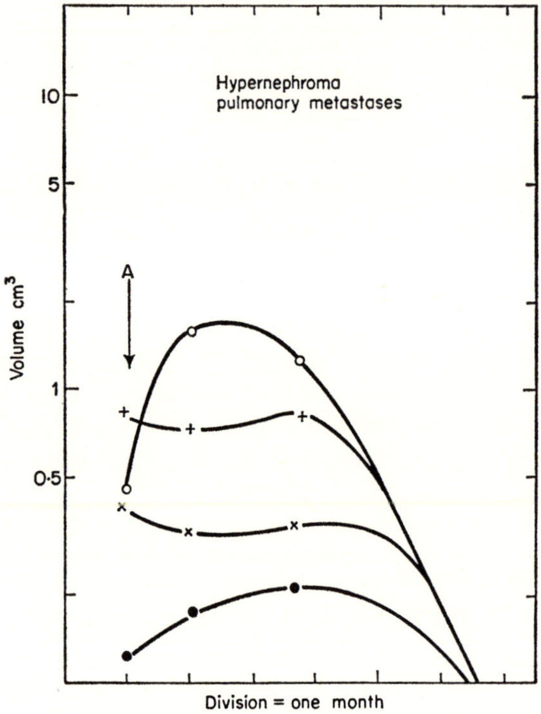

FIG. 2. Spontaneous regression of four pulmonary metastases in a case of hypernephroma after surgical removal of the primary tumour, indicating an effect of a host factor.

after radiation damage, compared with slow-growing tissues. A sufficiently long interval must therefore be maintained between the fractions of radiotherapy, and which allows the normal tissues time to recovery. The rationale of modern fractionization regimes is to take advantage of any differences in cell kinetics between normal and malignant tissues. The effect of different fractionization methods on tumours can be studied from the growth curves.

It is easy to follow the effect of radiotherapy and chemotherapy on the size of pulmonary tumours from roentgenograms and growth curves. An interesting study is the growth rate after palliative treatment. There are indications that, after reduction in the size of a metastases, the growth rate resumes at the same constant rate as before treatment (Breur 1965). There are no indications as yet showing a change in growth rate following treatment.

From the reduction in volume, the percentage of surviving cells for the given dose can be calculated. The surviving fraction from which the tumour can regrow can also be calculated. Such studies may be helpful for clinical radiotherapy in the same sense as experimental cell survival curve studies.

IV. Summary

Lung tissue forms an exceptionally good environment for tumour growth studies. It does not imply any geometrical bounds with respect to growth, it is fairly homogenous and provides good conditions for nutrition.

There are evidences for exponential growth of tumours and metastases of the lung. The problems and concepts to which the study of exponentially growing tumours will have relevance, are discussed. The role of such studies for clinical radiotherapy is indicated.

REFERENCES

Brenner, M. W., Holsti, L. R. and Perttala, Y. (1967). *Br. J. Cancer*. (In press.)

Breur, K. (1965). Growth Rate and Radiosensitivity of Human Tumours (in Dutch with English summary). Mouton, Haag.

Collins, V. P., Loeffler, R. K. and Tivey, H. (1956). *Am. J. Roentg.* 76, 988–1000.

Laird, A. K. (1964). *Br. J. Cancer* 18, 490–502.

Laird, A. K. (1965). *Br. J. Cancer* 19, 278–291.

Mendelsohn, M. L. (1963). *In* "Cell Proliferation" (L. F. Lamerton and R. J. M. Fry eds.), pp. 190–210. Blackwell, Oxford.

Mottram, J. C. (1935). *J. Path. Bact.* 40, 407–414.

Schwartz, M. (1961). *Cancer* 14, 1272–1294.

Spratt, J. S. and Spratt, T. L. (1964). *Ann. Surg.* 159, 161–171.

Welin, S., Youker, J. and Spratt, J. S. (1963). *Am. J. Roentg.* 90, 673–687.

Variation in the Response of Different Cells within a Tumour to Treatment with X-rays or Alkylating Substances

L. RÉVÉSZ, B. LITTBRAND and H. MODIG

Department of Tumour Biology, Karolinska Institute Medical School, Stockholm, Sweden

I. Introduction

In numerous studies during recent years, various lines of evidence have been obtained on the intercellular variation of different metabolic, nutritional, chromosomal, antigenic, etc. characteristics in tumours. The results are consistent with the view that neoplasms must be regarded as a genetically heterogeneous population of cells, in which interclonal variation represents a fundamental attribute, and evolution follows through the competition of various, individual population components (Harris 1964). Since many of the cellular properties which were found to vary, represent also factors of importance in determining sensitivity to X-rays or radiomimetic substances, it is conceivable that intercellular variation will also affect the extent of the damage which ensues when tumours are subjected to treatment with these agents. The observation on the differential radiosensitivity of selected sublines derived from experimental tumours is in line with this consideration (Glas and Révész 1963). At present, insufficient information is available concerning the inherent sensitivity variations in neoplastic cell populations and the sequential remodelling of the population structure which can be expected to occur on the basis of a heterogeneous response to treatment. Certain cell types which, under ordinary conditions, are present in small numbers, may have a selective value in connection with some particular treatment. Such a situation results conceivably in a shift in the balance between various, competing genotypes and this, in turn, can be of decisive importance with regard to the subsequent growth of the population. In the following, I shall report on the series of experiments which were performed in an extension of some of our previous studies (Révész, Glas, Hilding 1963, Révész, Bergstrand, Modig 1963, Révész and Littbrand 1964) with the purpose to gain information on the differential cellular response in tumours to treatment, and to investigate the factors which may govern it. Increased knowledge in this respect may contribute to a better understanding of the different interacting mechanisms which determine the outcome of various therapeutic measures.

II. Clonal Sublines of a Serially Irradiated Neoplastic Cell Population

In a first approach to the problem, experiments were conducted which aimed at an investigation of the cellular composition in a tumour line, subjected to serial irradiation by sublethal X-ray doses. Considering the mutagenic effect of X-rays, heterogeneity in the composition of the cellular population could be expected to be greatly increased in this line. By isolation of single cells, a number of clones were established, and studied with respect to radiosensitivity, certain chromosomal characteristics and concentration of free and protein-bound sulphydryl groups. The study of the chromosomal pattern and sulphydryl content was preferred due to the radiobiological significance of these properties, and the particular suitability of chromosomes as selective markers. It was thought that differences between the clones with regard to these characteristics would, to a great extent, reflect the intercellular variation in the parental population.

The parental population, denoted $SELD/_{23}$, was developed by fortnightly, intraperitoneal transfers in mice of 2×10^6 ELD ascites cells, irradiated before each passage serially with 2000 R *in vitro* under anoxic conditions. From a highly diluted suspension of this tumour, single cells were isolated each of which was subsequently injected intraperitoneally into a one-day-old mouse in order to permit proliferation in the absence of host immunological reactions. A series of clonal populations developed in about 20% of the inoculated animals after a period between 3 and 5 weeks. None of the clones, used in this study, was subsequently carried through more than 10 serial, intraperitoneal transfers, the accumulation of animal passages having been avoided by frequent storing of the material in a frozen tumour bank.

III. Variation in Chromosome Pattern

Chromosomes were studied in aceto-orcein squash preparations after treatment with colchicin. Figure 1 shows the frequency of cells with the six most commonly occurring chromosome numbers, as present in the original ELD, its serially irradiated line $SELD/_{23}$, in 19 clones (clone a to t) derived from the latter, and in two SELD populations ($SELD/_{23/32}$ and $SELD/_{23/42}$) which were transplanted serially for 32 and 42 transfer generations, respectively, without further irradiation. In $SELD/_{23}$, re-arrangement and increased heterogeneity of the population, as well as an alteration in the modal chromosome number from 46 to 45 can be noted, in comparison to ELD. The selective value of cells with 45 chromosomes after serial irradiation is apparent also in the prevalence of this cell type in SELD/$_{23/32}$ and SELD/$_{23/42}$, as well as in 13 of the 19 clones. As indicated by the composition of the population in the 6 other clones (clone a, k, l, m, n and t) in which cells either with chromosome numbers 44, 46, 48 or more than 70 are the principal contributors to the growth, some heteroploid components of SELD/$_{23}$ other than its modal type are also capable of continued proliferation and producing cell lineages of their own.

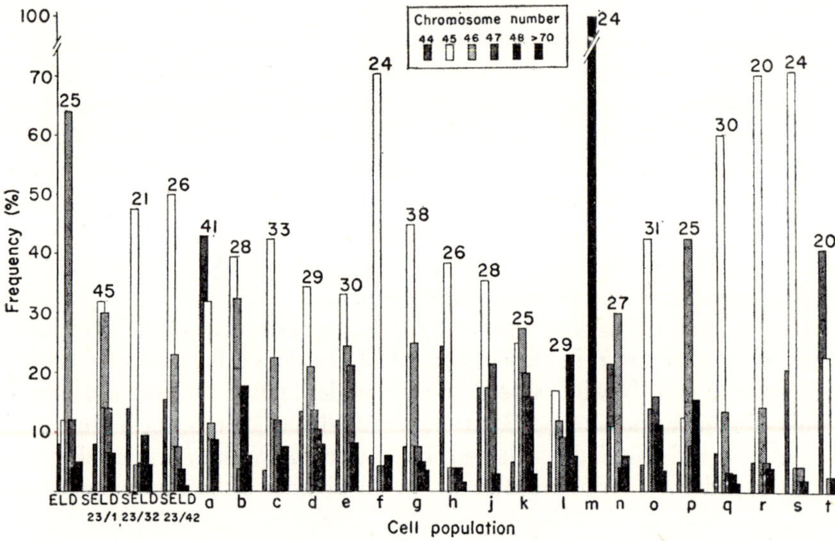

FIG. 1. Frequency distribution of cells with six different chromosome numbers in populations of ELD, three different transplant generations of its serially irradiated SELD subline, and in nineteen clones, derived from the first generation of the latter. The figures above columns indicate the number of cells in which exact chromosome counts were made.

Rapidly developing heterogeneity shortly after cloning is a striking characteristic of all the populations. This heterogeneity is of a variable degree since, as can be noted, the modal cell type accounts for more than 70% of the population in some of the clones, but less than 30% in some other. Cells with a spread of the chromosome numbers around the double modal value occurred in all cases indicating that endo-reduplication (Levan and Hauschka 1953) is a regularly occurring process and a constant source of the chromosomal variation. With a single exception (clone m), the frequency of cells with double modal chromosome number was not more than a few per cent suggesting that this cell type has a proliferative disadvantage in comparison with others. This phenomenon can be explained by the fact that cellular mass rather than cell number is the factor which limits proliferation in a mixed population of cells which have different ploidy and, consequently, different volumes (Hauschka *et al.* 1957). According to this consideration, clone *m*, deficient entirely in components with low chromosome numbers, must have originated from a single polyploid cell. Since the frequency of such cells is about 5% in the parental population, the occurrence of one polyploid clone among the 19 investigated is in agreement with the statistical expectation of the chance selection of one polyploid cell in connection with the random cloning procedure used.

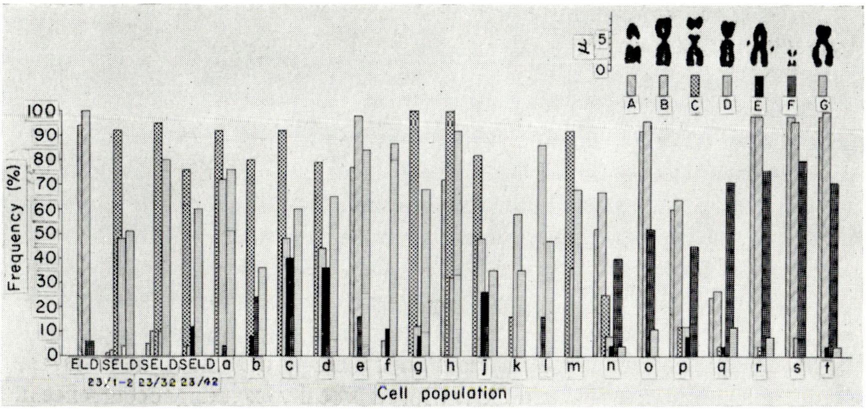

FIG. 2. Frequency distribution of five marker chromosomes (A–G) in the populations presented in Fig. 1.

Many chromosomes showed distinct alterations from the morphology observed in normal mouse cells. Among the structurally changed chromosomes which occurred most commonly, seven "marker" types were chosen for an analysis of their frequency distribution in the different populations. Three of the markers were present in the ELD and SELD/$_{23}$ karyotype and, in various combinations, also in the clones. These included a subtelocentric chromosome with secondary constrictions on both chromatides (A), a large metacentric (B) and a minute metacentric (F). The four other types included a metacentric chromosome with secondary constrictions on both chromatides (C), a submetacentric (D), one with ring-formation (E) and one subtelocentric with incomplete constrictions of the chromatides (G). These markers were present, in various combinations, in all populations except ELD. Considering the remarkable stability of the ELD karyotype through more than a decennium (Bayreuther 1952), the markers C, D, E and G can be considered as representing re-arrangements induced by radiation and not any unspecific variant formation.

Figure 2 shows the frequency distribution of the 5 markers as established by scoring 50 to 100 well spread metaphase plates in 2 to 5 separate preparations made of the respective lines. A salient feature is that each one of the populations is characterized by a particular distribution of various marker combinations. This is in a general agreement with observations made by Ruddle (Ruddle 1961) who showed that cell lines cultured *in vitro* can be uniformly labelled with unique marker chromosomes produced by exposure to X-radiation followed by clonal isolation of survivors. It is remarkable that, although not more than 3 of any of the specific markers were found together in a metaphase cell, nevertheless a combination of more than this number occurred in the population of several clones. This finding can be explained by assuming that the cloned cell carried a certain information which controlled the development of a particular chromosomal variation in subsequent generations. Further analysis of the data is in progress to investigate the feasibility of this supposition by attempting to establish the possibility that they represent closely similar but different accidental variations. Preliminary results of such studies concerning the B, C and E markers indicate that there does not exist any significant difference ($p > 0{\cdot}05$) in the relationship between the length of some particular chromatid portions of these markers, measured in the different clones. This strongly suggests that B, C and E markers are, indeed, of the same type in all cases. Some other peculiarities of the marker distribution, such as the regularly combined incident of A and B markers, the frequent combination of C and G, or the reciprocally excluding occurrence in the same clone of the A-B pair together with the C marker, are also subject to continued studies, and can be expected to elucidate the origin of these specific chromosomal variations.

IV. Radiosensitivity Variations

The radiosensitivity of the clones was estimated according to various criteria. In the tests *in vivo*, inhibition of population growth was used as an index. Irradiated cells were grown intraperitoneally in the mouse within Millipore diffusion chambers with an average pore size of $0{\cdot}45\ \mu$, and the cell number was determined at various intervals. This technique, described in detail previously (Glas and Révész 1963), permits examination of the proliferation of the graft in the absence of various cell-mediated host influences. Figure 3 shows the chamber growth of 8×10^6 cells from ELD, $SELD/_{23}$ and 11 of the clones after treatment with 2000 R *in vitro* in a medium devoid of oxygen due to the cellular metabolism. It can be seen that $SELD/_{23}$ and 8 of the clones (c, d, e, g, j, k, l, and m) reached a population size 3 to 4 times larger than the inoculum within about 25 days, while the growth of ELD and 3 other clones (a, b and h) was completely inhibited. If not irradiated, no major differences were found in the growth of these lines.

One of the more (h) and two of the less sensitive (g and l) clones were chosen together with the parental line for a study of the response to radiation with 450 or 300 R in an oxygenated medium. As has been reported previously, (Révész, Glas, Hilding 1963) only a slight difference was found in the growth pattern of the different lines after such treatment, and the size of their chamber population was of a similar magnitude during the whole growth period lasting for 20 days.

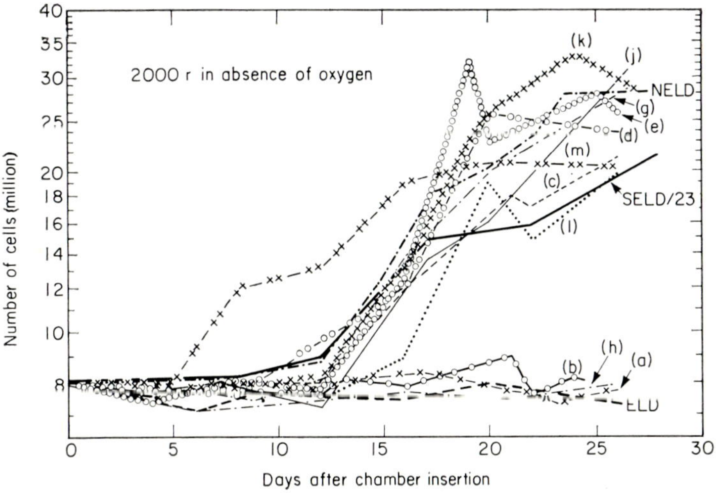

FIG. 3. Growth in Millipore chamber, inserted intraperitoneally in the mouse, of ELD, SELD eleven clones derived from the latter, and of NELD, derived from ELD by sublethal treatment with nitrogen mustard. The cells were irradiated with 2000 r *in vitro* in an oxygen-free medium before implantation.

One clone (a) which was sensitive, and two other (f and g) which were resistant in the *in vivo* tests, were selected, together with ELD, for essays using growth inhibition *in vitro* as the criterion of the radiation effect. A series of *in vitro* cultures of the respective lines, grown for 4 days after the explantation of 10^4 cells, were irradiated with either 450 R, delivered in an oxygen atmosphere, or with 1100 R in argon containing an oxygen impurity of less than 3 p.p.m. A description of the methods used has been published previously (Révész and Littbrand 1964). Following irradiation, the cultures were re-incubated for another period of 3, 7 or 10 days after which the population size was determined by bringing the cells into suspension with trypsin and counting their number with an electronic particle counter. Radiation induced growth inhibition was expressed as the ratio between the size of the irradiated population and the non-irradiated controls in the same experiment. The results presented in Fig. 4 indicate that, after the anoxic treatment, the growth of ELD and clone a was inhibited to a significantly larger degree (P varying between 0·01 and 0·05) than clone f or g. Conversely, after irradiation in oxygen, the growth of all four lines appears to be inhibited to a closely similar extent, with only statistically insignificant random variations in population sizes.

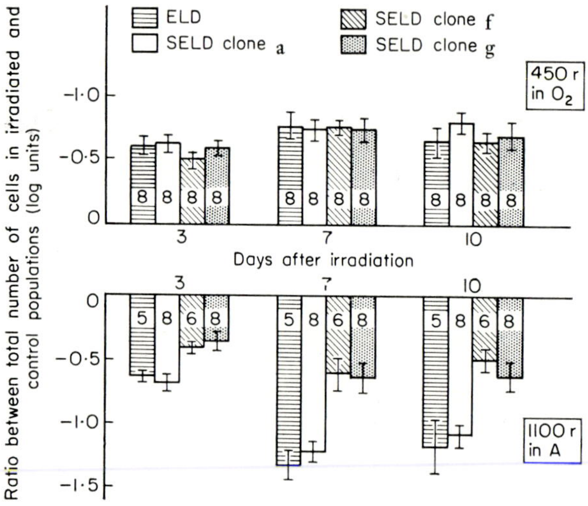

FIG. 4. Relationship between the size of irradiated and control cultures *in vitro* of ELD and three SELD clones after 3, 7 and 10 days of growth; subsequent to treatment with either 450 r in oxygenor 1100 r in argon. Columns illustrate the numerical value of the mean ratio of separate determinations, the number of which is indicated. For each of the means, the 95 % confidence interval is also shown.

Table I summarizes the pertaining data of some earlier (Révész and Littbrand 1964) and recent tests in which reproductive integrity, indicated by clonal survival of cells *in vitro*, was used as the index of sensitivity. When ELD, clones a and g were treated under oxygenated conditions, the value of D_0, indicating the dose which reduces the proportion of surviving cells by 1/e, differed only insignificantly from each other in the different lines. In contrast, the sensitivity of clone g can be calculated to be 24% less than ELD or clone *a* ($P<0·05$) when irradiation was made under anoxic conditions. The extrapolation number of the survival curves to 0 r was similar in all three lines, and had values close to 2 and 1 after oxygenated and anoxic treatments, respectively. The possible significance of these observations in indicating a qualitative, in addition to a quantitative, difference in the aerobic and anaerobic radiation injury had been discussed in other recent publications (Révész and Littbrand 1964, Littbrand and Révész 1964).

In summary, the results of different tests, applying different criteria in assessing the effect of radiation, are concordant in indicating that differences in radiosensitivity *do* exist between clonal sublines derived from a common, serially pre-irradiated ascites cell population. A mutation-selection process, effected by the serial irradiation, is a conceivable mechanism for the accumulation of radio-resistant cells in such a population. The sensitivity differences are, however, demonstrable only after irradiation under anoxic conditions,

TABLE I. Extrapolation number to (0 r) and the reciprocal of the slope constant (D_0) of the survival curve in vitro of ELD and two clones of SELD/23, after irradiation in oxygen or oxygen-free argon. Means are presented together with the 95 per cent confidence interval in parenthesis, calculated by statistical analysis of the survival data.

Cell line	D_0(r)		Extrapolation number	
	Treatment in oxygen	Treatment in argon	Treatment in oxygen	Treatment in argon
ELD	91 (83–101)	235 (222–251)	2·2 (1·5–3·4)	0·9 (0·7–1·3)
clone a	100 (95–106)	236 (226–248)	1·8 (1·5–2·1)	1·1 (0·9–1·5)
clone g	92 (87–98)	292 (275–310)	2·3 (1·8–3·0)	1·1 (0·9–1·4)

and are masked if exposure to radiation takes place in an atmosphere of oxygen. An oxygen effect resulting in a qualitatively different radiation damage, super-imposed upon the injury which develops in anoxia, has been considered to explain this phenomenon (Révész and Littbrand 1964). Another explanation for the absence of clones exhibiting differential sensitivities after treatment in oxygen can lie in the fact that exposure to serial irradiation was made, as a rule, under anaerobic conditions, and the selective pressure of this treatment could, therefore, act only on anoxic cells. It has been reported (Bhaskaran and Dittrich 1963) that when serial exposure was made routinely in an oxygen atmosphere, a cell line was developed which exhibited resistance to radiation also under aerobic conditions.

Considering the radio-sensitivity of the clones in relation to their chromosome pattern, it appears that several clones were identical in sensitivity although they varied in stem-line number. On the other hand, other clones varied in sensitivity to radiation, yet had identical modal chromosome number. This finding is consistent with our previous conclusion that the relation between chromosome number and radio-sensitivity may be wholly fortuitous, unless the chromosome sets compared are exact multiples of each other (Révész, Glas, Hilding 1963).

V. Variation in Sensitivity to Nitrogen Mustard

With the object of studying possible sensitivity variations to alkylating agents, in a series of preliminary experiments, the response of one radiation resistant (clone g) and one sensitive clone (clone a) was tested to treatment

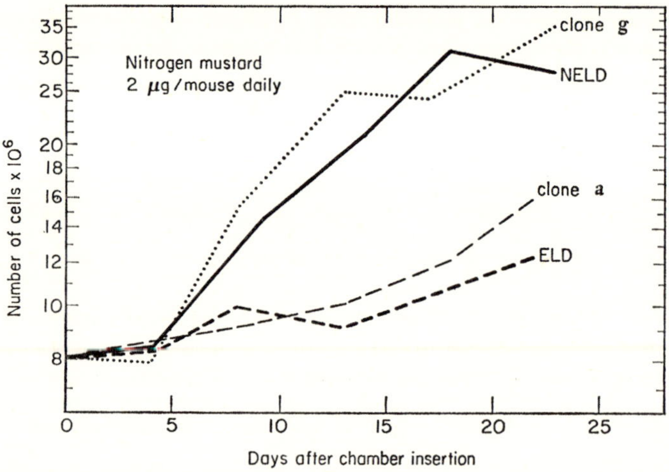

FIG. 5. Growth in Millipore chamber of ELD, its nitrogen mustard pretreated NELD subline, a radiation resistant (a) and a radiation sensitive (g) clone derived from SELD. Mice carrying the chambers were treated with 2 μg nitrogen mustard daily.

with nitrogen mustard (HN_2). Cells of these clones were grown in the Millipore chambers, referred to above, and implanted in the peritoneal cavity of groups of mice which were then treated by daily intraperitoneal injections of HN_2 in doses varying between 0·2 and 20 μg. The chamber growth of the clones in mice which received 2 μg HN_2 daily, are presented in Fig. 5. For comparison, the growth pattern of ELD exposed to a similar treatment is also illustrated together with another line, denoted NELD. This latter was derived by 24 serial transfers of ELD in mice treated with daily doses of 10 to 20 μg HN_2 and, in agreement with other observations made after the prolonged exposure of tumours to this agent (Wheeler 1963), it developed an increased resistance to the growth inhibitory effect of HN_2 treatment.

It can be seen that the growth of clone a was inhibited to an extent very similar to ELD. On the other hand, clone g exhibited a resistance which resembled NELD. When the growth of this latter line was studied after exposure to X-rays, a pattern similar to the radiation resistant clones was observed (Fig. 3). The cross resistance to X-radiation and HN_2, indicated by these results, is in agreement with findings with bacteria (Greenberg *et al.* 1961) and with another ascites tumour (Sakurai *et al.* 1960) but is at variance with observations reported with another tumour line (Levis 1963). That the occurrence of cross resistance may vary from case to case and that no generalization can be made is suggested also by the results of the experiments in which we used myleran: the growth of the radio-resistant clone g and j was, in this case, affected to an extent similar to the radiation and HN_2 sensitive ELD, used as a contol. This finding is consistent also with the proposal (Alexander and Mikulski 1961) that different initial chemical reactions may be responsible for the effect on growth of HN_2 and myleran.

VI. Variation in Cellular Sulphydryl Level

Considering the possible role of the intrinsic sulphydryl content in determining cellular sensitivity to radiation (Révész, Bergstrand, Modig 1963, Bacq 1965) and to nitrogen mustard (Calcutt and Connors 1963), chemical

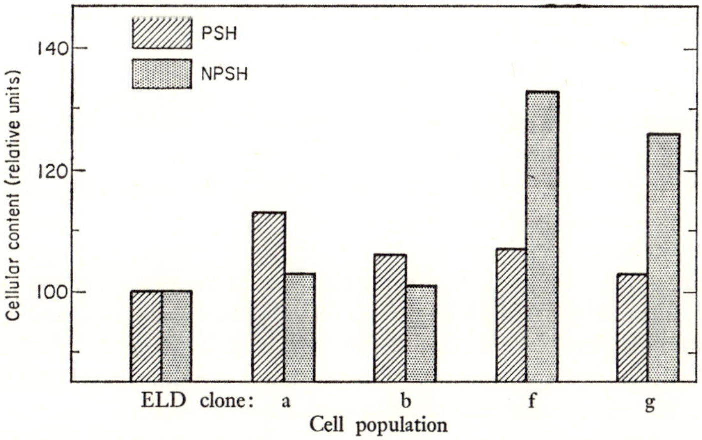

FIG. 6. Cellular content of protein-bound (P–SH) and non-protein bound (NP–SH) sulphydryl groups in two radiation resistant (f and g) and two radiation sensitive (a and b) clones, relative to ELD.

measurements were made of the protein bound (P—SH) and non-protein bound (NP—SH) concentration of sulphydryl groups, on the one hand, in clones which were found to be highly sensitive to radiation (clones a and b) or to HN_2 (clone a) and, on the other hand, in clones which showed an increased resistance to radiation (clones f and g), as well as to HN_2 (clone g). The relative cellular concentrations found by measurements using the techniques described previously (Révész, Bergstrand, Modig 1963) in combination with the method of Benesch *et al.* (1955) are presented in Fig. 6. Compared to ELD, P—SH was slightly increased in all cases. The NP—SH concentration was closely similar to ELD in the sensitive clones, but was significantly larger in the resistant ones. It can be noted that, in agreement with previous cytochemical measurements (Caspersson and Révész 1963), the 25% increase of NP—SH concentration relative to P—SH in clone *g* corresponded remarkably with the decrease of the radiation sensitivity of this line in relation to ELD (cf. the values of D_0 above in Table I).

The differing P—SH and NP—SH concentration in the clones suggest that there may exist a considerable variation in the metabolism of these compounds in the original population. Since increased NP—SH concentration relative to P—SH appears to be associated with resistance to both nitrogen mustard and X-rays, a variation of the sulphydryl level in neoplastic populations may obviously play a considerable role for the development of therapy resistant cell lines. In current clinical trials, biopsy material obtained from different tumours are being investigated with respect to the variation of sulphydryl content in order to elucidate this possibility.

VII. Selective Growth of Surviving, Resistant Cells

In a series of model experiments, we studied the possibility of a continued proliferation of cells which, due to their increased resistance to treatment, survive in a population, sensitive for the rest. In previous investigations (Révész 1956) in which a small number of viable neoplastic cells were randomly admixed with heavily irradiated (HR) cells of the same original tumour and then transplanted *jointly* to highly inbred, immuno-genetically compatible mice, we found that the viable part of the inoculum did not only survive, but was even enhanced in its growth by the lethally damaged portion. In cases when immunogenetic differences existed between the host and the transplant, the development of a host resistance was enhanced by HR cells, and tumour growth was eventually completely inhibited.

In an extension of these studies, in recent experiments viable tumour cells were inoculated in groups of highly inbred, genetically competent mice in a number which was, in itself, insufficient for the inoculation to result in a successful grafting. Suspensions of HR cells of the same tumour (10^7 in 0·5 ml medium) or sesame oil (0·5 ml) was deposited at the inoculation site *after* varying periods. Controls included, on the one hand, groups of mice which were inoculated with viable cells in the same subminimal number as the

TABLE II. Development of tumours from small inocula of spontaneous mammary carcinoma or methylcholantrene induced sarcoma cells in an immunogenetically competent host at a site treated with heavily irradiated cells or sesame oil at different periods after transplantation.

Tumour				No. tumour bearing mice/No. mice inoculated					
					Inoculation site treated with HR cells (H) or sesame oil (S)				
						Interval following inoculation			
Code	Mouse strain of origin	Recipient mouse genotype	No. cells inoculated	Untreated controls	Simultaneously with inoculation	2 weeks	3 weeks	4 weeks	6 weeks
First transplant generation spontaneous mammary carcinomas.									
S 99 D	DBA	(DBA × C3H)BC♂	10^4	4/8	H 4/4		H 4/4		
S 100 D	DBA	(DBA × C3H)BC♂	10^4	3/12	H 5/6				
S 100 D	DBA	(DBA × C3H)BC♂	10^4	0/4	S 3/4	S 2/4	S 3/4		S 1/4
S 103 D	DBA	(DBA × C3H)F_1♂	10^3	0/4	S 3/8	S 2/8			
			10^4	2/8	S 4/8	S 3/8			
S 104 D	DBA	(DBA × C3H)BC♂	10^3	0/6			H 2/4		H 3/6
			10^4	6/6			H 6/6		S 3/6
S 104 D	DBA	(DBA × C3H)BC♂	10^3	0/8		S 1/8		S 3/8	
			10^4	2/8		S 6/8		S 6/7	
S 104 D	DBA	(DBA × C3H)BC♂	10^4	2/5		S 6/6			S 6/6
S 100 A	A	(A × C3H)F_1♂	10^3	0/8				S 3/8	
			10^4	7/8				S 7/8	
S 101 A	A	(A × C3H)F_1♂	10^3	1/6		S 3/7		S 4/8	
			10^4	3/7		S 8/8		S 8/8	
First transplant generation methylcholanthrene induced sarcomas.									
MDC	DBA	(DBA × C3H)BC♂	10^3	0/11	S 3/6	S 0/6		S 1/6	S 0/6
			10^4	3/11	S 4/6	S 2/6		S 0/5	S 0/6
MAE	A	(A × C3H)F_1♂	10^3	0/6	0/6	S 0/6		S 0/6	
			10^4	2/6	3/6	S 1/6		S 2/6	

experimental groups but did not receive any treatment at the inoculation site; and, on the other hand, groups of mice which were inoculated with cells in a sufficient number for the progressive growth of the transplant. All inoculations were made subcutaneously in the interscapular region which was subsequently examined for tumour growth at weekly intervals during the life-time of the animal or at least during 8 months. A summary of the tests performed so far is presented in Table II. The results indicate that when cells from the first transplant generation of six different, spontaneously arisen mammary carcinomas were used, the subminimal inoculum was not destroyed in the respective hosts but could, in the majority of cases, remain in a latent state with maintained capacity for progressive growth upon the sesame oil or HR cell stimulus up to 4 to 6 weeks. In contrast, subminimal inocula from two methyl-cholantrene induced tumours, supposedly antigenic to the hosts (Hellström and Möller 1965) did not appear to have a similar capacity for latent survival. When the inocula consisted of large number of cells, a high incidence of tumour development occurred in all groups.

Population density in controlling cell proliferation in general, and tumour growth in particular, has been the subject of extensive studies (Harris 1964). The data in Table II suggest that sesame oil, conceivably by provoking some local, stimulatory host-reaction (e.g. an inflammatory reaction which may improve nutritional conditions) and HR cells, probably by a similar mechanism and/or by a feeder effect (Révész 1956) can compensate for the lack of a certain population density, necessary for progressive growth. This effect of sesame oil and HR cells may be comparable to the effect of substances which can promote the growth of cells with neoplastic alterations already induced by treatment with chemical carcinogens, but which are present in only a small number. In the absence of a promoting stimulus a small transplanted cell colony is still not destroyed but may survive in the host in a "dormant", non-proliferating state for some time. This is in keeping with the conclusions of Fisher and Fisher (1963) who found that a small inoculum could start multiplication weeks later upon delayed treatment with tissue homogenate, but is at variance with the results of Wallace (1965) who failed to demonstrate such an effect with HR cells.

Acknowledgments

The authors wish to acknowledge the valuable contributions to various phases of these investigations by Drs. H. Bergstrand, U. Glas and G. Hilding. This work was supported by grants from the Swedish Cancer Society, Swedish Atomic Research Council and Public Health Service Research Grant No. CA 06633–03S1 from the National Cancer Institute.

REFERENCES

Alexander, P. and Mikulski, Z. B. (1961). *Biochem. Pharmac.* **5**, 275–282.
Bacq, Z. M. (1965). "Chemical Protection against ionizing radiation." Charles C. Thomas, Springfield.
Benesch, R. E., Lardy, H. A. and Benesch, R. (1955). *J. biol. Chem.* **216**, 663–676.
Bhaskaran, S. and Dittrich, W. (1963). *Strahlentherapie* **122**, 270–278.
Calcutt, G. and Connors, T. A. (1963). *Biochem. Pharmac.* **12**, 839–845.
Caspersson, O. and Révész, L. (1963). *Nature* **199**, 153–155.
Fisher, B. and Fisher, E. R. (1963). *Cancer Res.* **23**, 1651–1657.
Glas, U. and Révész, L. (1963). *Int. J. Radiat. Biol.* **6**, 69–80.
Greenberg, J., Mandell, J. D. and Woody, P. L. A. (1961). *Cancer Chemother Rep.* **11**, 51–56.
Harris, M. (1964). "Cell Culture and Somatic Variation." Holt, Rinehart and Winston, New York.
Hauschka, T. S., Grinnell, S. T., Révész, L. and Klein, G. (1957). *J. natn. Cancer Inst.* **19**, 13–32.
Hellström, K. E. and Möller, G. (1965). *Prog. Allergy* **9**, 158–245.
Levan, A. and Hauschka, T. S. (1953). *J. natn. Cancer Inst.* **14**, 1–43.
Levis, A. G. (1963). *Nature* **198**, 498–499.
Littbrand, B. and Révész, L. (1964). *Nature* **203**, 889–891.
Révész, L. (1956). *Nature* **178**, 1391–1392.
Révész, L., Glas, U. and Hilding, G. (1963). *Nature* **198**, 260–261.
Révész, L., Bergstrand, H. and Modig, H. (1963). *Nature* **198**, 1275–1277.
Révész, L. and Littbrand, B. (1964). *Nature* **203**, 742–744.
Ruddle, F. H. (1961). *Cancer Res.* **21**, 885–894.
Sakurai, Y., Imamura, H., Moriwaki, A., Isaka, H. and Ishidate, M. (1960). *Gann* **51** (Suppl.) 35–36.
Wallace, A. C. (1965). *Cancer Res.* **25**, 355–357.
Wheeler, G. P. (1963). *Cancer Res.* **23**, 1334–1349.

DISCUSSION SUMMARY

Chairman: Professor Sir Alexander Haddow

The long discussion following Professor Alexander's paper was opened by Dr. Kekomäki who pointed out that it is strange, in the view put forward in the paper, why thymectomized animals are not dying of cancer. Professor Alexander agreed that this is a very pertinent point. He mentioned, however, that there are some possible outlets from this difficulty; one would be that the very important immune reaction to guard against cancer may be one of the earliest ones to be established. Therefore, even neonatal thymectomy may fail to abolish this reaction. Dr. Petersen then referred to some supporting evidence to Professor Alexander's contention. He mentioned, for example, that they have on record many patients with various types of immunological deficiences who have died of malignancies.

Professor Weiss stressed the fact that somatic mutation in terms of chromosomal re-arrangement cannot be the only way of establishing a permanent deviant cell line. If that were the case, we would not have all the thousands of differentiated strains in the body which all have the same genome. Clearly the extragenic environment can perpetuate itself by calling on the genome to reproduce the same kind of combination.

Some arguments against the view put forward by Professor Alexander were then put forward by Professor Wrba and Dr. Wasastjerna. Professor Wrba mentioned that Feldman and Rehovot have shown that two tumours induced in the same way in one animal may have different immunological characteristics. Dr. Wasastjerna, in turn, referred to a study where it had been found that immunization of the recipient animal against the tumour *increases* the percentage of takes. Professor Alexander replied that he was not conversant with the last experiment but this would of course be incompatible with his viewpoint.

It was then pointed out by Dr. Glinos that to his knowledge there had not been one substantiated case of spontaneous transformation of adult human tissues, although it was possible to transform human cells with viruses. Professor Alexander answered that one can only detect spontaneous transformation to malignancy if the human cells are transplanted into situations where there is no homograft reaction against the cells—when this is done amnion cells can readily be shown to transform *in vitro*.

In answer to Professor Grundmann's question about the effects on tumour when anti-tumour lymphocytes were grown in a millipore chamber, Professor Alexander said that there is a temporary regression in growth. There was thus an indirect effect, but the mechanism was, as yet, obscure. However, an interesting hypothesis had been put forward by Hall who assumed that the effective lymphocytes equalled "messenger cells": they pass the message to those cells, whatever they are, which make the antibody.

Professor Claude pointed out that one must differentiate between malignant cells and malignant growth. He wondered whether it would also be worth considering the role played by the connective tissue. In tissue culture cells grew more easily when they have got rid of connective tissue; this might also be an important factor in the resistance against tumour growth. Professor Alexander explained that this had also impressed him in his autografting experiments. They have interpreted the finding that the primary tumour recurs but that the autograft is rejected in terms of tumour architecture. The immune reaction can eliminate a graft of isolated cells but it is less effective against residues of tumours with a supporting stroma.

The discussion was continued by Dr. Koss who expressed his differing opinion to the views presented by Professor Smithers. Dr. Koss pointed out that in a way any similarity between cancer as a clinical disease and cancer as a biological process is remote and accidental. Dr. Koss mentioned that they had three models of human carcinogenesis, which have been rather thoroughly investigated and which give us some clues as to the actual sequence of events. In the first two, breast and uterine cervix, it is possible to identify a very early stage of cancer, called carcinoma *in situ*. This morphologically identifiable process lasts for a great many years, and it may or may not go on to invasive human cancer; the cells are often tetraploid and resistant to radiation. What makes lesion *in situ* progress to human cancer, is not known. In the third case, that of urinary bladder, the sequence of events was indeed a striking one. Exposure to carcinogen produces a transient inflammatory reaction; this is followed by a clinically and cytologically normal stage which may last for as long as 30 years. Thereupon follows a cytologically abnormal and clinically normal stage lasting for as long as 10 years, and then comes as the terminal event the clinical cancer. Dr. Koss pointed out that one very basic fact of life, and the secret of carcinogenesis, lies in the first line: what is happening to the cells after carcinogen exposure and how do they manage to maintain their functional and morphological integrity for 30 years without necessarily giving rise to cancer. By the time clinical cancer sets in it is too late to learn anything about the process of carcinogenesis.

Professor Smithers answered that it is possible to look at the problem the other way around, and to say that the slow development of tumours is a clinical story: if one removes the incitant factor of the early stages, the tumours will regress. Furthermore, histologically one can find no point at which a change has suddenly taken place between the preparation for a "clinical cancer" and its final demonstration.

The discussion was summarized by the Chairman who mentioned that one thing that really impressed him in these meetings was the inclusion of the differentiation process. It was not possible to discuss regulation and cell growth for very long without bringing in the very fundamental matters of differentiation and dedifferentiation. This is one of the key features of the cancer problem.

The Chairman then pointed out that a very intriguing suggestion was made by Professor Goss that cell division and high differentiation were compatible when the end products had been removed. This would help one to dispel what has always been a great puzzle and anomaly. There are indications that cells may well be susceptible to redifferentiation, and there are also hints of specific RNA effects inducing a change of that kind.

The Chairman then said that many workers have been carrying out a systematical comparison between so called cancer cell of a given type and the so-called corresponding normal cell. Most of the work has been negative, and it is beginning to be appreciated that probably this is a false comparison. It could be that the true comparison should be with an embryonic cell not only growing fast but invading. It was clear that we must go on with tumour experiments, but at the same time we must think. There were men who worked far too hard, measuring all the time, most of the measurements being true but irrelevant. They would do far better to go to the South Sea Islands and sit under a tree and think for six months.

Section VIII

General Discussion

Editorial comment

To achieve a wide biological view of the problem of the cellular control of growth, representatives of botany, biochemistry, embryology, and microbiology were invited to the symposium. Their contributions to the general and especially private discussions were found important. Of these participants Professor Larsen has submitted a report representing the points of view of a plant physiologist. This has been accepted in the publication with gratitude.

The importance of methodological aspects became apparent many times during the discussions. Professor Iversen has prepared a report dealing with this subject.

Reflections by a Plant Physiologist at a Symposium on Growth Control in Animals

P. LARSEN

Botanical Laboratory, University of Bergen, Bergen, Norway

ALTHOUGH the pattern of growth in animals differs from that in plants in many respects, there are still enough similarities between growth in the two kingdoms to make it seem worth while for a worker in one of the fields to consider the theories, techniques, and results of workers in the other.

I. Definition of Growth

Many laymen and some botanists regard any visible change during the development of a plant as growth, and in some languages the word for "growth" is derived from the word for "plant". In plant physiology, however, it has been proved practical to adopt a much narrower definition of growth: Growth is an active increase in volume. By "active" is meant that the increase is due to life processes in the plant, so that for instance reversible imbibitional swelling and reversible turgor extension are excluded. In a sprouting potato tuber, growth in this sense is restricted to the shoots. The tuber and shoots as a whole may at the same time shrink in volume, due to transpiration, and (in the dark) loose dry matter on account of respiration.

Growth as defined here is of course the *net increase* in volume, a fact that needs emphasis only when both losses and increases occur in the chosen time interval. A general definition also covering this case, has been formulated by Weiss and Kavanau (1957, p. 3) and adopted by other zoophysiologists: "Growth, as the measured gain of organic mass of a circumscribed living system, is the net balance of mass produced and retained (incremental component) over mass destroyed and otherwise lost (decremental component) within the given boundaries." It might be interesting to examine the applicability of this definition to plant growth.

A plant physiologist would want to stress that "organic mass" must include the often very large vacuoles in the living cells of higher plants (example: a water melon with a water content of 95%). Seeds, for instance of pea or cress, germinating in the dark and with access to only air and water, will produce seedlings with a volume many times as large as that of the soaked seeds, but with a lower total content of organic dry matter, because some organic matter has been lost by respiration. There is no doubt that such seedlings have grown, but it can perhaps be disputed whether their "organic mass" has increased or not. In such cases, therefore, plant physiologists prefer to base their definition of growth on the increase in volume. Usually, however, the loss of organic matter cannot be attributed to a loss in volume, and the plant physiologist can thus make little use of a distinction between incremental and decremental components of growth.

There are situations, on the other hand, where such a distinction is useful in the analysis of growth also in plants. One example, dynamically analogous to the growth of the epidermis in animals, is the formation of secondary cortex ("borke", the outer layers of what is popularly known as bark) on woody plants. One or several layers of living cells produce cells which differentiate into cork cells. The cork thus formed may later be shed, sometimes together with other tissues external to it. Another example is the root cap, which protects the root meristem as the growing root tip makes its way among the soil particles. A meristem produces a few layers of slightly differentiated root cap cells which are later sloughed or rubbed off as the root tip proceeds, but are continuously replaced by new cells.

The preceding remarks concerned multicellular systems. A notion of negative growth, however, also turns up when analyzing the growth of individual cells. At the moment a new cell wall has been formed, the parent cell can be said to have undergone negative growth. The so-called *initial cells* in many lower plants thus exhibit alternatingly positive and negative growth while still retaining their identity. In other cases it is difficult, impossible, or even meaningless to attempt a distinction within newly formed cells between parent and daughter cells. In certain meristems, however, where cell divisions occur in waves, and in synchronized cultures of unicellular organisms, the obvious fact that negative growth of individual cells has occurred is reflected in the rhythmic course of the curves resulting from plotting the average cell size against time. When studying this kind of growth, however, it usually seems preferable to shift the emphasis from the individual cells to the tissue or the culture as a whole.

II. Growth in Plants

Growth in plants is of two kinds: embryonic growth, and cell elongation.

Embryonic growth is characterized mainly by increase in the mass of protoplasm. Although mitosis and cell division do not *per se* constitute an increase in volume, these processes are traditionally included in the term embryonic growth. This kind of growth occurs in the embryo, in meristems, and in the cambium. The terms meristematic growth and cambial growth can be used where appropriate, but as a general term, embryonic growth is to be preferred. By stretching "embryonic" to mean "juvenile" one can speak of embryonic growth in a meristem and a cambium as well as in microorganisms.

The quantitatively most conspicuous raw materials for embryonic growth (as distinct from cell elongation) are nitrogen and phosphorous, required together with carbohydrates and other materials for the synthesis of nuclear and cytoplasmic constituents.

Nitrogen and phosphorous are rather scarce in the surroundings of plants, and an ordinary seed plant would not attain any great size if its total increase in volume depended exclusively on embryonic growth. The most conspicuous volume increases in higher plants are caused by *cell elongation*: the embryonic cell increases—up to several hundred percent—in length. (Growth in other dimensions may occur, for instance in the epidermis cells in leaves and in the parenchyma of fleshy fruits. This growth has the same characteristics as cell elongation and will not be treated separately.) Although cell elongation is accompanied by some synthesis of new cytoplasm, the quantitatively dominating raw materials for cell elongation are carbohydrates and water. The carbohydrates are used for the formation of cell wall material: cellulose and pectic substances (and later lignin). Water is the dominant constituent of the cell sap, which as a vacuole takes up most of the new space formed by cell elongation. Rapid growth in plants is caused by cell elongation on basis of the relatively cheap raw materials water and carbohydrates.

Cell elongation is regulated by auxins, gibberellins, various growth inhibitors, and the enzymatically controlled equilibrium among these regulators. Cell elongation in plants seems to have little in common with most growth processes in animals. It may, however, be pointed out that the problem of why the embryonic cells, or some of them, start elongating instead of undergoing new divisions may be of general biological significance. During the development of a plant, some or all of its embryonic tissues may normally enter a period of dormancy during which neither embryonic growth nor cell elongation takes place even though the external conditions seem favourable for growth. Release from dormancy can often be achieved by external stimuli such as temporary cold treatment or changes in photoperiod, by application of chemical stimulants, or by removal of inhibitors.

Students of growth in both animals and plants are searching for growth regulating factors. For such work are needed biological tests and suitable preparations to test. As regards regulators of cell elongation in plants, available tests are numerous and relatively simple. They have been helpful in identifying such plant hormones as the auxins and the gibberellins and to indicate the existence of natural growth inhibitors. When it comes to mitosis, cell division and the growth of the cytoplasm, on the other hand, the direct tests are usually time consuming. One has to wait for several weeks before tissue cultures can be weighed or the number of mitoses in paraffin block sections have been counted. Once the pioneer work with such direct tests has been successfully completed, however, simpler (although perhaps less specific) tests may sometimes be devised on the basis of the chemical or biochemical properties of purified or crystallized, active compounds (example: cytokinins; see below).

Some of the results obtained in studies on the regulation of embryonic growth in plants may be of interest in animal physiology and will be briefly presented below.

III. Cytokinins

Certain plant tissues exhibit potentially unlimited growth on a basal medium without the addition of specific growth regulators (White 1939) or with auxin alone; others do not. One tissue in which the cells do not

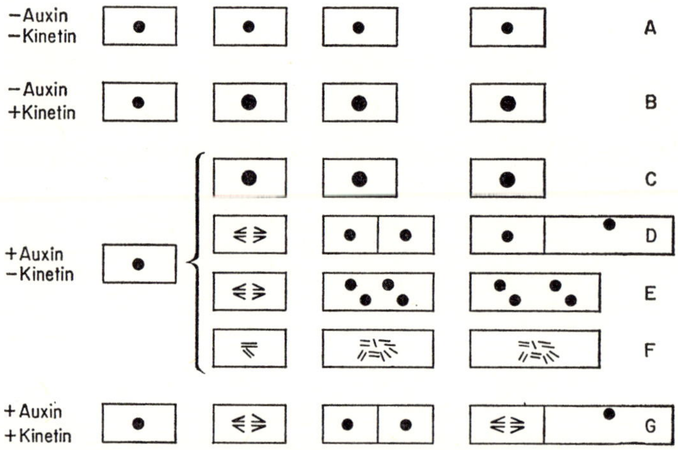

FIG. 1. Effects of kinetin and auxin applications on mitoses and cell divisions in tobacco pith tissue. The scheme indicates successive stages of the fate of an individual cell.—A no change in the nucleus.—B and C slight increase in DNA content.—D a few mitoses and cell divisions.—E a few mitoses, but no cell divisions, polynucleate cells.—F formation of polyploid nuclei, no cell divisions.—G normal mitoses and cell divisions.—Growth will continue only in G.—(Summary of observations by Naylor *et al* 1954, and by Patau *et al* 1957).

divide on such a medium is the pith of tobacco stem. In a search for compounds that would induce cell divisions in this tissue, Skoog and his co-workers at the University of Wisconsin, Madison, U.S.A., discovered that an artifact produced by autoclaving samples of commercial DNA possessed the desired property. The compound was isolated (Miller *et al.* 1955) and proved to be 6-(furfurylamino) purine (an adenine derivative), which was later synthesized. It was named kinetin because it stimulates cell division (cytokinesis). Kinetin can induce cell divisions in tobacco pith tissue only if auxin, for instance indole-3-acetic acid (IAA), is also present in suitable concentration (Skoog and Miller 1957). Figure 1 shows schematically the fate of various pith cells as responses to auxin and kinetin treatment. Addition of auxin alone can give rise to different types of development among which the formation of polynucleate cells and polyploid nuclei are notable. Continuous embryonic growth (and elongation of some cells) is obtained only if both auxin and kinetin are given at suitable concentrations (optima around 2 mg and 0·2 mg, respectively, per litre).

Several other adenine derivatives show similar activity, usually to a smaller degree (Miller 1961). Only one such compound has been identified as naturally occurring, viz. zeatin. This is 6-(4-hydroxy-3-methylbut-2-enyl) aminopurine. Zeatin was isolated from immature maize kernels (Zea Mays). Seventy kilograms of maize kernels yielded 1 mg of pure, crystalline zeatin, more active than kinetin (D. S. Letham, Division of Scientific and Industrial Research, Auckland, New Zealand; Letham *et al.* 1964). Other plant materials have yielded preparations with similar properties, for instance plums (Letham and Bollard 1961) and malt, i.e. germinating barley (van der Kerk *et al.* 1964). Skoog and his group (Rogozinska *et al.* 1965) have recently isolated a similar material from the so-called blanching water from a pea cannery. Peas are treated with 90 to 96°C hot water before being canned. A quantity of 6900 litres of blanching water, used to blanch 25 tons of green peas, yielded 99 kg of solid material which was worked up in the process of isolating the active compound. Of this powder, 14 kg yielded 47 mg of highly active material, which, however, is not yet quite pure, and has not yet been chemically identified. It does seem to be a purine derivative and is not identical with kinetin.

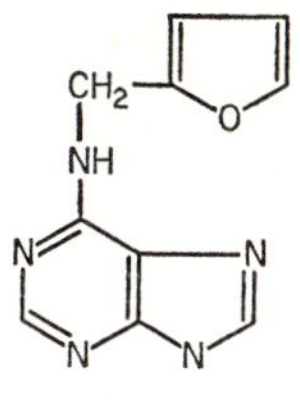
Kinetin

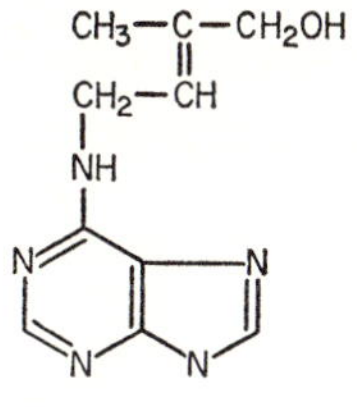
Zeatin

After several adenine derivatives were found to have the same properties as kinetin in bio-assays, a generic name for this group of compounds was sought. The term kinins was proposed by Miller *et al.* in 1956. It turned out, however, that the term kinins had already been employed by zoo-physiologists (Jaques and Schachter 1954) to designate compounds that acted primarily on smooth muscles (making them contract or relax, depending on organ). These compounds, as far as they have been identified, are peptides, containing 9 or 10 individual amino acids in a definite order (example: bradykinin). So the substances originally studied by Skoog and his group should not be called kinins. Other names have been proposed instead: "cytomins" (K. V. Thimann, mentioned by Kefford 1963), "phytokinins" (Dedolph *et al.* 1963) and "cytokinins" (Skoog *et al.* 1965). Although the term "phytokinins" has been adopted to some extent (Dedolph 1965), this term will be misleading if the same compounds are found in animals and turn out to be active also in stimulating cell division in animal cells. So the term *cytokinins* will be used here.

The cytokinins also influence other processes than embryonic growth. The ratio between the concentration of kinetin and auxin is important for the differentiation in callus tissues (formation of roots or shoots). Cytokinins stimulate the growth in area of excised leaf disks by increasing the size of the individual cells, and they will stimulate the elongation of the tips of very young coleoptiles (about 2 mm long) of cereal grain seedlings. A remarkable property of cytokinins is their ability to delay the ageing of excised leaves. Leaves treated with kinetin remain green much longer than untreated controls, and spraying or infiltration of a portion of a leaf with kinetin will not only prevent the otherwise unavoidable breakdown of protein in that region, but promote RNA synthesis and even attract amino acids from other parts of the leaf and promote protein synthesis. By the use of ^{14}C-labelled compounds it has been shown that both natural and unnatural amino acids are attracted to the kinetin-rich regions (Mothes 1964). Some of these properties of the cytokinins have been used as a basis for bio-assays which are less time-consuming than the direct tests for cytokinin activity (involving determination of embryonic growth, DNA synthesis, or the counting of cells or mitoses). The stimulating effect of kinetin on the elongation of very young coleoptile tips can, for instance, be assessed in 24 hours. Tests based on this effect are, however, less specific than the direct tests on tissue cultures; and the possible interference of such regulators as auxin or gibberellin should therefore be controlled.

The isolation of zeatin is just one specific example of the occurrence of growth stimulators in the endosperm of immature seeds. In the immature coconut, the endosperm consists of a solid part and a liquid part. The latter is a convenient material for physiological and biochemical studies and has been intensively investigated by several workers, among them F. C. Steward and his co-workers at Cornell University, Ithaca, New York (Steward and Shantz 1959, Steward 1963, Steward *et al.* 1964).

The maize endosperm material owes only part of its activity to zeatin, and coconut milk can also be divided in several fractions, which, however, show little activity when tested separately. Steward, working primarily on carrot cell cultures, distinguishes (1) highly active materials that specifically induce cell division and (2) a neutral fraction that acts as a kind of synergist or catalyst for the active fraction. The growth response to coconut milk is greatly enhanced if (3) some "reduced" (organic) nitrogen is added.

The activity of materials in group (1) might be called cytokinin activity, although this term is not used by Steward. Various compounds in the active fraction of coconut milk have been isolated (one of them is 1,3-diphenylurea), but none of them accounts for the entire activity of whole coconut milk. Among the synergists may be mentioned inositols (myo-inositol and scyllo-inositol) and sugar alcohols (sorbitol and mannitol). The reduced nitrogen (group 3) is most conveniently supplied as casein hydrolysate or individual amino acids.

Since auxin is required for normal cell divisions, and since kinetin stimulates cell enlargement in leaves and young coleoptile tips, there appears to be no sharp distinction between regulators of embryonic growth and of cell enlargement (cell elongation). The effect of application of a given growth regulator will evidently depend on the amounts of this and other growth regulators already present in the tissue.

It would be interesting to learn if compounds classed as cytokinins or their synergists also act on animal cells—perhaps in combination with various other factors. Likewise, preparations active on animal cells should also be tested on plant cells. (It may be noted in this connection that the kinetin preparations first tested by Miller *et al.* were originally obtained by autoclaving DNA from herring sperm and calf thymus.) The configuration of the active part of the molecules which stimulate (or retard) cell division may be rather similar in the two kingdoms, although the carrier part of the compounds may differ widely in view of the very different transport mechanisms in animals and plants and the presence of the cellulose wall in plant cells.

IV. Crown Gall

As another example of common interest to students of growth in animal and plant cells may be taken the crown gall disease (Klein and Link 1955, Braun and Stonier 1958). Crown gall tumours can be called forth in a plant stem by inoculating a wound with a culture of *Agrobacterium tumefaciens*. The largest and most rapidly growing tumours are produced if the inoculation is postponed till 48 to 72 hours after the wound was made. In this period of the wound healing cycle the host cells show the highest metabolic activity, and at about 60 hours normal cell divisions begin in both infected and uninfected wounds. If the inoculation is postponed further, progressively less vigorous tumours result. So the host cells must be in a certain "conditioned" state in order to be most responsive. In some plants secondary tumours will develop at some distance from the primary tumour. The secondary tumours are normally bacteria-free, and bacteria-free tumour tissue can also be obtained from primary tumours, for instance, by isolating parts of them from cultures on synthetic media. The bacteria are capable of changing the host cells in such a way that they produce all the factors necessary for cell division and enlargement.

Bacteria-free bits of tumour tissue, whether taken directly from a host or from a tissue culture, can grow as tumours when grafted onto a host plant. It seems that it is the tumour tissue that grows, whereas the tumour cells do not convert host tissue into tumours. Extracts of actively growing tumour cells will induce cell division and enlargement in tobacco pith tissue cultures in much the same way as coconut milk or a combination of IAA and kinetin, but the pith cells do not become autonomous with respect to growth factors.

The bacteria are said to produce a "tumour inducing principle" (TIP) when they act on conditioned host cells. The nature of this TIP is unknown, but current hypotheses suggest that either a special kind of DNA or a virus is an essential part of the TIP. A high concentration of DNA has been demonstrated in tumours, but also a protein with a very high molecular weight. Tumour cells are frequently polyploid.

The elucidation of the fundamental principles of tumour induction in the crown gall disease must be of equal interest to plant and animal pathologists, and the disease has actually attracted considerable attention by students of cancer in animals and man. But aside from their pathological importance, the peculiarities of tumour formation and tumour growth also serve to underline the rather precise regulation of growth which exists in the normal plant and its organs and tissues, a regulation which at least in principle must be common to all living things and is the basis for naming them organisms.

Zoo-physiologists and plant physiologists share a common interest in clarifying the general principles underlying the regulation of growth. As far as this regulation is effected by chemical entities, the identification of these rests first on the availability of biological tests. These must be provided by the physiologists themselves. The isolation of growth regulators seems so far to have been successful mainly from materials that can be obtained in quantities of kilograms, tens of kilograms, or more. So an investigation of such materials would seem profitable. Past experience also indicates that the co-operation with industrial firms is clearly advantageous, and that the co-operation with competent biochemists is indispensable.

REFERENCES

Braun, A. C. and Stonier, T. (1958). "Morphology and Physiology of Plant Tumors." Protoplasmalogia. X. Pathologie des Protoplasmas. 5a. 93 pp. Springer. Wien.
Dedolph, R. R. (1965). *Science* 149, 658.
Dedolph, R. R., Wittwer, S. H. and MacLean, D. C. (1963). *Scient. Am.* **208**, 13.
Jaques, R. and Schachter, M. (1954). *Br. J. Pharmac. Chemother.* 9, 53–58.
Kefford, N. P. (1963). *Science* **142**, 1495–1505.
Kerk, G. J. M. van der, Eyk, G. W. van and Weber, J. A. (1964). *Chem. Weekbl.* No. 14, Deel 60, 185–194.
Klein, R. M. and Link, G. K. K. (1955). *Q. Rev. Biol.* **30**, 207–277.
Letham, D. S. and Bollard, E. G. (1961). *Nature* **191**, 1119–1120.
Letham, D. S., Shannon, J. S. and McDonald, I. R. (1964). *Proc. chem. Soc.* 230–231.
Miller, C. O. (1961). *A. Rev. Pl. Physiol.* **12**, 395–408.
Miller, C. O., Skoog, F., Okumura, F. S., von Saltza, M. H. and Strong, F. M. (1956). *J. Am. chem. Soc.* 78, 1375–1380.
Miller, C. O., Skoog, F., von Saltza, M. H. and Strong, F. M. (1955). *J. Am. chem. Soc.* 77, 1392.
Mothes, K. (1964). *In* "Régulateurs naturels de la croissance végétale". C.N.R.S., Paris, pp. 131–139.
Naylor, J., Sander, G. and Skoog, F. (1954). *Physiologia Pl.* 7, 25–29.
Patau, K., Das, N. K. and Skoog, F. (1957). *Physiologia. Pl.* **10**, 949–966.
Rogozinska, J. H., Helgeson, J. P., Skoog, F., Lipton, H. and Strong, F. M. (1965). *Pl. Physiol.* **40**, 469–476.
Skoog, F. and Miller, C. O. (1957). *In* "Symposium of the Society of Experimental Biology" XI. The Biological Action of Growth Substances, pp. 118–131.
Skoog, F., Strong, F. M. and Miller, C. O. (1965). *Science* **148**, 532–533.
Steward, F. C. (1963). *Scient. Am.* **209**, 104–113 and 156.
Steward, F. C. with Mapes, M. O., Kent, A. E. and Holsten, R. D. (1964). *Science* **143**, 20–27.
Steward, F. C. and Shantz, E. M. (1959). *A. Rev. Pl. Physiol.* **10**, 379–404.
Weiss, P. and Kavanau, J. L. (1957). *J. gen. Physiol.* **41**, 1–47.
White, P. R. (1939). *Am. J. Bot.* **26**, 59–64.

Some Methodological Remarks Concerning the Measurements of Growth

O. H. IVERSEN

Institute for General and Experimental Pathology, University of Oslo, Rikshospitalet, Oslo, Norway

IT is important that the general discussion of this conference also considers some methodological questions regarding the measuring of growth. We often have to measure the rate of cell renewal, and for this we can use different methods:

I. Mitosis Counts

It will be evident for all of us that the mitotic index (or the mitotic count) is not a measure of the rate of cell proliferation, unless we know the mitotic duration. To measure the *mitotic rate*, we must rely upon a stathmokinetic method, Colchicine or Colcemid being mostly used. This is not a perfect method, but used properly it can give an estimate of the mitotic rate. By proper use is meant, that one must follow the increase in mitotic count at different time intervals after the injection of Colcemid or Colchicine. One has to have values 1, 2, 3 and 4 hours after the injection, and use only the first straight line increase in the mitotic count. For different tissues and different rates of proliferation the Colcemid or Colchicine effect may fade off after 4, 5 or 6 hours. This has to be investigated in each case. The tangent of the angle between the straight part of the line and the abcissa is the rate of entry of cells into mitosis, which can be accepted as an estimate of the rate of cellular proliferation.

One has also to measure the mitotic count without using Colcemid. We then have *Mitotic count* (*without Colcemid*) = *Mitotic rate* (*with Colcemid*) × *Mitotic duration.*

Mitotic rate is here defined as the number of cells dividing per time unit per given number of cells. With this method it is possible to find also the mitotic duration.

Many recent works have shown that the mitotic duration may vary widely. Generally it seems to be short when the proliferation rate is high, and long when the proliferation rate is low. There are many evidences now pointing to the fact that the high mitotic count in some tumours is mainly due to long mitotic duration, and not to a high growth rate.

II. ^{3}H-Thymidine Technique

The measuring of the number of labelled cells after injection of tritiated thymidine and autoradiography is open to the same type of critics as is the mitotic count. The labelling index is, however, a little better than the mitotic count, because the time for DNA synthesis is longer than the mitotic duration. But one can do two injections with one hour interval (if G_2 is longer than one hour), and in this way measure the increase in number of labelled cells during one hour. In this way one obtains a measure of the proliferation rate as expressed by the rate of cells entering into DNA synthesizing phase per hour per given number of cells. The time for DNA synthesis can be calculated by this method, but also by determining the first wave of labelled mitoses.

Grain counts may be a help, but it is very time consuming and has many pitfalls, for instance the very great problem of standard section-thickness. The measurement of the total radioactivity per mg tissue can be used, but the incorporation value of tritiated thymidine is not *eo ipso* a measure of the proliferation rate. At this conference we have seen autoradiographs from thymus showing a few grains over rapidly dividing cells, and many grains over slowly dividing cells. It is thus evident that the incorporation value must be used with the proper critics.

III. Increase in Weight or Cell Number

The measurement of the increase in weight or cell number can be a good method, and should be used as a complementary method whenever possible. More complicated approaches using labelled precursors of protein, RNA, etc., can be used for this purpose. But the methods have their limitations, mainly depending on the fact that regularly there is also a *cell loss* from organs and tissues. Increase in mass or number of cells is the net gain of cell renewal over cell loss:

$$\text{Mass} = \int_0^t (\text{renewal})\,\mathrm{d}t - \int_0^t (\text{loss})\,\mathrm{d}t$$

IV. Measuring of Cell Loss

This is sometimes possible by using a radioactive tracer. If it is possible, one should calculate the rate of cell loss from any tissue measured.

V. The Normal Circadian Variations

Circadian variations in the cell renewal process are very important. Dr. Echave Llanos demonstrated this for us in an indisputable manner. We have to work with standardized patterns of feeding, light and darkness, and we must only compare results of experiments performed at the same time of the day. It is important to know the rhythm of the animals used.

VI. Homogeneous Population or Not?

We must be sure that we work with a homogeneous population which do not change in relative numbers during the experimental period. If it changes, we must determine the qualitative and quantitative types of changes. We must know the size of the generative pool of cells, and the size of the differentiating cell pool.

VII. Self-maintaining or Transit Populations

The importance of knowing what type of cell population we work with was pointed to by Dr. L. G. Lajtha, and need not be repeated here.

VIII. The Effect of Extracts

Time seems ripe for us to seriously consider the necessity of taking biochemists into our field, so that we can fractionate our extracts and find out what we are working with. Before this is done, no real progress can be made. When I mentioned this point to Dr. Echave Llanos, he remarked: "This is not very important, this is necessary!"

CONCLUDING REMARKS

Professor Paul Weiss

We have had $2\frac{1}{2}$ days of extremely successful and pretty well focused discussion about the problem of growth control. Growth and control were treated separately in some detail, but the main emphasis was on growth control, and the material lends itself to summarizing, mostly of the blank spots on the map of knowledge that had come out in these meetings.

It is very important to create a basis now, not for any dogmatic or final decisions which might handicap further research so badly needed, but rather for emphasizing essential points where inconsistencies and incongruities between the different points of view and the different partial approaches to this central problem call for further investigation.

I hope my comments might preclude some further ambiguities such as have already crept in, by anticipating some of them. I will eliminate the terms "differentiated" and "undifferentiated" because, after all, any later cell has had a long ancestry of earlier cells. So, what is called "differentiation", on a broader scale, is of course the whole process of changes leading from some beginning in the egg after its division, through a continuous series of generations, to a more specialized terminal stage. It definitely must not be confined to just the terminal transformation of a "blast" to a form by which the character or type of cell is identified. This last phase of "expression" is a very interesting phenomenon in itself—"cyto-differentiation", as I call it—but it does not bear on the basic problem of differentiation, which is the divergence of self-perpetuating properties among cell strains of originally common origins.

The point I would like to stress as the result of the meeting is, first, that I feel sure that everybody has become cognizant of the need for exercising greater care in defining the ground rules for his work. I have grown up close to the field of physics, which has in its units of centimeters, grams, seconds, and now even quanta, an invariable common and universal mode of communication which all physicists must use and which means the same to all of them. By contrast, we find more than a dozen different meanings for "differentiation", and each of us adheres to our particular brand of its meaning. Is it surprising then to find so many inconsistencies and contradictions in the literature dealing with "differentiation"? The position is similar in the field of "growth". Biology has not reached the stage yet where it has terms of universally accepted meaning for unequivocal communication. I think everybody has become aware of this truth in our meeting.

The second observation is that the reference frames themselves have been so vague. I therefore have been greatly impressed by the growing efforts to define precisely the systems we are working with, including time and nature of what happens and the conditions and properties of the test objects. Animals and tissues behave differently, at various times and circumstances. I have witnessed here the beginnings of a stricter and more disciplined definition. Let us continue this trend. As I said earlier, the young people should not be drawn only into the field of "molecular" biology, where most dramatic basic facts have been discovered and, for a while, the operation will be essentially a "mopping-up" process. There is far more urgent work to be done in the areas we have discussed, where so much basic matter is wholly unknown. I will come back to this presently.

The past lack of definitions of our reference system reminds me of a story about Till Eulenspiegel, the jester. There was an enemy invasion of his town and the people wanted to save their church bell; but the city fathers did not know how to hide it and asked him for advice. So he said: "Well, we have a lake here, so let's take the bell out and sink it, and when the war is over we go and raise it again". So, they took the bell out in a boat and went to the middle of the lake with great aplomb and said: "But how are we going to locate it again once the war is over?" He answered: "That is very simple, you just mark the place by making a notch in the side of the boat where you drop it overboard", which they did, and—well, you can imagine the rest of the story. I have seen signs of similar folly in our own literature, I think our meetings here have brought about a great deal of clarification.

I am sure, however, that there are far more positive aspects worth pointing out that may prove acceptable after this meeting, pending proof to the contrary. We must bear in mind that when we talk about growth, we refer to *heterogeneous* systems. We have learned as children in school that we cannot add up apples and pears without depriving them of their specific characters: two apples and four pears just make six bodies. When it comes to growing systems, you must, as stressed by Dr. Iversen, give a complete account of all the separate items—how much, how long, how much is going in, and how much is coming out—and it is the balance sheet of all these component, but highly diverse, contributory events that we are measuring as a result.

This brings me to a very important matter, namely that we are still not talking about the events or the processes themselves that are going on in the cells and in the organism, but we are talking about our observations and we then proceed to project the limitations of our observational or technical powers into the objects we are dealing with. This is true whether we talk of malignancy, of cancer, of thresholds or any of those things. These are properties of our observational acuity or of the methods available for detection, discrimination and identification. But we must guard against inferring from marked discontinuities in our observations that there are corresponding discontinuities in the properties of the system we are studying. As a physicist, I have learned to respect the distinction between the observer and his object, which he calls "operationalism"; the biologist, particularly in this field of growth control, will gradually have to acquire the same operational discipline of thinking.

So far I have spoken of the great uncertainties, but some certainties have also emerged. One of them seems to be that there is a definite humoral feedback relation in growth control. I think we will never be at the point where we may ascribe growth control to one single monopoly or factor, assuming that everything else follows from that. I accept Professor Haddow's maxim that, basically, nature is simple, but we have to enucleate that simplicity by hard and conclusive work and prove it, and not just try to get it by mental over-simplifications, as we are apt to do. If growth control is simple, it must appear to be simple *after* demonstration; yet, for the time being, this meeting has shown, to the contrary, that no single mechanism of growth control or of feedback system can be conceded an explanatory monopoly.

As in development in general, we are faced here, not with the old-time linear, mechanistic cause-effect chain, where when one cuts one link, all the rest drops out. We are dealing with network systems, where one can destroy some part and the rest will still produce something which to the observer looks like a perfectly good likeness of what we would normally see if that particular thread of the fabric had not been severed. We are dealing with multiple-mesh fabrics with numerous pathways connecting any two points, so that when one connection fails, there are lots of bypasses left to keep communication going undisrupted. Anybody who studies development knows this for a fact and it is about time to have it more broadly introduced into our concepts of growth. Simplification will presumably consist in finding general rules for network behaviour, rather than finding a single linear cause-effect relation chain that could be considered to be *the* responsible growth control mechanism.

This brings me nearer to the problem of growth control in answer to "functional needs"—which has been raised—that is, whether or not functional activity as such, or some chemical feedback system related to it, is keeping the system in a steady-state equilibrium. In the first place, the problem cannot be phrased in terms of an "either-or" proposition. Secondly, "functional need" has still to express itself through some mechanism. How does it work? That "functional load" has a controlling effect, is true of many tissues. Yet, it is not universally true. I give you one example. Norman Robbins, one of my students, has shown, in line with Beidler, that the sensory cells in the papillae of taste buds of the tongue have an extremely short life. Their life expectancy in many mammals is no more than about a week. They are constantly reproduced at the periphery of the bud, then move toward the centre, where they are shed. At the same time, nerve connections are constantly made, broken and re-made. That is a very quick turnover of the cell population. Now, exactly the same thing goes on even after subcutaneous transplantation of the taste buds into foreign locations in body cavities, where certainly the "functional load" of "tasting" something cannot be invoked to explain the high rate of growth. I could give many more examples of this kind, just as I can cite more cases of the other kind, like the muscles of the heart, which were mentioned as proof for the role of functional load. So again I would say the meeting has shown once more that we must not yet generalize; just keep on working.

It has also shown that there is this distinct dichotomy between a cell deciding either to make another cell like it by cell division or to go into manufacturing a specialized cell product; for instance, a thyroblast, to grow and give rise to two thyroblasts or to go into operation and form colloid and thyroxin. This can serve as a model for the progressive differentiation of cell lines, too: a cell may go on to replicate itself by proliferation or it may be switched into an alternate course and from then on proliferate according to the new pattern. Some of its progeny may then undergo a further switch, and so forth. This, to my mind, occurs in steps of dichotomies, and the whole series of developmental changes from the egg on is effected by such "binary decisions". The fact that the germinal cells of the epidermis have this bipotentiality, for instance, is shown by the experiments on their metaplasia by Fell and Mellanby and our own: depending on the vitamin A level, a basal cell of the chick epidermis can be thrown either into a keratinizing track or into a mucus-producing track, which are the only two courses open to this cell, and it must take either one or the other.

Now if you want to look into the molecular basis of such switching and have it related to the genic system, you must bear in mind that the genic system is entrapped in its nuclear and cytoplasmic cages. These have been modified progressively during ontogeny by the antecedent series of such dichotomizing interactions, confronting thus the genome inside with ever new intracellular environments. It seems, as Professor Haddow hinted, that there is gradually emerging a branching pattern of specialization of cell strains, where two equivalent cells, which are to all purposes identical because they come from the same mother cell, are going one into Course A and the other into Course B, establishing each a cell colony which breeds true from here on, one according to Type A and the other according to Type B. They retain the new differential despite the fact that they all have the same genome. It is becoming more and more likely now that this happens concurrently, and perhaps instrumentally linked, with an interaction in which the RNA system is branched into an A type and a B type RNA. The evidence is not very conclusive as yet but it is emerging. If this is true then we would have to assume that there are as many RNA types as there are cell types; and there are far more cell types than meets the eye under the microscope.

This brings me to the next point which has come out in these sessions. The outlined dichotomy is a strict alternative for the cell; the two possible courses in each case are mutually exclusive. When we speak of "differentiation"—"differentiated" cancers, and so on—we ought to consider whether we are talking about the character of a given cell, that is, its intrinsic properties, or about the particular incidental way in which it manifests its character in a given setting. The overt manifestations, being secondary reaction products, are too variable to be relied upon as diagnostic criteria of cell character. We consequently are not surprised to find that cells can lose a lot of these overt signals without losing their specific characters. Indeed, cell lines can also change in character, but they do not lose the differentials that distinguish one cell strain or cell line from another. This is what is wrong with the universal term "differentiation". The differentials are not extinguished, and when we say that a cell "dedifferentiates" and later "redifferentiates", this does not mean that it can now yield again any and all the cell forms that could have been formed in earlier stages of development from its precursor or ancestor cell. In other words, there is such a confusion in the whole literature on these terms that I heartily endorse the suggestion of Professor Haddow to have a separate symposium on these problems of differentation, so that there might emerge better orientation and guidance, just as this conference has produced orientation and guidance on humoral growth control.

That there is humoral inhibition of growth seems pretty clear, but its nature is not definable as yet. That there is some sort of self-inhibition and that some of it comes from the growth process of a particular organ itself, is likewise clear. That is to say, the organ system itself is involved in throttling or controlling its own total growth regardless of contiguity of its mass, that is, regardless of whether the mass of that organ or system is in one piece or is dispersed, as it is in the case of the blood system and in the cases of multiple organs, paired organs, and the like. But I am not going to impose my own ideas about this upon you; that is not my function here, as I am merely to summarize the results of the meeting.

The only problem which has not been properly discussed, although the word has been used many times, is the question of *specificity*.

This, I find, has been the major deficiency of the meeting—deficiency, rather than defect. We have heard many times the question: "Is this or that 'specific', or is it an 'unspecific' influence?" But such a categorical distinction does not exist. There are degrees of specificity and without defining them, it will be difficult to deal with the subject.

When we talk about degrees of specificity, the problem of mutual recognition, of self-recognition and of alienation among cells comes to mind, which was introduced earlier. I have a lot of practical experimental data relevant to it, and so have others. They refer to cell surface configurations, molecular conformations, surface antibodies (such as Prof. Alexander brought up yesterday), surface changes as a result of homograft reaction, and so on. These are important problems, but to me the immune reaction is merely a model of a much broader biological principle, namely, that of complementary matching configurations. The conformation of matching cells need not be confined to structural matching only—the static jigsaw-puzzle kind of thing—but should include functional mutualisms, that is, relationship in the physiology between neighbouring cells so that each can live harmoniously with the other one, because what one produces is useful, or at least not harmful, to the other one, and vice versa: they are mutually dependent and attuned. Many examples of such systems are demonstrable in the body. In that view, cell "alienation"—a cell becoming emancipated from its neighbours, as perhaps in the case of metastasis, or as Prof. Alexander pointed out, being somehow made different—does not have to be visualized in terms of loss of solid interdigitation or of changes in molecular surface topography, but rather as loss of physiological interrelations which had harmonized the cell with its environment: with its basement membrane, its surrounding cells, its liquid medium (if it is an epithelium), and so on.

One thing that has never been mentioned is that organ specificity does not have to be viewed as restricted to the epithelial tissues of a particular organ, but presumably extends to the stroma and the vascular bed, so that the capillary system of a particular organ, as well as its stroma, has as a result of embryonic interactions, become just as specific and characteristic for that particular organ as the epithelial components. This possibility has to be taken into account when we study an organ-specific or site-specific or tissue-specific effect. We have to make sure that a given "organ-specific" immunological effect, for instance, might not perhaps have exerted its primary action on the capillary bed or the stroma of that particular organ and that only secondarily, as a result of some rather unspecific consequences, like an anaemic condition or a hyperaemic condition, the more conspicuous change arises by which we finally identify the terminal result.

I call this to your attention because you notice that there has not been any mention of this possibility, and it is a very real one. It is very real because you know it, for instance, from hormone reactions: the male sex hormone in the rooster, for example, circulating diffusely, dilates primarily and specifically the capillary bed of the comb; growth of the comb then follows secondarily. We could cite many similar cases also for the smooth musculature. The vascular tree, I would maintain, is differentiated into as many qualitatively specified branches as there are organs that it supplies. And a lot of the effects that we record on the organs operate primarily on that local vascular target. The same differential specialization is also true of the mesenchyme and connective tissue.

In other words, for the time being, the problem of the specificity of growth control, as Professor Haddow says, is still enlarging, and the more we learn, the more it will at first continue to enlarge. I am convinced that, in the end, things will fall into place and make sense. If they do not at present, we must not try to squeeze them into an inadequate partial and, by definition, incomplete picture. This is the essential conclusion I have drawn from this meeting. I hope I have not offended anybody in pointing out some of the deficiencies, for I have been a partner in the process of causing these deficiencies or retaining them.

I would like then to conclude on this very happy and cheerful note, the best note on which we can close: to look forward optimistically to the future, encouraged by the marked resolution of problems that has taken place in this meeting, and retaining our cheerful, optimistic, forward-looking and friendly spirit, so that we may not waste energy or divert energy from useful constructive experimentation to fighting a sort of religious warfare between opposing, but illusory, monopolistic doctrines. I thank all of you for your contributions and for the wonderful spirit in which this meeting has been carried on, and I particularly want to thank our hosts for the remarkable opportunity they have provided and which I will extoll more explicitly at a later occasion. I say with a deep sense of responsibility and humility that I, for one, have found it exceedingly worthwhile to take time off at a very bad time of the academic year to come here, and not only hear the comments but meet all of you in person. I feel assured that you feel equally well rewarded. Thank you very much.

AUTHOR INDEX

(Numbers in italics refer to pages in the References at the end of each paper)

A

B

C

D

E

F

G

I

J

K

N

O

P

T

U

V

W

SUBJECT INDEX

A

B

D

E

F

G

H

M

N

O

P

R

S

T

U

V

X